Texts in Theoretical Computer Science
An EATCS Series

T0238637

Editors: W. Brauer J. Hromkovič G. Rozenberg A. Salomaa
On behalf of the European Association
for Theoretical Computer Science (EATCS)

Subir Bandyopadhyay

Dissemination of Information in Optical Networks

From Technology to Algorithms

In Cooperation with Ralf Klasing

With 95 Figures and 4 Tables

 Springer

Author

Subir Bandyopadhyay
School of Computer Science
University of Windsor
401 Sunset Avenue
Windsor
Ontario N9B 3P4, Canada
subir@uwindsor.ca

Series Editors

Prof. Dr. Wilfried Brauer
Institut für Informatik der TUM
Boltzmannstr. 3
85748 Garching, Germany
brauer@informatik.tu-muenchen.de

Prof. Dr. Juraj Hromkovič
ETH Zentrum
Department of Computer Science
Swiss Federal Institute of Technology
8092 Zürich, Switzerland
juraj.hromkovic@inf.ethz.ch

Prof. Dr. Grzegorz Rozenberg
Leiden Institute of Advanced
Computer Science
University of Leiden
Niels Bohrweg 1
2333 CA Leiden, The Netherlands
rozenber@liacs.nl

Prof. Dr. Arto Salomaa
Turku Centre of
Computer Science
Lemminkäisenkatu 14 A
20520 Turku, Finland
asalomaa@utu.fi

ISBN: 978-3-642-09197-1   e-ISBN: 978-3-540-72875-7

Texts in Theoretical Computer Science. An EATCS Series. ISSN 1862-4499

ACM Computing Classification (2008): B.4, C.2, F.2, G.2

© 2008 Springer-Verlag Berlin Heidelberg
Softcover reprint of the hardcover 1st edition 2008

Cover Design: KünkelLopka GmbH Heidelberg

Printed on acid-free paper

9 8 7 6 5 4 3 2 1

springer.com

This book is dedicated to the memory of my parents Sushobhan and Nandini and my grandparents Krishnadhan, Dhirabati, Atulyadhan, Nirabati, Sudhansu, and Usha.

Preface

For the last twenty years, communication between computers has been one of the most important and all-pervasive features of the computer revolution. Early computer networks used copper wires as the physical medium for communication. Important limitations of copper as a medium include its relatively high attenuation, susceptibility to malicious attacks, and electromagnetic interference. The tremendous growth of the Internet and the World Wide Web has made high-bandwidth computer communication a crucial and strategic infrastructure. Deregulation of the telephone industry and the dramatic increase of data, as opposed to voice, traffic over communication networks have also spurred the deployment of high-speed networks. The rapid growth of optical networks is primarily due to the inherent high speed and the reliability of optical communication. First-generation optical networks simply replaced copper wires by optical cables, to take advantage of the higher bandwidth of optical communication, with switching and other network operations handled, as before, by electronics. The speed of electronic processing is the bottleneck for such networks. In second-generation optical networks, the routing, switching, and many other network operations are done at the optical level. Second-generation optical networks are being increasingly deployed to meet the ever increasing demand for high speed backbone networks.

This book is the second in a series of books on communication networks. The first book, entitled "Dissemination of Information in Communication Network", by Juraj Hromkovič, Ralf Klasing, Andrzej Pelc, Peter Ružička, and Walter Unger, dealt with classical direct communication between connected pairs of nodes of a communication network. This book primarily deals with second-generation optical networks and is intended to be a textbook for graduate students, as well as a monograph, surveying the main areas of research in second-generation optical networks. Every effort has been made to create a self-contained book that stresses the fundamentals, in depth and in detail. In particular, the mathematical programs, often used in optimizing optical networks, have been explained carefully so that the readers can feel confident in developing similar formulations. Interesting problems have been posed as exercises. Appendices on topics such as linear programming and network flow programming have been included to help readers unfamiliar with these areas. Students will find it convenient to test their understanding of the subject by solving problems given in each chapter. The bibliographic notes section in each chapter gives a comprehensive review of work in the area. Students interested in pursuing further studies in a subtopic should find the bibliographic notes helpful in narrowing down their literature search.

Professor Dr. Juraj Hromkovič, ETH, Zurich, invited me to write this book. I would like to take this opportunity to thank him for his unstinting help and generous advice throughout this period. I also very much enjoyed the warm hospitality I enjoyed during my trips to Zurich. Dr. Ralf Klasing, CNRS, LaBRI, Université Bordeaux, wrote most of Chapter 4. He took enormous pains to proofread the book several times and made many significant improvements to it. I am most grateful to him for his advice, and have learned a lot from his meticulous approach to this project. Drs. Arunita Jaekel and Yash Aneja of the University of Windsor read the book carefully and made numerous useful suggestions for improvements. Dr Jaekel was also very helpful in finalizing the bibliographic notes sections in different chapters. The help I received from my graduate students was very useful. In particular, I would like to mention Ataul Bari, Abul Kalam, Vic Ho, Quazi Rahman, A. K. M. Aktaruzzaman, and Delwar Faruque for their help and suggestions.

This book would not have been possible without the constant support, patience, encouragement and understanding of my family members Bharati, Avik, Anjali, Rupa, Prasenjit, and Nayan.

Windsor, Canada   Subir Bandyopadhyay
July 2007

Contents

1

Introduction to Optical Networks

Computer communication started with copper wire as the medium for carrying electrical signals encoding the data to be communicated from one computer to another. Copper as a medium of communication has a number of limitations and, in the last two decades, enormous progress has been made in using alternative media for communication. This section reviews the main developments in the area of communication using optical signals with optical fibers as the medium of communication [30, 88, 89, 172, 222, 286, 324, 370].

1.1 What Is an Optical Network

Optical fibers are essentially very thin glass cylinders or filaments which carry signals in the form of light (optical signals). An optical network connects computers (or any other device which can generate or store data in electronic form) using optical fibers. To facilitate data communication, an optical network also includes other optical devices to generate optical (electrical) signals from electrical (respectively optical) data, to restore optical signals after it propagates through fibers, and to route optical signals through the network. These devices will be briefly reviewed later on in Chapter 2. Optical networks have found widespread use because the bandwidth of such networks using current technology is 50 tera-bits per second. In other words, it is theoretically possible to send 50×10^{12} bits per second using a single fiber. First-generation optical networks simply replaced copper wires with optical fibers. However, there are the following important differences between copper and fiber as communication media:

- optical devices are much more expensive compared to electronic devices, so it is important to optimize the use of optical network resources,
- a number of optical signals at different carrier wavelengths may be simultaneously carried by the same fiber,

- the speed at which optical signals may be communicated is far greater than the speed at which data can be processed by electronic circuits.

Second-generation optical networks [80, 286, 324, 370] take into account these differences and the recent developments in optical devices and network technologies. The technology of using multiple optical signals on the same fiber is called *wavelength division multiplexing* (WDM). In this book, the focus is on second-generation WDM networks, with arbitrary topologies, also known as *mesh* networks.

1.1.1 Important Advantages of WDM Optical Networks

As mentioned already, a key advantage of optical technology is speed. Internet traffic has been increasing rapidly since the early 1990s, doubling every six months. The increasing volume of business-to-business communication is another reason for the rapid increase in the demand for additional speed. New applications for data transfer are also a notable development. An example is *video-on-demand*, where users can select a video to be downloaded from some central server. Other important widespread applications include downloading music and exchanging digital pictures. There are other important advantages of optical technology, as follows [286]:

- *Low signal attenuation.* As a signal propagates through fibers, the signal strength goes down at a low rate (0.2 db/km). This means that the number of optical amplifiers needed is relatively small.
- *Low signal distortion.* As a signal is sent along a fiber optic network, it degrades with respect to shape and phase. Signal regenerators are needed to restore the shape and timing. Low signal distortion means that signal regeneration is needed infrequently.
- *Low power requirement.*
- *Low material usage.*
- *Small space requirements.*
- *Low cost.*

1.1.2 Key Terminology in WDM Optical Networks

In an optical network, all possible sources or destinations of data (typically computers) are called *end nodes*. Another important component of an optical network is the *optical router*. Every router node has a number of input (output) fibers, each carrying one or more incoming (outgoing) optical signals. The purpose of a router is to direct each incoming optical signal to an appropriate outgoing fiber. An example of a simplified model of a second-generation optical network is shown in Figure 1.1, where a circle represents an end node, a rectangle represents a router node and a directed line represents a fiber. These fibers are unidirectional and the arrow on the line gives the direction in which optical signals can flow. Such a diagram is called a *physical topology* since it shows the major physical components of the network. A *lightpath* is an optical connection from one end node to another, used to carry data in the form of encoded optical signals. Such a lightpath always starts from an end node, traverses a number of fibers and router nodes, and ends in another end node. Figure 1.2 shows a number of lightpaths using the physical topology in Figure 1.1. The lightpath L_1, shown using a dashed line, starts from end node E_1, passes through router nodes R_1, R_2, R_3, and terminates in end node E_3. The lightpath L_2, shown using a dotted line, starts from end node E_2, passes through router nodes R_2, R_3, R_4, and terminates in end node E_4. The other lightpaths have routes through the physical topology, as shown in Figure 1.2. Since the lightpaths determine which end nodes can directly communicate with each other, once the lightpaths are set up, the physical topology is irrelevant for determining a strategy for communication. It is convenient to view the lightpaths as edges of a directed graph[1] G_L where the nodes of G_L are the end nodes of the physical topology. Such a graph is called the *logical topology* (some researchers use the term *virtual topology*) of an optical network and the edges of such a graph are called *logical edges*. The logical topology corresponding to the lightpaths shown in Figure 1.2 is shown in Figure 1.3.

To make a distinction between an edge in a physical topology, representing a fiber, and an edge in the logical topology, representing a lightpath, the following notation will be used.

An edge from node N_x to node N_y in the physical topology will be shown by a single arrow $N_x \rightarrow N_y$. Here N_x (N_y) may be either an end node or a router node. An edge from end node E_i to end node E_j, representing a lightpath from E_i to E_j, in the logical topology will be depicted by a double arrow $E_i \Rightarrow E_j$.

It should be noted that the actual route of a lightpath through the physical topology is irrelevant as far as the logical topology is concerned. For instance, using the same physical topology shown in Figure 1.1, in Figure 1.4 lightpaths L_1 and L_2 have routes $E_1 \rightarrow R_1 \rightarrow R_4 \rightarrow R_3 \rightarrow E_3$ and $E_2 \rightarrow R_2 \rightarrow R_1 \rightarrow R_4 \rightarrow E_4$ respectively. The logical topology remains the same as that shown in Figure 1.3.

For a given set of end nodes, different physical topologies, representing different connections between end nodes and router nodes could give the same logical topology. For instance, if the physical topology, shown in Figure 1.5, is used, it is quite simple to set up lightpaths L_1 from end node E_1 to end

---

[1] In many networks, whenever there is a directed edge from end node E_i to end node E_j, there is also a directed edge from end node E_j to end node E_i. In such networks, an undirected edge may be used between end nodes, replacing a pair of directed edges, for simplicity.

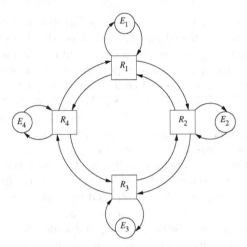

Fig. 1.1. The physical topology of a typical WDM network with four end nodes $E_1, \ldots, E_4$ and four routers $R_1, \ldots, R_4$

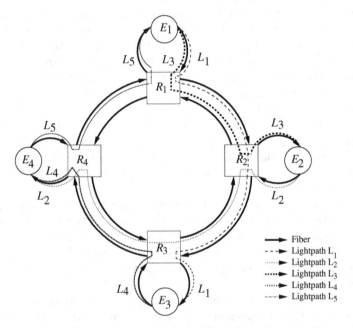

Fig. 1.2. Some lightpaths on the physical topology shown in Figure 1.1

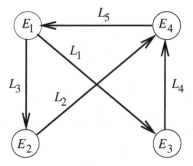

Fig. 1.3. Logical topology G_L corresponding to the lightpaths shown in Figure 1.2

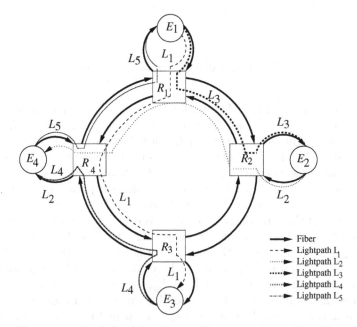

Fig. 1.4. New routes for lightpaths L_1 and L_2 on the physical topology shown in Figure 1.1

node E_3, ... , lightpath L_5 from end node E_4 to end node E_1 giving exactly the logical topology shown in Figure 1.3.

Exercise 1.1. Show a routing for the lightpaths $L_1, \ldots, L_5$ through the physical topology shown in Figure 1.5. □

1.1.3 Data Communication in a WDM Optical Network

If there is a lightpath from end node E_x to end node E_y, then this lightpath is an obvious candidate (but not the only candidate) for data communication

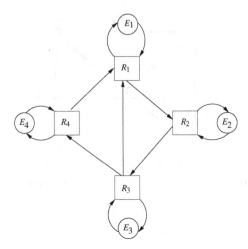

Fig. 1.5. A different physical topology

from E_x to E_y. If there is no lightpath from E_x to E_y, then a communication from E_x to E_y is possible only if there are end nodes E_{i_1}, E_{i_2}, ..., E_{i_m}, such that there is a lightpath from E_x to E_{i_1}, a lightpath from E_{i_1} to E_{i_2}, ..., a lightpath from E_{i_m} to E_y. End node E_x has to communicate the data to E_{i_1}, E_{i_1} then communicates the data to E_{i_2}, ..., E_{i_m} finally communicates the data to E_y. It is convenient to use the notion of a directed path (or dipath), in the logical topology G_L, from one end node to another, for such a communication. A directed path through a logical topology is called a *logical path*. The logical path used by the above communication from E_x to E_y is $E_x \Rightarrow E_{i_1} \Rightarrow E_{i_2} \Rightarrow \ldots \Rightarrow E_{i_m} \Rightarrow E_y$. In general, there may be a number of logical paths between any pair of end nodes. If data has to be communicated through an optical network from source E_x to destination E_y, one or more appropriate logical paths from E_x to E_y have to be selected for this communication.

For example, in the logical topology shown in Figure 1.3, to send data from end node E_1 to end node E_3, lightpath L_1 corresponding to the logical edge $E_1 \Rightarrow E_3$ is the only choice. To send data from end node E_1 to end node E_4, one possible logical path is $E_1 \Rightarrow E_2 \Rightarrow E_4$. If this logical path is used, the lightpath L_3 will be used to send the data from the source end node E_1 to the intermediate end node E_2. The lightpath L_2 will be used to send data from the intermediate end node E_2 to the destination end node E_4. In this process, the data in the source end node E_1 is converted into optical form and is sent to end node E_2 using lightpath L_3. At end node E_2 the data is extracted from the lightpath L_3 so that the data is again in electronic form. Then the data is again converted into optical form and is communicated from end node E_2 to end node E_4 using lightpath L_2 where the data is again converted back to electronic form.

Exercise 1.2. Consider the logical topology shown in Figure 1.6.

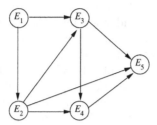

Fig. 1.6. The logical topology of a typical WDM network

i) Find all the logical paths from end node E_1 to end node E_5.
ii) In this logical topology, is it possible to communicate from any end node to any other end node? If not, then is it possible to add one lightpath so that any end node can communicate with any other end node?

□

1.2 Categorizations of WDM Networks

WDM networks may be categorized in a number of different ways. Three important ways of categorization of WDM networks are described below.

1.2.1 Broadcasting Networks and Wavelength-Routed Networks

WDM networks may be broadly classified as *wavelength-routed* networks and *broadcast-and-select* networks.

In a *wavelength-routed network*, the wavelength of the optical signal and the fiber it is using determine the subsequent path used (hence the name wavelength-routed) by the signal. Since each optical signal is sent along a specified path and not broadcast to all nodes in the network, the power requirement of such a network is lower than that of a broadcast-and-select network. This type of network may contain a large number of end nodes but is more complex and expensive than a broadcast-and-select network. The network shown in Figure 1.2 is a wavelength-routed network since the end nodes communicate using lightpaths which are routed from their sources to their respective destinations based on their wavelengths. For instance, the routers R_1, R_2, or R_3 have been set up in such a way that, when signals using the wavelength of lightpath L_1 are received by router nodes R_1, R_2, and R_3, they are sent forward to router node R_2, router node R_3, and end node E_3 respectively. The same is true for all the lightpaths in this network.

In a *broadcast-and-select network* for unicast[2] communication, the source end node selects an appropriate wavelength λ_p and broadcasts the data to be transmitted to all end nodes in the network using the wavelength λ_p. The receiver at the destination end node must be tuned to the wavelength λ_p while the receivers at all other end nodes are tuned to wavelengths different from λ_p. The net result is that the data is detected and processed only at the destination node. A typical broadcast-and-select network is shown in Figure 1.7. In this figure, each end node has one transmitter and one receiver. For convenience, the transmitters in each end node are shown on one side and the receivers on the other side. Each end node has a connection to a device called a *passive star* (to be reviewed in Section 2.3.2). The ith end node E_i has a transmitter t_i and a receiver r_i, and generates a signal using wavelength λ_i, for all $i, 1 \leq i \leq 4$. The receivers in the end nodes are tuned, as shown by the wavelength above the rectangle depicting each end node. As shown in the figure, the passive star receives signals using wavelengths $\lambda_1, \lambda_2, \lambda_3, \lambda_4$ from end nodes E_1 to E_4 and broadcasts all these signals to all the receivers, r_1, r_2, r_3, and r_4 in the corresponding end nodes. Only the receiver at end node E_1, being tuned to wavelength λ_3, selects this signal. Similarly, each of the remaining receivers only selects one signal. A broadcast-and-select network is simple and easy to implement but the size of the network is limited due to the requirement that the signal has to be broadcast to all end nodes. This type of network will be discussed in more detail in Chapter 6.

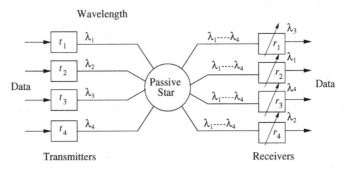

Fig. 1.7. A broadcast-and-select network

1.2.2 Static and Dynamic Lightpath Allocation

There are two approaches for deciding a strategy for data communication in a wavelength-routed network. The more standard approach is to set up lightpaths on a semipermanent basis so that, once the lightpaths are set up to

---

[2] Unicast communication means communication from one source to only one destination.

handle the expected volume of data between the ordered pairs of end nodes, the lightpaths will continue to exist for a relatively long period of time (weeks or months). This approach is called *static lightpath allocation*. When the communication pattern changes sufficiently, the existing lightpaths will be taken down and new lightpaths will be set up to handle the changes in traffic. In the second approach, called *dynamic lightpath allocation*, lightpaths are set up on demand and, when a communication is over, the corresponding lightpath is taken down (i.e., the lightpath no longer remains operational).

1.2.3 Single-hop and Multi-hop WDM Networks

In order to make the most economical use of a network, maximizing the throughput of the network is important. The cost of devices for optoelectronic conversion is high, and the speed of processing in electronic circuits is much less than the speed at which optical signals propagate through fibers. Minimizing the number of stages of electronic processing of data in WDM networks is therefore a crucial objective. It is highly desirable that the data be kept in optical form from the time it leaves the source end node until it reaches its destination end node. In a *single-hop* network, all data communication involves a path length of one logical edge. In other words, exactly one lightpath is involved in each communication. Single-hop networks are also called *all-optical networks*, since the communication is always in the optical domain [81, 178, 286, 322, 324].

In a network with $\mathcal{N}_E$ end nodes, the number of end node pairs is $\mathcal{N}_E \cdot (\mathcal{N}_E - 1)$ so that the number of lightpaths, and hence the amount of optical resources needed in a single-hop optical network, becomes uneconomical when $\mathcal{N}_E$ is relatively large. In a *multi-hop* network, [1, 157, 236, 286, 324], some data communication involves more than one lightpath. In the network shown in Figure 1.1, with lightpaths shown in Figure 1.2, the communication from end node E_1 to end node E_4 involves two lightpaths (either L_3 and L_2 or L_1 and L_4). The network is therefore a multi-hop network. If, in the network shown in Figure 1.1, there were lightpaths between every ordered pair of end nodes, then it would become a single-hop network.

1.3 Important Problems in WDM Networks and Solution Approaches

Some important problems in the area of WDM network design that have been investigated are as follows:

- How to set up a lightpath, or a set of lightpaths, to make optimum use of the network resources.
- How to best define the logical topology of a multi-hop network.

- Given a logical topology, what is the optimum strategy to handle all the requests for data communication in the network.
- How to handle faults in the network.

The problems mentioned above are (mathematical) optimization problems [377], and are often tackled using *linear program* formulations or using *combinatorial optimization*. The idea of a *linear program*[3] (LP) is to optimize some objective function, subject to linear constraints of a number of decision variables [377] appearing in the problem formulation. If, in a problem formulation, some of the variables are constrained to have nonnegative integer values and the others are *continuous* variables, each capable of having any nonnegative value, the formulation is called a *mixed integer linear program* (MILP) [377]. In most of the optical network design problems considered in this book, a mixed integer linear program is needed for an exact solution. If some of these integer variables are required to have a value of either 0 or 1, such variables are called *binary* variables. A major problem with the MILP formulations is the fact that they are NP-complete. This means that, for optical networks, as the size of the network grows, the time required to solve the MILP formulation grows exponentially. Even for small problems, the actual time to solve them can be unacceptably large. For this reason, heuristics are often used to solve the problems within a reasonable amount of time, giving solutions that are "good" but not necessarily optimal. In this book, heuristics as well as MILP formulations will be covered.

1.4 A Typical Problem in Multi-hop Wavelength-Routed Network Design

As an informal introduction to the above topics, the problems in designing a multi-hop wavelength-routed network using static lightpath allocation are given below. The design takes into account the following:

- the physical topology of the network,
- the optical hardware available at each end node which determines how many lightpaths may originate from or end at that end node,
- the characteristics of the fibers, which, for example, determine how many lightpaths may be allowed in a single fiber,
- the traffic requirements between each pair of end nodes.

It is convenient to represent the traffic requirements in the form of a matrix $T = [t(i, j)]$, often called a *traffic matrix*. The entry $t(i, j)$ in row i and column j of traffic matrix T denotes the amount of traffic from end node E_i to E_j, $i \neq j$. The unit may be specified in a number of ways. Some popular units for specifying data communication are

---

[3] A review of the simplex method to solve an LP appears in Appendix 1.

- megabits/second (Mbps),
- gigabits/second (Gbps),
- the amount of data that may be carried by a single lightpath,
- the signal rate, using the *Optical Carrier level* notation (OC-n), where the base rate (OC-1) is 51.84 Mbps and OC-n means $n \times 51.84$ Mbps.

A typical traffic matrix T, for the network in Figure 1.1, is shown below. Here the capacity of a lightpath is used as the unit.

$$
T = \begin{bmatrix}
0.00 & 0.30 & 0.10 & 0.30 \\
0.20 & 0.00 & 0.30 & 0.20 \\
0.35 & 0.10 & 0.00 & 0.30 \\
0.00 & 0.20 & 0.10 & 0.00
\end{bmatrix}
$$

The entry $t(1,4)$ is 0.30, meaning that the amount of data to be communicated from end node E_1 to end node E_4 is 0.30 units. If a single lightpath can carry data at the rate of 10.0 Gbps[4] (i.e., OC-192), the expected data communication rate from end node E_1 to end node E_4 is 3.0 Gbps.

Designing a multi-hop network is complex due to the following considerations:

- In general, there are numerous choices for the routes of each of the lightpaths through the physical topology.
- There are restrictions on the properties of lightpaths. For instance, as noted above, the number of lightpaths allowed on a fiber is limited by the characteristics of the fibers.
- In general, a very large number of logical topologies may be mapped to the same physical topology. Given a set of traffic requirements, possibly in the form of a traffic matrix T, some logical topologies are better in the sense of utilizing network resources[5] better.
- There may be multiple logical paths between a given pair of end nodes. The traffic between such a pair of end nodes may be shared by some or all of these logical paths. For example, in the logical topology shown in Figure 1.3 there are two logical paths ($E_1 \Rightarrow E_2 \Rightarrow E_4$ and $E_1 \Rightarrow E_3 \Rightarrow E_4$) between end nodes E_1 and E_4. In traffic matrix T, the traffic between end nodes E_1 and E_4 is 3.0 Gbps. Part of this traffic may be routed using the logical path $E_1 \Rightarrow E_2 \Rightarrow E_4$ and the rest using the logical path $E_1 \Rightarrow E_3 \Rightarrow E_4$.

The objective of the design is to optimize network performance. A very useful metric for this is the traffic on the logical edge carrying the maximum

---

[4] The amount of data that can be carried by a lightpath depends on the technology used and is likely to increase in the future. Unless otherwise stated, it is assumed that all lightpaths in a network have the same capacity for data communication (i.e., 10.0 Gbps in this example).

[5] The question of what are the "network resources" and what is meant by "better utilization" will be discussed in later chapters.

traffic. For reasons that will be discussed later, an important objective is to minimize this traffic. One technique to solve such optimization problems is *linear programming*. To specify this problem, integer variables are needed. A detailed discussion of this problem appears later on, but, informally, any ordered pair of end nodes may have a lightpath between them and each lightpath may choose any valid path, in the physical topology, from its source to its destination. The presence (or absence) of a lightpath requires a 0/1 variable and the use of a particular edge in the physical topology in a given lightpath also requires a 0/1 variable. In such a representation, for nontrivial networks, the number of integer variables becomes extremely large and the problem becomes computationally intractable.

To address this situation, it may be necessary to use techniques that are faster but may give suboptimal solutions. For example, it is possible to substantially simplify the problem by decoupling the problem of finding a logical topology from the problem of routing through the logical topology to minimize the traffic on the logical edge carrying the maximum traffic in the network, but at the expense of potentially worse solutions. The problem of finding a logical topology may be considerably simplified by using a heuristic.[6] In this case, the problem of routing through the logical topology reduces to that of a *multi-commodity network flow problem* [4] (to be discussed in detail in Section 7.3) that can be specified using straightforward linear programming techniques. Even so, it is interesting to note that each nonzero entry in the traffic matrix corresponds to a commodity in the multi-commodity network flow model. Since most pairs of end nodes are expected to have some traffic between them, the number of such nonzero entries in a traffic matrix is expected to be $O(\mathcal{N}_E^2)$, where $\mathcal{N}_E$ is the number of end nodes in the network. In other words, the number of commodities increases rapidly with $\mathcal{N}_E$. For example, if $\mathcal{N}_E = 30$, the number of nonzero entries in the traffic matrix (commodities in the multi-commodity network flow model) is almost 900. This poses major computational challenges requiring appropriate operations research techniques that will be discussed later on in Section 7.3.

Exercise 1.3. Consider the logical topology shown in Figure 1.3 and the traffic matrix T given above. You must show how each entry in the traffic matrix $t(i,j), t(i,j) \neq 0$, for all $i, j, 1 \leq i, j \leq 4$, may be communicated from its source E_i to its destination E_j. When determining the strategy to communicate traffic $t(i,j)$, you are permitted to split the data into smaller parts. For example, for the traffic $t(1,4) = 0.3$ from E_1 to E_4, there are two logical paths to consider: $E_1 \Rightarrow E_2 \Rightarrow E_4$ and $E_1 \Rightarrow E_3 \Rightarrow E_4$. Part of the total traffic of 3 Gbps from E_1 to E_4 may be sent using the logical path $E_1 \Rightarrow E_2 \Rightarrow E_4$ and the rest using the path $E_1 \Rightarrow E_3 \Rightarrow E_4$. In general, each lightpath will be used to carry traffic from multiple source-destination pairs. For instance, the lightpath from E_1 to E_2 will also be used to carry the traffic

---

[6] One problem with many of the heuristics is that there is no guarantee about the quality of the solution.

from E_1 to E_2. Make sure that the traffic on each lightpath is less than 1.0, the capacity of the lightpath. Indicate how much traffic is being carried by each lightpath.   □

Exercise 1.4. Consider again the logical topology shown in Figure 1.3 and the traffic matrix $T_1 = [t_1(i,j)]$ given below.

$$T_1 = \begin{bmatrix} 0.00 & 0.30 & 0.10 & 0.50 \\ 0.20 & 0.00 & 0.30 & 0.20 \\ 0.35 & 0.10 & 0.00 & 0.30 \\ 0.30 & 0.20 & 0.10 & 0.00 \end{bmatrix}$$

Is it possible to ensure that each entry in the traffic matrix $t_1(i,j), t_1(i,j) \neq 0$, for all $i, j, 1 \leq i, j \leq 4$, may be communicated from its source E_i to its destination E_j on this logical topology?

If traffic T_1 cannot be handled, can you add one lightpath to the network (in other words, add one logical edge to the logical topology shown in Figure 1.3) so that this traffic may be handled?

The solution is not unique and any logical edge which would give a valid routing such that no logical edge carries a total traffic of more than 1.0 is a valid solution.   □

1.5 Structure of the Book

In Chapter 2, major optical devices will be briefly summarized. The purpose of this chapter is to familiarize the readers with the terminology used in describing optical hardware and its purpose. Readers interested in more details on how the optical hardware works will be directed to relevant books and papers.

Chapter 3 gives an overview of different types of WDM networks and some important problems in WDM network design. This chapter gives more details about broadcast-and-select networks, wavelength-routed networks, single-hop and multi-hop networks, the route and wavelength assignment (RWA) problem, networks with static and dynamic lightpath allocation, and networks with wavelength conversion.

Chapters 4 and 5 cover different ways of setting up lightpaths. In Chapter 4, the RWA problem is modeled as a graph-coloring problem, thus allowing tools from graph theory to be applied. The hardness of the problem is established, and bounds for general digraphs and general requests are derived. The RWA problem for specific types of requests and for specific networks is also considered.

In Chapter 5, a number of techniques for route and wavelength assignment (RWA) in different types of networks are considered. This includes networks with and without wavelength converters, networks with static and dynamic lightpath allocation, and ring networks. For static lightpath allocation, since the lightpaths, once established, will persist for a considerable time, it is worthwhile to spend some time in finding the "best" allocation. For this purpose MILPs have been proposed. Heuristic solutions for handling this problem within a reasonable time have also been suggested. It is important to find a quick RWA in a lightpath allocation scenario. Some heuristics and distributed algorithms for dynamic RWA are also covered in this chapter.

Chapter 6 describes a logical topology design suitable for broadcast-and-select networks. This topology is interesting in that additional nodes can be added relatively easily, with only small changes to the network.

Chapter 7 covers the design of the logical topology to handle a set of traffic requirements. This problem has been discussed briefly in Section 1.4. MILP formulations are proposed to solve the problem. Such a formulation is computationally intractable except for very small networks. Heuristic solutions are also proposed to tackle this problem. Traffic routing over large logical topologies is a problem in itself and needs special techniques.

In view of the fact that lightpaths typically carry 10 Gbps each (and may carry even 40 Gbps each in the future) and cover all the continents, it is easy to see that optical networks are quite vulnerable to faults. The most common fault is due to fiber cuts, for instance, a fiber may be accidentally cut during earth-moving operations. Chapter 8 discusses different types of faults and how to handle them. The problems are complicated and solutions have been proposed using MILP and heuristics.

The capacity of a lightpath is much larger than a typical request for data communication from an individual user. Chapter 9 discusses *traffic grooming* techniques for efficiently utilizing the large data communication capability of lightpaths to handle many requests for data communication at relatively lower rates.

2

Introduction to Optical Technology

This chapter contains a brief overview of some of the major components used in WDM networks. The terms filters, switches, multiplexers, and demultiplexers have been traditionally used to describe electrical circuits for data communication. These devices deal with electrical signals in analog or digital form. Devices with similar functions that work with optical signals have been developed in recent years for use in optical networks.

The discussions below describe what each device does, so that the reader can follow the design issues discussed in subsequent chapters. For details on how optical devices work, excellent coverage is available in [324]. When describing individual devices, a number of other citations have been included to help interested readers get more information about the devices and how they work.

The topics covered below include

    i) optical fibers — the medium of communication,

    ii) optical transmitters to generate optical signals for communication,

    iii) optical receivers to detect optical signals for conversion to electronic form, and

    iv) other optical devices to allow propagation of optical signals from their sources to their respective destinations in a way that helps maximize the utilization of the fiber-optic networks.

2.1 Optical Fiber

An optical fiber (Figure 2.1) consists of a cylindrical *core* of silica (Figure 2.2), with a refractive index[1] μ_1, surrounded by cylindrical *cladding* of silica with a lower refractive index μ_2 [172, 176, 190, 191, 222, 223, 286, 324].

---

[1] The refractive index μ of an optical medium is the ratio of the speed of light in vacuum to the speed of light in that medium.

The *buffer* surrounding the cladding encapsulates the fiber for mechanical isolation, protection from physical damage, etc.

The idea of optical communication using a fiber is to send the optical signal at an angle of incidence greater than the critical angle $\sin^{-1} \mu_2/\mu_1$. If light passing through one optical medium with refractive index μ_1 meets another optical medium with a lower refractive index μ_2, at an angle of incidence greater than the critical angle, total internal reflection takes place, where all the light is reflected back into the medium with refractive index μ_1. Optical signals propagate through the core using a series of total internal reflections (Figure 2.3).

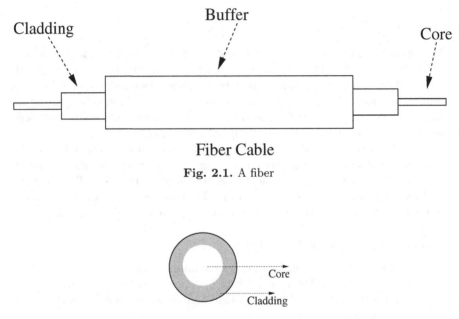

Fig. 2.1. A fiber

Fig. 2.2. The cross-section of a fiber

The attenuation of an optical signal propagating through a fiber is acceptably low in the wavelength band of 1450 to 1650 nanometers (nm) so that this band has been used for optical communication. The interval 1530 - 1565 nm is called the *C-band*. Optical devices for this band are available and the C-band has been widely used in WDM networks. Other bands such as the *L-band* from 1565 to 1625 nm will be available in the near future. The availability of different wavelength bands will increase the rate of data communication that can be handled by existing networks.

In a WDM network, in general, a fiber carries a number of optical signals in the band being used. These optical signals must obviously be at different carrier wavelengths. It is convenient to visualize the available bandwidth (which

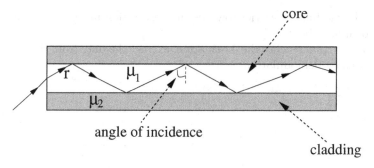

Fig. 2.3. Optical signal propagation through a fiber, using total internal reflections

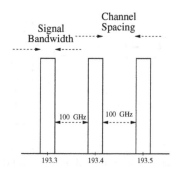

Fig. 2.4. Signal bandwidth and channel spacing

today is the C-band) as a set of *channels*. Each optical signal is allotted a distinct channel so that each channel has sufficient bandwidth to accommodate the modulated signal. In order to avoid interference between different optical signals, each channel is separated from the other channels by a certain minimum bandwidth called *channel spacing* (Figure 2.4). It is typical to have a channel bandwidth of 10 GHz and a channel spacing of 100 GHz in current networks. This means that the C-band can accommodate up to 80 channels, each having a bandwidth of 10 GHz. Shorter channel spacing (25 GHz) will lead to as many as 200 channels in the C-band alone.

2.2 Optical Communication Fundamentals

The main idea for communicating data from its source to the intended destination is as follows:

- use an optical carrier signal at some wavelength in the band of 1450 to 1650 nm,
- at the source of the data, modulate the carrier with the data to be communicated,

- send the modulated carrier towards the destination using a path involving one or more fibers,
- when the signal reaches the destination, extract the data from the incoming signal using demodulation.

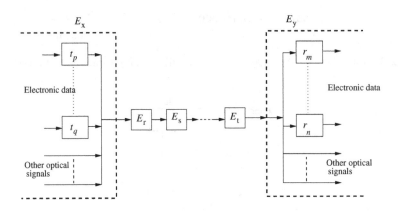

Fig. 2.5. Data communication in a WDM network

Figure 2.5 is a highly simplified diagram showing a part of a wavelength-routed WDM network. In the figure, only the transmission side of end node E_x and the receiving side of end node E_y are shown. End node E_x (E_y) has a number of transmitters (receivers), $t_p, \ldots, t_q$ ($r_m, \ldots, r_n$), tuned to wavelengths $\lambda_p, \ldots, \lambda_q$ ($\lambda_m, \ldots, \lambda_n$) corresponding to channels $c_p, \ldots, c_q$ ($c_m, \ldots, c_n$) respectively. Here $c_p, \ldots, c_q$ and $c_m, \ldots, c_n$ have at least one channel c_i in common. A simplified explanation of how the data may be communicated from E_x to E_y using channel c_i, on a network with only one incoming fiber to each end node and one outgoing fiber to each end node, is given below.

In end node E_x, the data to be communicated to end node E_y is in electronic form. This data is the input to transmitter t_i and is converted to optical signals using channel c_i. Other data in end node E_x for communication to other end nodes are similarly converted, in end node E_x, to optical signals using distinct channels from $c_p, \ldots, c_q$, each different from c_i. All these optical signals generated by transmitters $t_p, \ldots, t_q$ within E_x are combined with other optical signals coming from other end nodes that use E_x as an intermediate node in their respective paths to their destinations. These optical signals from other end nodes must use channels $c_k, \ldots, c_l$ distinct from $c_p, \ldots, c_q$. The output of end node E_x consists of optical signals at channels $c_p, \ldots, c_q, c_k, \ldots, c_l$

carried by the outgoing fiber from E_x to E_r, an intermediate[2] end node for the data to be sent from E_x to E_y using channel c_i. The node E_r simply passes on the signal using channel c_i to the next intermediate end node E_s. Continuing in this way, the optical signal using c_i is routed through a number of intermediate end nodes, $E_r, E_s, \ldots, E_t$, in its progress to the destination E_y. Intermediate end nodes E_r, E_s, ..., E_t are just like E_x. Each end node has the following tasks:

- receive optical signals from the preceding node,
- separate the different optical signals on the input fiber,
- extract, from incoming signals, optical signal(s) encoding data intended for itself, and convert data intended for itself to electronic form,
- forward, without electronic processing, other incoming signals (including the signal using channel c_i) intended for other end nodes,
- convert, to optical signal(s), electronic data that have to be communicated to other end node(s),
- combine the optical signals generated within the end node with the incoming signals that are using the end node as an intermediate end node, and send the combined optical signals using the outgoing fiber from the end node.

End node E_t sends its output to E_y. At end node E_y, the same tasks outlined above are carried out. Signals intended for end node E_y are converted to electronic form using receivers. In particular, the signal using channel c_i is converted by the receiver r_i (one of the receivers $r_m, \ldots, r_n$, tuned to wavelength λ_i, corresponding to channel c_i) to electronic form. This concludes the process of communication from end node E_x to end node E_y using channel c_i. In Figure 2.5, each node has only one fiber carrying incoming signals and one fiber carrying outgoing signals. This was done for simplicity. In mesh networks, each node (router node or end node) is connected to a number of fibers carrying incoming (outgoing) signals from (to) a number of nodes.

To accomplish such communication, an optical network must include optical and electronic devices that

- generate, at the source of the data, optical signals at some carrier wavelength in the band of 1450 to 1650 nm, modulated by the data to be transmitted.
- combine a number of optical signals and feed the combined signals to the outgoing fiber. Some of these optical signals are generated within the node itself and the remaining signals are passing through this node.
- route the incoming optical signals from the incoming fibers to the outgoing fibers.

---

[2] In this highly simplified description, the network has no router nodes. In Section 3.2 a more detailed model where the intermediate nodes may be router nodes as well will be discussed.

- overcome the effects of attenuation and distortion as the optical signal propagates through fibers.
- extract, at each end node, one or more modulated optical signal(s) from the set of all the optical signals carried by the incoming fiber or fibers. These extracted signals contain data intended for that end node.
- generate the data, in electronic form, from the modulated optical signal, when the signal reaches its destination.

A brief review of these devices is given in the following section.

2.3 Optical Devices

2.3.1 Optical Transmitters, Modulators, and Receivers

The most important type of optical signal generators are *lasers* [112, 355]. Laser stands for Light Amplification by Stimulated Emission of Radiation. An important characteristic of lasers is that the light output is *coherent* so that all the photons are in the same phase. The result of this property is that the light is very intense and tightly focussed. For wide-area WDM networks, laser output power is a very important consideration. It is also necessary that the generated signal have a low spectral width and wavelength stability, meaning that the signal produced has a wavelength which can be tightly controlled to ensure that the channel bandwidth is not exceeded. Semiconductor lasers are widely used due to their small size and ease of fabrication [324].

Present-day optical networks [176, 286, 324] use fixed wavelength transmitters where the wavelength of the output optical carrier signal is fixed. The number of channels on a fiber is increasing rapidly with improvements in optical technology. In a network with hundreds of channels, the use of fixed wavelength transmitters creates difficulties. If a particular end node should generate only, say, five optical signals, it makes sense to have only five transmitters. If five fixed wavelength transmitters are used, the wavelengths of these five lightpaths are fixed for all time. The other alternative of having hundreds of transmitters, one for each possible carrier wavelength for each of the fibers carrying outgoing signals from that end node, is prohibitively expensive. It is also necessary to have some spare transmitters to take care of failures. The issue of the wavelengths of these spare transmitters is also a problem if fixed wavelength transmitters are used. It is highly desirable that tunable lasers [324] are used in the future. Such transmitters are available in the laboratory and will be soon available for commercial applications. Ideally such transmitters should be able to change the transmission wavelength quickly. Available transmitters in a laboratory setting use current injection, temperature control, or mechanical or micro-electro-mechanical tuning. One problem is the time needed to change the carrier wavelength.

For low data rate and communication over short distances, light emitting diodes are also used as low-cost transmitters where power requirements are relatively low.

Modulation is needed to send data using an optical carrier generated by a transmitter [3, 29, 324]. Modulation is a process of converting data in electronic form to encode an optical signal. A common scheme is *on-off keying* (OOK) which sends a bit by turning the light off (on) if the bit is a 0 (1). A second scheme is to use subcarrier modulation where the data modulates a microwave signal and this microwave signal then modulates the optical signal. The advantage is that several sources of data may modulate different microwave signals which may be combined. The combined microwave signals then modulate the optical signal [324].

As the name implies, a *receiver* receives or detects at the destination of data communication optical signals at some designated carrier wavelength and extracts the data sent from the source so that the data is once again in electronic form [3, 29, 324]. In doing so, the signal has to be detected using a photo detector, the resulting electrical signal amplified, the noise minimized, and the distortion reduced; and then the bits may be extracted from the signals.

2.3.2 Optical Couplers

A *directional coupler* has n inputs and m outputs. This device generates m outputs, where each output is a linear combination of the n input signals, each input signal having a distinct carrier wavelength [176, 286, 324]. The simplest coupler has two inputs and two outputs (Figure 2.6) and is called a 2×2 *coupler*. Such a device has an associated parameter, α, called the *coupling ratio*. A 2×2 coupler, as shown in Figure 2.6, has inputs I_1 and I_2 having powers $\mathcal{P}_1$ and $\mathcal{P}_2$ and generates outputs O_1 and O_2.

There are two versions of optical couplers — *wavelength-independent* and *wavelength-selective*. In a wavelength-independent coupler, the value of α is independent of the wavelengths of the optical signals. If the coupler shown in Figure 2.6 is wavelength-independent, the power output O_1 (O_2) is the signal I_1 (I_2) at a power level $\alpha \mathcal{P}_1$ combined with signal I_2 (I_1) at a power level $(1 - \alpha)\mathcal{P}_2$. The total power output from O_1 (O_2) is therefore $\alpha \mathcal{P}_1 + (1 - \alpha)\mathcal{P}_2$ $((1 - \alpha)\mathcal{P}_1 + \alpha \mathcal{P}_2)$. A very common scheme is to have $\alpha = 0.5$, resulting in the so-called 3db 2×2 coupler. In a 3db 2×2 coupler, the input signals are evenly distributed to the outputs so that both of the input signals will be communicated, at half their power, on both of the outputs. This idea can be extended, so that, using 3db 2×2 couplers as building blocks, it is possible to build larger couplers where the power from each input is distributed equally to all the outputs. Such a coupler is called a *passive star coupler* (PSC) since it may be fabricated using only fibers — a passive device which is relatively inexpensive. A passive star coupler with eight inputs and outputs is shown in

Figure 2.7. Passive star couplers with n_{OC} inputs and outputs ($n_{OC} \times n_{OC}$ PSC) are important devices for the broadcast-and-select type of networks that will be discussed in Chapter 6.

If the coupler shown in Figure 2.6 is wavelength dependent, the coupling ratio depends on the wavelength of the signal. For example, let the input I_1 of the coupler shown in Figure 2.6 be connected to a fiber carrying two signals, one at wavelength λ_1, one at wavelength λ_2; let input I_2 be unused; and let $\alpha_1 = 1$ ($\alpha_2 = 0$) for wavelength λ_1 (λ_2). It may be readily verified that the output O_1 (O_2) will carry only the signal at wavelength λ_1 (λ_2). In other words, the device can be used to split the two signals at two carrier wavelengths, carried by the fiber carrying the input to I_1, into two separate outputs, each only carrying signals using a single carrier wavelength. Other, more complex, uses of this idea are possible [324].

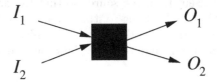

Fig. 2.6. A 2×2 coupler

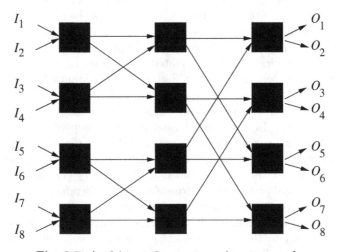

Fig. 2.7. An 8-input, 8-output passive star coupler

2.3.3 Optical Filters and Switches

As stated earlier, a fiber in a WDM network carries, in general, a number n of optical signals using channel c_i dedicated to the ith signal, for all $i, 1 \leq i \leq n$, where the value of n cannot exceed the maximum number of channels, n_{ch}, permitted by the fiber used in the network. Filters and switches [176, 324] allow routing of signals through fibers as needed and for extracting optical signals at the destination.

Fig. 2.8. An optical filter

The purpose of a filter is to separate a signal from the other signals on the same fiber. Figure 2.8 shows an optical filter that has, as input, a fiber carrying signals on channels c_1, c_2, c_3, and c_4. This particular filter has been designed in such a way that it extracts the signal that uses channel c_3. This type of device is very useful, for instance, to extract an optical signal when it reaches its destination for its conversion to an electrical signal using a receiver and a demodulator.

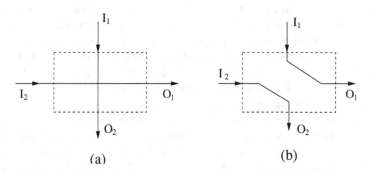

Fig. 2.9. Two modes of a 2×2 switch

The simplest type of an optical switch has two inputs and two outputs and is called a 2×2 *switch* (Figure 2.9). Any of the inputs I_1 and I_2 can be connected to any of the outputs O_1 and O_2 so that, by proper adjustment of the switch, the signals on input I_1 may be sent either through output O_1 or through output O_2. Input I_2 will be sent to the remaining output. Using the 2×2 switch as the building block, very large switches, with as many as

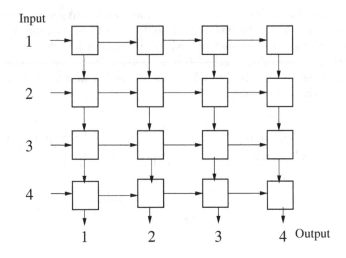

Fig. 2.10. A crossbar network

several thousand inputs and outputs, are possible. It is important to reduce the number of 2×2 switches in any path from the input port to the output port to minimize the effects of signal degradation due to effects such as crosstalk and attenuation. It is also important to have *nonblocking* networks where signals from any input may be routed to any unused output. A nonblocking network is called *wide-sense nonblocking* if any unused input can be connected to an unused output without rerouting any of the existing connections [324]. A very simple switch is a *crossbar* [324], which is nonblocking (Figure 2.10). The maximum number of switches encountered by a signal in an $n_c \times n_c$ crossbar is $2n_c - 1$, which is unacceptable when n_c is large.

Many other switching architectures (e.g., Clos [94], Beneš [41], Spanke-Beneš [363]) have been proposed with different advantages and disadvantages. An example of a Beneš network is shown in Figure 2.11. In an $n_c \times n_c$ Beneš network, each path goes through $(2 \log_2 n_c - 1)$ 2×2 switches, which is very attractive. It is a nonblocking network but not a wide-sense nonblocking.

2.3.4 Multiplexers, Demultiplexers, and Cross-connect Switches

A *multiplexer* (MUX) has a number of inputs, each carrying signals using a distinct channel. A multiplexer generates an output that combines all the signals [324]. This is very useful when optical signals from a number of transmitters,

Input   Output

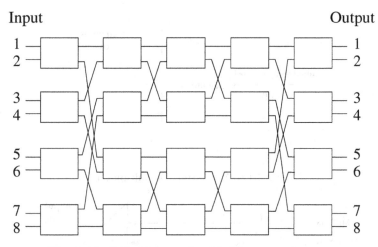

Fig. 2.11. A Beneš network with 8 inputs and outputs

each using a different channel, are to be combined for transmission on a fiber. Figure 2.12 shows a multiplexer with four inputs.

A *demultiplexer* (DEMUX) serves the opposite purpose — its input is a fiber carrying, say, m_{DM} optical signals, with the ith signal using channel c_i. A demultiplexer has, at least, m_{DM} outputs, with the ith output carrying the optical signal using channel c_i, for all $i, 1 \leq i \leq m_{DM}$. Figure 2.13 shows a demultiplexer with four outputs.

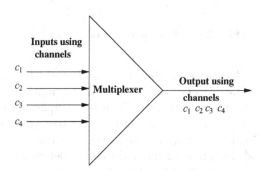

Fig. 2.12. A 4-input Multiplexer

The purpose of an *optical cross-connect switch* (OXC) is to route optical signals [324]. A $k_{OC} \times k_{OC}$ OXC has k_{OC} inputs and k_{OC} outputs. If the ith input to the OXC is carrying signal s_j^i using channel c_j, then the signal s_j^i may be routed to the pth output provided that no other signal using channel c_j is routed to the pth output, for all $i, j, 1 \leq i \leq k_{OC}, 1 \leq j \leq n_{ch}$. Figure 2.14 shows a 2×2 cross-connect switch where n_{ch}, the maximum

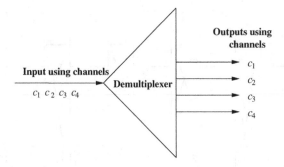

Fig. 2.13. A 4-output Demultiplexer

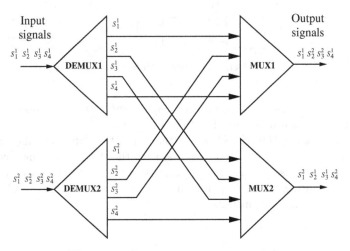

Fig. 2.14. A static cross-connect switch

allowed number of channels on a fiber, is 4. As shown in Figure 2.14, an OXC consists of k_{OC} demultiplexers (multiplexers) on the input (output) side, each connected to an input (output) fiber. Each demultiplexer (multiplexer) has n_{ch} outputs (inputs), with the ith output (input) carrying the signal that uses channel c_i, for all i, $1 \leq i \leq n_{ch}$. The connections between the outputs of the demultiplexers and the inputs of the multiplexers determine exactly how the signals on the input fibers will be routed to the output fibers.

Cross-connect switches may be categorized as *static* or *dynamic* [324]. The cross-connect switch shown in Figure 2.14 is a static device since the connections between the outputs of the demultiplexers and the inputs of the multiplexers are fixed. The routing achieved by this switch cannot be changed, so the routes taken by the lightpaths in a network using such switches are determined at the time the network is set up.

The routing performed by a dynamic cross-connect switch may be modified by changing the connections between the outputs of the demultiplexers and

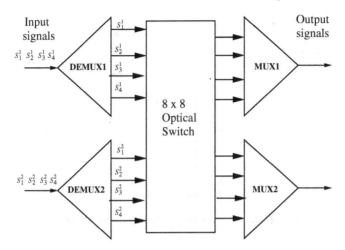

Fig. 2.15. A dynamic optical cross-connect switch

the inputs of the multiplexers. This may be accomplished by inserting a switch between the demultiplexer and the multiplexer (Figure 2.15). By changing the settings of individual 2×2 switches, the connections between the outputs of the demultiplexers and the inputs of the multiplexers may be altered as needed. An excellent coverage of these topics appears in [303].

2.3.5 Add-Drop Multiplexers and Optical Line Terminals

Add-drop multiplexers (ADMs) are very useful in optical networks [324]. An ADM is essentially a pair consisting of a multiplexer and a demultiplexer where some of the outputs from the demultiplexer are not connected to any of the inputs of the multiplexer. Each output of the demultiplexer not connected to an input of the multiplexer is connected to a receiver. Each input to the multiplexer which is not connected to an output of the demultiplexer is connected to the output of a transmitter/modulator. An ADM with an input carrying four signals is shown in Figure 2.16. As shown in Figure 2.16, signals using channels c_2 and c_3 are "dropped", meaning that these signals are not sent to the multiplexer. These optical signals will be converted to electrical signals for use in the end node containing this ADM. This node also creates two optical signals using channels c_2 and c_3 which are also inputs to the multiplexer that generates the output.

Reconfigurable ADMs are also possible. The idea is the same as that used in dynamic cross-connect switches, and uses switches to determine which wavelengths are to be dropped from which incoming fiber and which wavelengths are to be added to which outgoing fiber. Add-drop multiplexers using optical devices are now available and are called *Optical Add/Drop Multiplexers* (OADM).

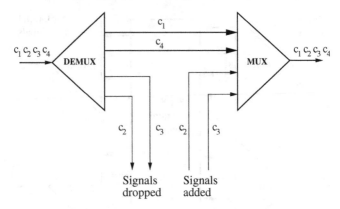

Fig. 2.16. An optical add-drop multiplexer

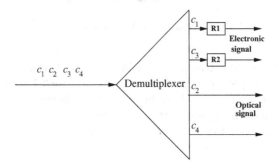

Fig. 2.17. An optical line terminal to receive data

The term *Optical Line Terminal* (OLT) is sometimes used to describe a multiplexer (demultiplexer) coupled with a transmitter (receiver) [324]. The end nodes use an OLT to

- extract data, in electronic form, from optical signals on an incoming fiber or
- generate optical signals from data in electronic form, to be transmitted on an outgoing fiber.

In order to extract data from optical signals, a demultiplexer first separates optical signals at different carrier wavelengths from the incoming fiber. Then a receiver/demodulater extracts the data in electronic form. Such an OLT has one input port, connected to a fiber carrying incoming signals. The outputs of the demultiplexer are either directly connected to client applications or connected to a number of receiver/demodulaters. The output of each receiver/demodulater generates data in electronic form. Figure 2.17 shows an OLT to receive data where the incoming signals use channels c_1, c_2, c_3, and c_4. The demultiplexer separates the signals. The signals using channels c_1 and c_3 are fed to receivers r_1 and r_2 generating data in electronic form.

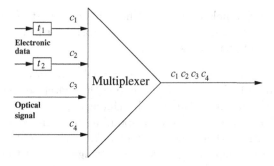

Fig. 2.18. An optical line terminal to generate optical signal

To generate the optical signal from data in electronic form, the data is used to modulate an optical signal at a carrier wavelength generated by a transmitter. Typically, an OLT to generate optical signals has a number of input ports, each generating a modulated carrier at a different wavelength. In addition there may be some client supplied optical signals. A multiplexer combines all these optical signals and the output of the multiplexer is connected to some outgoing fiber. Figure 2.18 shows an OLT where the incoming client signals use channels c_3 and c_4. Transmitters t_1 and t_2 take electronic data from two sources and convert them to optical signals using channels c_1 and c_2. The multiplexer combines the signals using channels c_1, c_2, c_3, and c_4. The output of the multiplexer is connected to a fiber carrying outgoing optical signals.

2.3.6 Wavelength Converters

A wavelength converter (WC) is a device that allows the carrier wavelength of an optical signal to be changed [286, 324]. As shown in Figure 2.19, a WC has an input carrying incoming signals and an output carrying the same number of outgoing signals. Each outgoing signal encodes the same data as an incoming signal but the carrier wavelength of an outgoing signal is allowed to be different from that of the corresponding incoming signal.

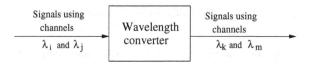

Fig. 2.19. A block diagram of a wavelength converter

Figure 2.19 shows a typical wavelength converter. The input of the WC is connected to a fiber carrying two signals using carrier wavelengths λ_i and λ_j

corresponding to channels c_i and c_j. The WC converts the signal at carrier wavelength λ_i (λ_j) to a signal using a carrier wavelength λ_k (λ_m).

WCs may be categorized as follows:

- *full wavelength conversion* where the channel c_j of an incoming signal can be any channel, $1 \leq j \leq n_{ch}$. By changing the switch settings, the channel c_j used by the input signal may be changed to any other channel, say c_k, $1 \leq k \leq n_{ch}$, used by the corresponding outgoing signal. Figure 2.20 shows a situation where the input of a WC are signals using channels c_1, c_2, c_3, and c_4. Depending on switch settings, each of these signals may be converted to any other channel.

- *fixed wavelength conversion* where the channel c_j of an incoming signal can be any channel, $1 \leq j \leq n_{ch}$, as in the previous case, but the scheme for wavelength translation is fixed. Figure 2.21 shows a situation where the input of a WC are signals using channels c_1, c_2, c_3, and c_4. The conversion scheme is fixed, so that c_1 is always converted to c_2, c_2 to c_1, c_3 to c_4, and c_4 to c_3.

- *limited wavelength conversion* where, for a given channel number c_j of an incoming signal, the corresponding channel number c_k for the outgoing signal is a member of a predefined subset of the set $\{1, 2, \ldots, n_{ch}\}$ [349, 423]. Figure 2.22 shows a situation where the inputs of a WC are signals using channels c_1, c_2, c_3, and c_4. Depending on switch settings, c_1 can be converted to either c_1 or c_2, c_2 can be converted to any channel in the set $\{c_1, c_2, c_3\}$, etc.

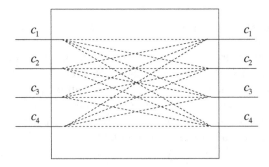

Fig. 2.20. Full wavelength conversion

While a significant amount of theoretical work has been carried out on designing networks with wavelength converters, all-optical full wavelength converters are still not feasible. Available technology does allow wavelength conversion using optoelectronic conversion. This means that a WC essentially converts the incoming signals to electronic signals and then converts the electronic signals to optical signals at any desired wavelength. However, such

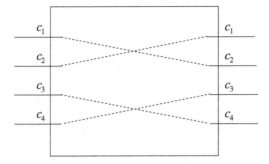

Fig. 2.21. Fixed wavelength conversion

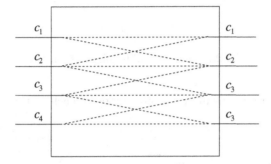

Fig. 2.22. Limited wavelength conversion

conversion is inherently slow since it happens at electronic speed. Some theoretical research work on limited wavelength conversion has been carried out. It has been demonstrated that it is technologically feasible to convert an incoming signal using some channel c_i to an outgoing signal using channel c_j, if $|i - j| \leq k$ where k is a small constant.

Exercise 2.1. Consider the physical topology of the network given in Figure 2.23. (This figure also appears in Chapter 1.) The problem is to have a single-hop network so that each end node in the network has a lightpath to all other end nodes in the network. Using the devices discussed in this chapter, except the wavelength converters, indicate how many wavelengths are needed. Show the design of any one end node and one router node.   □

Exercise 2.2. Consider the dynamic optical cross-connect switch shown in the network given in Figure 2.15. Let the cross-connect switch be a Beneš network[3] (see Figure 2.11) so that the output signals are as shown in Figure 2.24. The problem is to set up the switches in the Beneš network.

---

[3] useful websites for a tutorial introduction to the Beneš network —
http://www.dcs.gla.ac.uk/fp/workshops/fpw96/Sheeran.pdf,
http://www.essex.ac.uk/ese/staff/davidh/swtheory.pdf

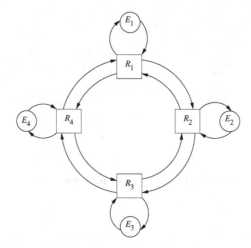

Fig. 2.23. The physical topology of a network

Draw a diagram showing a possible mode of each of the switches in the Beneš network. □

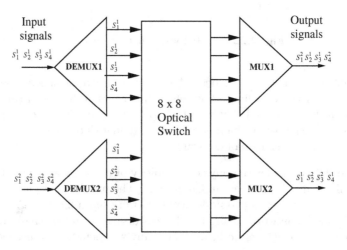

Fig. 2.24. A router node with an 8×8 Beneš network implementing a given routing

Exercise 2.3. Consider again the network shown in Figure 2.23. Let each router node in Figure 2.23 have a WC providing limited wavelength conversion as defined in Figure 2.22.

If a lightpath, from end node E_1 to end node E_4 uses channel number c_1 on the fiber $E_1 \rightarrow R_1$ and the route $E_1 \rightarrow R_1 \rightarrow R_2 \rightarrow R_3 \rightarrow R_4 \rightarrow E_4$

through the physical topology, what channel numbers can this lightpath use on the edges $R_1 \rightarrow R_2$, $R_2 \rightarrow R_3$, and $R_3 \rightarrow R_4$?   □

2.4 Bibliographic Notes

In this chapter, a brief description of some optical components has been given. A vast literature of work in this area is available. For an introduction to the topic, references [175, 190, 191, 221, 326] are useful. A good coverage of optical hardware, in the context of WDM network design, is available in [285, 288, 324, 358]. More detailed and technical descriptions of optical components are available in [113, 114, 171, 173, 174, 176, 253, 299, 382, 391]. A review of the emerging technologies in optical hardware appears in [255]. Wavelength converter technology has been reviewed in [122, 429]. Another implementation of limited wavelength translation appears in [350].

3

WDM Network Design

This chapter contains an overview of different types of WDM networks and fundamental problems in designing such networks. The simplest type of WDM networks, from the technological point of view, is the broadcast-and-select network. Broadcast-and-select networks are suitable for smaller networks, typically local area networks. For larger networks, wavelength-routed networks are more appropriate since wavelength routing uses optical signals that are sent from the source to the destination through a fairly small number of intermediate nodes without broadcasting to all the nodes in the network. Second-generation optical networks are wavelength-routed networks that have the potential to be more cost effective than first-generation networks. In designing wavelength-routed networks, major problems include determining which end nodes will be connected by a lightpath, how each lightpath should be routed through the physical topology, and effective use of wavelength converters in improving the throughput of optical networks.

3.1 Broadcast-and-Select Networks

A broadcast-and-select network [286] is typically a local network with a relatively small number, say $\mathcal{N}_E$, of end nodes, each end node equipped with one or more transmitters and receivers. All the end nodes are connected to a passive star coupler (Figure 1.7). Such a network is capable of supporting both unicast and multicast[1] communication and may be either a single-hop or a multihop network. A simple situation is one where each end node has a single tunable transmitter and a single receiver with a tunable filter. If this is a single-hop network that supports unicast communication from end node E_s to end node E_d, for all s, d, $1 \leq s, d \leq \mathcal{N}_E$, every source-destination pair (E_s, E_d) is assigned a unique wavelength λ_{sd}. For unicast communication from

---

[1] Multicast communication is communication from one source to a set of destinations.

E_s to E_d, the transmitter in E_s as well as the receiver in E_d have to be tuned to wavelength λ_{sd}. Then, the transmitter in end node E_s broadcasts the data to be communicated to all end nodes in the network. Since only the receiver in end node E_d is tuned to wavelength λ_{sd}, only this receiver picks up (or receives) the data from end node E_s. For multicast communication from end node E_s to end nodes $E_{d_1}, E_{d_2}, \ldots, E_{d_m}$, for some $m \leq \mathcal{N}_E$, the transmitter in E_s has to be tuned to some wavelength, say λ. All the receivers in end nodes $E_{d_1}, E_{d_2}, \ldots, E_{d_m}$ also have to be tuned to the same wavelength λ. The receivers in all other end nodes must be tuned to wavelengths other than λ. No other end node must use λ for data transmission until this communication is over. At this stage, the transmitter in end node E_s can start transmitting. Only the nodes $E_{d_1}, E_{d_2}, \ldots, E_{d_m}$ will receive the data communicated by node E_s.

To support unicast, simultaneous communication in a network with $\mathcal{N}_E$ end nodes, where each end node has one tunable transmitter and one tunable receiver, at least $\mathcal{N}_E(\mathcal{N}_E - 1)$ distinct wavelengths are needed, since there may be simultaneous communication involving any number of the $\mathcal{N}_E(\mathcal{N}_E-1)$ possible source-destination pairs and each communication is broadcast to all the end nodes in the network and must involve distinct wavelengths. Such a network cannot handle many end nodes since the signal transmitted by the source of any communication has to be broadcast to all the end nodes in the network, so that the power requirements in a large network will be excessive. However, such a network does not require any expensive optical devices and involves only a passive star coupler (described in Section 2.3.2), a device that is relatively cheap and robust.

One of the difficulties in a single-hop broadcast-and-select network is that $\mathcal{N}_E(\mathcal{N}_E - 1)$, the number of wavelengths needed, increases rapidly with $\mathcal{N}_E$. Another difficulty is that tunable transmitters and receivers which can quickly change from one wavelength to another are not available. One solution to the latter problem is to equip each end node with $\mathcal{N}_E - 1$ fixed wavelength transmitters and receivers. However, it is expensive to put in $\mathcal{N}_E - 1$ transmitters and receivers in each end node.

These problems have been addressed by *multi-hop broadcast-and-select networks*. In a multi-hop broadcast-and-select network, each end node is allowed to communicate with a very limited number, say k_B ($k_B << \mathcal{N}_E$), of other end nodes. This can be easily achieved by equipping each end node with k_B fixed wavelength transmitters and receivers. By adjusting the wavelengths of the transmitters and receivers, the set of end nodes that any given end node can communicate with is fixed in advance. If end node E_p can communicate directly with end node E_q, then there exists a wavelength λ_{pq} (unique for all ordered pairs (E_p, E_q) where E_p can communicate with E_q) such that end node E_p (E_q) has a transmitter (receiver) tuned to wavelength λ_{pq}. To send data from E_p to E_q, end node E_p modulates the transmitter tuned to wavelength λ_{pq} with the data to be communicated and then broadcasts the data.

Only end node E_q, having a receiver tuned to λ_{pq}, will receive this signal. This receiver will convert the optical data to electronic form.

In multi-hop broadcast-and-select networks, communication from a source E_s to a destination E_d means that a number of other end nodes $E_1^{sd}, E_2^{sd}, \ldots,$ E_r^{sd} participate in the communication. The idea is that E_s will send the data to node E_1^{sd}, E_1^{sd} will send the data to E_2^{sd}, $\ldots$, E_r^{sd} will send the data to the destination E_d. To achieve this, it must be possible to find, for all source-destination pairs (E_s, E_d), a sequence of end nodes $E_1^{sd}, E_2^{sd}, \ldots, E_r^{sd}$ so that E_s can communicate with E_1^{sd}, E_1^{sd} can communicate with E_2^{sd}, $\ldots$, E_r^{sd} can communicate with E_d. To send data from E_s to E_d, $r+1$ separate communications are involved. At each intermediate end node $E_i^{sd}, 1 \le i \le r$, there is an opto-electronic conversion to convert the optical signals sent from the preceding end node into electronic form, followed by an electro-optical conversion to convert the electronic data into the optical signals for communication to the following end node. Since electronic speed is much less than optical speed, there is a delay involved at each intermediate node. It is therefore desirable to have a small value of r.

To capture these limitations, it is convenient to consider the logical topology, discussed earlier (Section 1.1.2), corresponding to the physical topology of a multi-hop broadcast-and-select network and the k_B lightpaths from each end node. The directed graph G_L representing the logical topology of a multi-hop broadcast and select network must be connected, so that each end node can communicate with any other end node. To ensure that the value of r is low for all values of E_s and E_d, it is necessary to ensure that the diameter [42] of graph G_L is as small as possible. It is also important to ensure that such a graph can be constructed for any number $\mathcal{N}_E$ of end nodes in the network. In Chapter 6, the design of such networks will be explored.

Exercise 3.1. Consider the logical topology shown in Figure 3.1. Show how a broadcast-and-select network having this logical topology may be used to communicate from E_4 to E_5 and from E_2 to E_3 using a minimum number of stages of opto-electronic and electro-optical conversions (in other words, the minimum number of hops). A formal way to find such a route is discussed in Appendix 2.   □

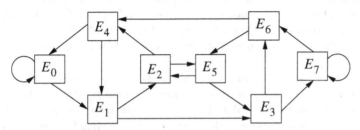

Fig. 3.1. The logical topology of a broadcast-and-select network

3.2 Wavelength-Routed Networks

When designing a WDM network, it is always possible for each lightpath to span only one fiber. In this type of a network, the fiber, carrying one or more optical signals, is simply replacing a copper link connecting two adjacent nodes in the network. The first-generation optical networks did exactly that. The development of OADM, OXC, and OLT have made possible second-generation optical networks where optical signals from end node E_x may traverse a number n, $n \geq 1$, of fibers to reach end node E_y without being converted to electrical signals at any intermediate end node in the route from E_x to E_y.

Figure 2.5 may be refined by including OADM, OXC, and OLT — optical devices reviewed in Chapter 2.

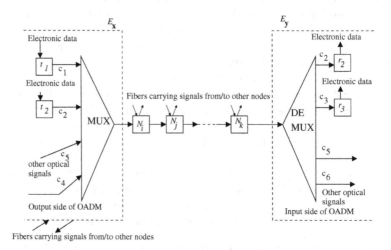

Fig. 3.2. Data communication in a WDM network

Figure 3.2 is a more detailed version of the part of a WDM network depicted in Figure 2.5 showing two typical end nodes E_x and E_y where E_x is communicating with E_y. In this diagram, end node E_x has two transmitters t_1 and t_2 tuned to wavelengths λ_1 and λ_2 (corresponding to channels c_1 and c_2) respectively. End node E_y has two receivers r_2 and r_3 tuned to wavelengths λ_2 and λ_3 (corresponding to channels c_2 and c_3) respectively. It is assumed in this discussion that no wavelength conversion takes place in this communication from E_x to E_y.

End node E_x may communicate data to E_y using channel c_2 as follows. In end node E_x, the data to be communicated to node E_y is in electronic form. Transmitter t_2 converts this data to optical signals using channel c_2. Similarly transmitter t_1 converts some other data to optical signals using channel c_1, for communication to some other end node. The multiplexer of an OADM within E_x combines these two optical signals using channels c_1 and c_2 with other optical signals using channels c_4 and c_5, coming from other nodes. These signals from other end nodes are using E_x as an intermediate end node. The output of the multiplexer, consisting of optical signals using channels c_1, c_2, c_4, and c_5, is carried by one of the outgoing fibers from E_x to the next node N_i. The node N_i may either be an end node (which can be a source or a destination of data) or be a router node whose only responsibility is to direct signals from different input fibers carrying incoming signals to appropriate output fibers carrying outgoing signals. Continuing in this way, the optical signal using c_2 is routed through a number of intermediate nodes $N_i, N_j, \ldots, N_k$ towards the destination E_y. Each intermediate node has a number of input ports connected to fibers carrying incoming signals from other end nodes or routers and a number of output ports carrying outgoing signals to other end nodes or routers. In Figure 3.3, a typical intermediate node N_r is shown. The node N_r is an end node where the OADM is dropping two optical signals at channels c_p and c_q. Node N_r is the destination of these two signals and the receivers r_p and r_q convert them to electronic data. The OADM in N_r is also adding two optical signals using channels c_i and c_j generated by the transmitters t_i and t_j. The intermediate node N_r is therefore sending two sets of data, possibly to two different destinations, and receiving two sets of data.

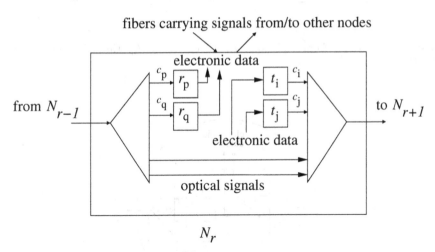

Fig. 3.3. An intermediate node in data communication from E_x to E_y

In this process, there may be optical amplifiers (not shown in Figure 3.2) to boost the optical signals to the desired level. At end node E_y, the demultiplexer part of an OADM extracts the signals using channels c_2 and c_3 from the signals carried by one of the fibers carrying incoming signals to E_y. At E_y, the signal using channel c_2 is converted by the receiver r_2 tuned to wavelength λ_2 (corresponding to channel c_2) to electronic form.

While OADMs are used for ring networks, in networks with a complex interconnection, it is convenient to use OXCs which allow efficient cross-connections involving many incoming and outgoing fibers, each carrying a large number of optical signals. In this scenario, a distinction is made between nodes which are used purely for cross-connections and the nodes which add or drop optical signals. To simplify the network model and to include this second type of node, it is assumed, from now on, that the end nodes will only have an OLT and that there is another kind of node, called *router node*, which only has an OXC (Figure 3.6). In this model, an end node will not have any cross-connecting capability while a router node only has cross-connecting capability. From now on, the ith node will be depicted by R_i if it is a router node.

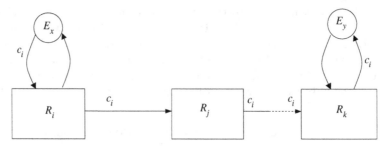

Fig. 3.4. A route from end node E_x to end node E_y through router nodes $R_i, R_j, \ldots, R_k$

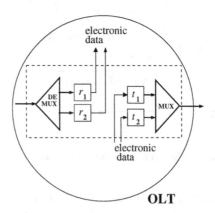

Fig. 3.5. A typical end node in a WDM network

The new model of a WDM network is shown in Figure 3.4. Each of the end nodes in Figure 3.4 has an OLT. A typical end node is shown in Figure 3.5 with two incoming optical signals decoded by receivers r_1 and r_2. As shown in

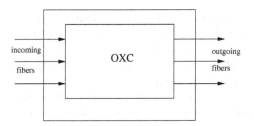

Fig. 3.6. A router node in a WDM network

Figure 3.4, each end node is connected to one or more router node(s), where each router node is an OXC as shown in Figure 3.6. In this model, a lightpath from a source E_x to a destination E_y, using the route $E_x \to R_i \to R_j \to \ldots \to R_k \to E_y$, starts from end node E_x, passes through router nodes R_i, R_j, ..., R_k and ends at end node E_y. There are two scenarios to consider in such wavelength-routed networks.

Scenario 1: In this case, no node in the network has a wavelength converter. This is the scenario considered in Figures 3.2 and 3.4. Here, a lightpath from a source E_x to a destination E_y, on all fibers in its route $E_x \to R_i \to R_j \to \ldots \to R_k \to E_y$, uses the same channel c_i. To do this, no other signal on the fibers $E_x \to R_i, R_i \to R_j, \ldots, R_k \to E_y$ is allowed to use the channel c_i. This is an important restriction and is called the *wavelength continuity constraint*, meaning that the carrier wavelength of a lightpath does not change from fiber to fiber. In a network satisfying the wavelength continuity constraint, when considering a route for a lightpath, some channel $c_i, 1 \leq i \leq n_{ch}$, must be available on every fiber on the route, where n_{ch} is the maximum number of channels each fiber in the network can accommodate.

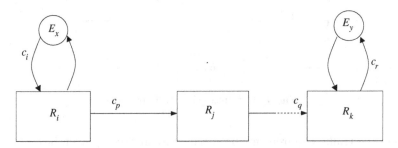

Fig. 3.7. A route from E_x to E_y in a wavelength-convertible network

Scenario 2: In this case, each node in the network has a full wavelength converter. Looking once again at the problem of communicating from E_x to E_y, using the route $E_x \to R_i \to \ldots \to R_k \to E_y$ (Figure 3.7), the channel used by the lightpath from E_x to E_y changes, from channel c_i on the fiber $E_x \to R_i$, to channel c_p on the fiber $R_i \to R_j$, ..., to channel c_q on the fiber to R_k, and to channel c_r on the fiber $R_k \to E_y$. Since there is a wavelength converter at node R_i, the incoming signal having a carrier wavelength λ_i may be converted to an outgoing signal having any carrier wavelength λ_p (i.e., the channel may be changed from channel c_i to any channel c_p), $1 \leq p \leq n_{ch}$, for transmission using the fiber R_i to R_j. To do this, no other signal on the fiber $R_i \to R_j$ must use the channel c_p. Since every node has a full wavelength converter, the wavelength continuity constraint is not needed. When implementing a new lightpath and a route is under consideration, it is only necessary to ensure that the total number of lightpaths on each fiber in the route is less than n_{ch}.

Other variations on Scenario 2 are possible, depending on the type of wavelength conversion available in the network. These variations include fixed wavelength conversion or limited wavelength conversion at every node in the network. Some networks have wavelength converters (full, fixed, or limited) in only a limited number of nodes.

3.2.1 Advantages of Second-Generation WDM Networks

There is considerable scope to reduce the cost of an optical network using OADMs and OXCs. To illustrate this, an example using a network with four end nodes E_1, E_2, E_3 and E_4 is shown in Figure 3.8. It is assumed that the wavelength continuity constraint is applicable in both networks.

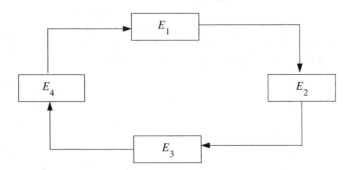

Fig. 3.8. A network with four nodes

Let the traffic requirements be specified by matrix T_1 given below.

$$T_1 = \begin{bmatrix} 0.0 & 1.0 & 2.0 & 2.0 \\ 2.0 & 0.0 & 2.0 & 1.0 \\ 2.0 & 2.0 & 0.0 & 2.0 \\ 1.0 & 3.0 & 1.0 & 0.0 \end{bmatrix}$$

Depending on whether or not OADMs and OXCs are available, two scenarios have been considered. In the first scenario OADMs and OXCs are not used. Each end node has an OLT as shown in Figure 3.9. In the second scenario OADMs are available in every node as shown in Figure 3.10. In Figure 3.10, only two transmitters and receivers are shown in each node. The actual number of transmitters (receivers) will depend on the number of lightpaths starting from (terminating at) a given node.

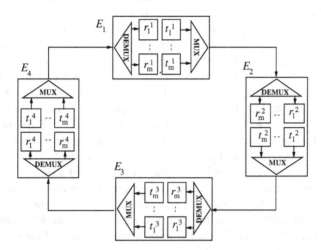

Fig. 3.9. A WDM network without any OADM or OXC

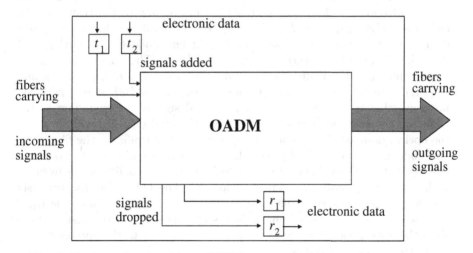

Fig. 3.10. An end node with an OADM, two transmitters, and two receivers

To analyze the two scenarios in detail, the cost of the hardware associated with the fiber $E_1 \rightarrow E_2$, consisting of the multiplexer part of E_1 and the demultiplexer part of E_2, will be examined below.

Scenario 1: The entire traffic from end node E_1 to all other end nodes (which requires a total of five lightpaths — 50 Gbps assuming that each lightpath can carry up to 10 Gbps) has to be communicated to end node E_2. In addition, all traffic that uses the edge from end node E_1 to node E_2, has to be received at end node E_1, converted from optical signals to electronic data, reconverted back to optical signals and then forwarded to end node E_2. For instance, the traffic from end node E_4 to end node E_2, requiring three lightpaths, must be received and converted to electronic form at end node E_1. Then, this has to be converted back to optical form and forwarded to end node E_2. Other traffic that has to be routed in this manner corresponds to the traffic for the ordered end node pairs (E_4, E_3) requiring one lightpath and (E_3, E_2) requiring two lightpaths. In summary, a total of 11 lightpaths must start from end node E_1 and terminate at end node E_2. To estimate the cost due to the traffic on the fiber from end node E_1 to end node E_2, on the output side of end node E_1, one OLT is needed to convert 11 sets of data into 11 streams of optical signals and multiplex these 11 lightpaths to be transmitted using the fiber from end node E_1 to end node E_2. Similarly, on the input side of node E_2, one OLT is needed to demultiplex 11 lightpaths and to convert each of these 11 streams into electronic form. The situations for the other end nodes are similar.

Scenario 2: In this case, the traffic from end node E_1 to all other end nodes (which requires a total of five lightpaths) have to be added, by using five transmitters to the other optical signals carried by the fiber from end node E_4 to end node E_1. The optical signals corresponding to the traffic for the end node pairs (E_4, E_3), (E_4, E_2), and (E_3, E_2) will remain in the optical domain for transmission to end node E_2. At end node E_2, an OADM will drop the signals whose destination is end node E_2. Thus the optical signals corresponding to the traffic for the end node pairs (E_1, E_2), (E_3, E_2), and (E_4, E_2) will be dropped at end node E_2 by the OADM at end node E_2 for conversion to electronic data. The optical signals corresponding to the traffic for the end node pairs (E_1, E_3), (E_1, E_4), and (E_4, E_3) will remain in the optical domain. To estimate the cost due to the traffic on the fiber from end node E_1 to end node E_2, an OLT is needed at end node E_1 to create five lightpaths. The OADM at end node E_1 combines six lightpaths from E_4 with the five lightpaths generated within E_1. The OADM at E_2 drops six wavelengths and passes on the remaining five wavelengths to end node E_3.

The cost of the second scenario is less than the cost of the first scenario since the optical switches are less expensive compared to the opto-electronic or electro-optical converters.

An important point to note is that, in the above example, the traffic between any pair of end nodes is a multiple of the capacity of a lightpath. When this is not true, the outcome of the cost analysis may be quite different. To illustrate this, T_2, another example of a traffic matrix for the same physical topology shown in Figure 3.8, is given below.

$$T_2 = \begin{bmatrix} 0.0 & 0.1 & 0.2 & 0.2 \\ 0.2 & 0.0 & 0.2 & 0.1 \\ 0.2 & 0.2 & 0.0 & 0.2 \\ 0.1 & 0.1 & 0.1 & 0.0 \end{bmatrix}$$

If OADMs are used in the same way as in the previous example, the hardware cost in the second scenario is the same as before. In the first scenario the traffic received by end node E_1 to be communicated to end node E_2 again is due to the pairs of end nodes (E_1, E_2), (E_3, E_2), (E_4, E_2), (E_1, E_3), (E_4, E_3), and (E_1, E_4) — a total of 0.9 units. Since this is less than 1.0, a single lightpath is sufficient to carry this traffic to end node E_2. The cost for the first scenario in this example, for the fiber from E_1 to end node E_2, is an OLT on the output side of E_1 to inject this signal into the fiber and an OLT on the input side of node E_2 to extract the electronic signal from the single lightpath. Since the unit of transmission at the optical signal level is one lightpath, the second scenario is not economical, in this example, since many lightpaths, each carrying a fraction of its capacity (typically 2.5 or 10 Gbps), are needed. In summary, there may very well be situations where the use of OADMs and OXCs are not justified. OADMs and OXCs do offer additional design choices and, depending on circumstances, give more economical networks. Optimal design of wavelength-routed networks is complicated because of the large number of possible ways to design the network.

3.2.2 Single-hop and Multi-hop Networks

The notion of single-hop and multi-hop networks has been introduced already in Chapter 1. When designing such networks, there are many factors to consider. For single-hop networks, the logical topology is already available — it is a completely connected graph. The issue for single-hop networks is how to implement all the lightpaths. This is called the route and wavelength assignment (RWA) problem that will be discussed in detail in the next section (Section 3.3) and in Chapters 4 and 5. For multi-hop networks, an optimal logical topology has to be determined and the RWA done for each logical edge. As mentioned already (in Section 1.1.2), in general, there are many logical topologies for the same physical topology. The logical topology must correspond to a connected graph so that it is possible to guarantee that any end node may communicate with any other end node. The logical topology must be such that the traffic between all end nodes may be handled without exceeding the data-carrying capacity of each lightpath (currently either 2.5 Gbps or 10 Gbps depending on the technology). There are limitations on the number of lightpaths starting from an end node or terminating at an end node, determined by the number of transmitters and receivers at that end node. Each fiber can support up to n_{ch} lightpaths. The issue of whether wavelength converters (discussed in Section 3.4) are available also has to be taken into account. Since the design problems are complicated, a lot of work has been done in this area. Chapters 4 and 5 deal with the RWA problem. Chapter 7 deals with the issue of logical topology design and also the issue of routing the traffic over a given logical topology.

3.3 Route and Wavelength Assignment Problem in WDM Networks

To establish a lightpath from end node E_x to end node E_y, its route through the physical topology and, for every fiber in its route, the channel assigned to the lightpath are important. As explained in Section 1.1.2, such a lightpath will be denoted by $E_x \Rightarrow E_y$. In networks where the wavelength continuity constraint is satisfied, the same channel will be assigned to a lightpath on every fiber in its route. In networks with full wavelength converters at each node, the channel used by a lightpath may vary from one fiber to another. Each fiber, as mentioned earlier, can support a maximum number n_{ch} of channels. The value of n_{ch} is determined by the fiber used and the available technology. This means that the routes assigned to the lightpaths should be such that the number of lightpaths sharing any fiber does not exceed n_{ch}. The channel assigned to lightpaths should be such that two lightpaths sharing a fiber are never assigned the same channel. This problem is called the *route and wavelength assignment problem* (RWA).

3.3.1 Static and Dynamic Lightpath Allocation

Two versions of the RWA problem have been considered by researchers. If the set of lightpaths to be set up is known in advance (either specified as such to the network designer or computed by solving some optimization problem), the problem is called the *off-line RWA* problem. In this formulation the traffic demand is known in advance and is not expected to change in the near future. This is also called *static lightpath allocation* since the lightpaths, once established, are not modified until the traffic pattern changes sufficiently to warrant a different set of lightpaths. The lightpaths in such a scheme exist for relatively long periods of time until it is decided to run the RWA algorithm with a new set of lightpaths that will replace the existing set of lightpaths to accommodate the changed traffic pattern.

The other version of the problem is to set up the lightpaths on demand and is called *online RWA* (also called *dynamic lightpath allocation*) problem [166, 343]. In this problem, requests for data communication are considered as and when they occur. Each request is to communicate, from a specified source E_x to a specified destination E_y, a certain amount of data (or equivalently, to communicate for a certain amount of time). When the request for communication arrives, the request has to be processed to determine, if possible, an optimum route and a channel on every edge of the route to service the request. In a scheme for online RWA, all existing lightpaths have to be considered when creating a new lightpath to support a new communication request. When the communication is over, all resources devoted to this communication have to be reclaimed for possible use in future communication. In other words, in dynamic allocation, lightpaths are set up when needed and are taken down when the communication is over.

A dynamic lightpath allocation scheme does not guarantee that the communication from a source E_x to a destination E_y will always be possible. If the conditions for establishing a lightpath are not satisfied, the communication will be blocked and may possibly be attempted again, after some delay, with the expectation that the conditions for establishing a lightpath are now satisfied. Clearly, in realistic situations, it is necessary to ensure that the probability that a communication will be blocked, called *blocking probability*, is very low.

In this section, the case of dynamic lightpath allocation in a network with no node equipped with a wavelength converter will be considered, so that every lightpath satisfies the wavelength continuity constraint. The case of a network with each node equipped with one (or more) wavelength converter(s) will be considered in Section 3.4. In response to a request for communication from a source E_x to a destination E_y in such a network, the RWA problem is to determine a route ρ_{xy} from E_x to E_y and a channel c_i such that the channel c_i is not currently in use in any edge on the route ρ_{xy} (Figure 3.4).

If the attempt for RWA is successful, the next step is to set up the lightpath using the route ρ_{xy} and channel c_i. The step involves reserving channel number c_i on every edge in this route ρ_{xy}. The OLTs at the source, the destination, and the OXCs in the router nodes on the selected route have to be set so that optical signals using the channel c_i are forwarded from the source E_x and are routed to the destination E_y. Communication starts when the OLTs in the source and the destination end node and the OXCs at all the intermediate router nodes have been set properly. When the communication is over, the network management software, in some way, must keep track of the fact that the communication has terminated so that the channels dedicated to this communication are marked as available for processing future requests.

Route and wavelength assignment is a complex problem and, for nontrivial networks, cannot be solved optimally in a reasonable time. Exact and approximate solutions of the RWA problem will be reviewed in Chapters 4 and 5.

3.4 Wavelength-Convertible Networks

A wavelength-convertible network has wavelength converters at some (or all)
nodes. It is easy to devise situations where a network, following the wavelength continuity constraint, fails to establish a new lightpath while the same
network, where nodes are equipped with wavelength converters, can succeed
in establishing the same lightpath. This means that wavelength-convertible
networks are capable of making better utilization of the network resources.
The following example illustrates the advantage of networks with wavelength
converters.

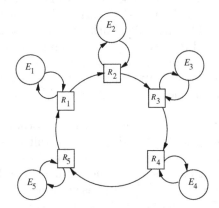

Fig. 3.11. A unidirectional ring network

A WDM all-optical unidirectional ring network is shown in Figure 3.11.
Each fiber in the network can support three channels c_1, c_2, and c_3. Dynamic lightpath allocation has been used in this network to establish four
lightpaths L_1, L_2, L_3, and L_4. Lightpath L_1 (respectively L_2, L_3, and L_4)
has been assigned channel c_1 (respectively c_2, c_1, and c_3) and route $E_1 \rightarrow$
$R_1 \rightarrow R_2 \rightarrow R_3 \rightarrow E_3$ (respectively $E_2 \rightarrow R_2 \rightarrow R_3 \rightarrow R_4 \rightarrow E_4$,
$E_3 \rightarrow R_3 \rightarrow R_4 \rightarrow R_5 \rightarrow E_5$, and $E_1 \rightarrow R_1 \rightarrow R_2 \rightarrow R_3 \rightarrow E_3$). This is
shown in Figure 3.12. In Figure 3.12, if lightpath L_i uses channel c_j on one
of the fibers in the route of the lightpath, a label $L_i(c_j)$ is shown next to the
fiber.

The problem of handling a request for a lightpath L_5 from E_3 to E_2 will
be discussed below, for two situations. In situation 1 (2) , no (each) router
node in the network has been equipped with a wavelength converter. In this
network there is only one route $E_3 \rightarrow R_3 \rightarrow R_4 \rightarrow R_5 \rightarrow R_1 \rightarrow R_2 \rightarrow E_2$.

Situation 1) The request cannot be allowed and the call must be blocked
    since there is no channel which is free on all fibers in the route $E_3 \rightarrow$
    $R_3 \rightarrow R_4 \rightarrow R_5 \rightarrow R_1 \rightarrow R_2 \rightarrow E_2$. This represents a wastage of network
    resources since unallocated channel(s) exist on all the fibers in the route

but the call is blocked since the same channel is not available in all the fibers in the route.

Situation 2) The request for a lightpath L_5 from E_3 to E_2 can be handled with the lightpath having channel c_3 on each fiber in the route $E_3 \rightarrow R_3 \rightarrow R_4 \rightarrow R_5 \rightarrow R_1$ and channel c_2 on the fibers in the route $R_1 \rightarrow R_2 \rightarrow E_2$. The situation after establishing the lightpath L_5 is shown in Figure 3.13.

This example shows that wavelength converters can reduce the blocking probability and improve the utilization of network resources.

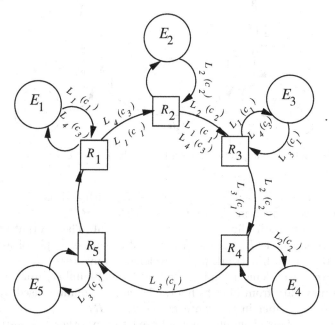

Fig. 3.12. A ring network with lightpaths L_1, L_2, L_3, and L_4

Clearly, the pattern of lightpaths already established determines whether or not a request for a new lightpath may be allowed. For instance, in situation 2, after the situation shown in Figure 3.13 has been established, if there is a request for a lightpath from E_1 to E_3, the request cannot be satisfied since all three channels c_1, c_2, and c_3 on the fiber $R_2 \rightarrow R_3$ have already been allocated.

In a mesh network, there are, in general, a number of paths between pairs of end nodes. If each router node in a mesh network is equipped with full wavelength converters, a dynamic lightpath allocation scheme, to establish a lightpath from end node E_x to end node E_y, will be unsuccessful only if, for each route from E_x to E_y, all n_{ch} channels, in at least one fiber in the route, have been already allocated to earlier requests for lightpath allocation. If there

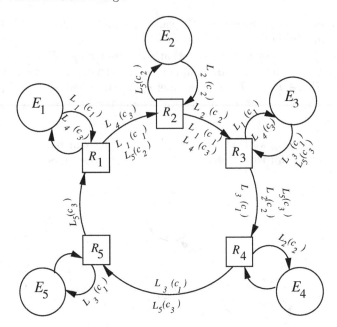

Fig. 3.13. A ring network with lightpaths L_1, L_2, L_3, L_4, and L_5

is any route $E_x \to R_i \to R_j \to \ldots \to R_k \to E_y$ with at least one unassigned channel on every fiber in the route $E_x \to R_i \to R_j \to \ldots \to R_k \to E_y$, the dynamic lightpath allocation will be successful. The lightpath from end node E_x to end node E_y shown in Figure 3.7 is an example of a lightpath established in a mesh network with wavelength converters.

The carrier wavelength (and hence the channel number) used by the new lightpath may vary from fiber to fiber. The channel number used by the new lightpath on any fiber in the route $E_x \to R_i \to R_j \to \ldots \to R_k \to E_y$ can simply be any channel number not allotted to any other lightpath on that fiber. In other words, in networks where each node is equipped with a full wavelength converter, for each lightpath to be established, only the route has to be determined such that every fiber in the route has at least one unutilized channel, and wavelength assignment is no longer a problem.

There has been some theoretical work on simplified models of all-optical networks using dynamic route and wavelength assignment [60, 230]. Barry and Humblet [32] investigated the effect of length of the route (i.e., the number of fibers in a lightpath), the switch size, and the number of fibers shared by lightpaths. A summary of the analysis in [32] on the effect of length of the route is given below.

Let end node E_x request a lightpath to end node E_y using a route involving H fibers and router nodes $R_1, R_2, \ldots, R_{H+1}$ as shown in Figure 3.14.

Each fiber, as mentioned before, can carry n_{ch} channels. It is assumed that node E_x (E_y) has a transmitter (receiver) available to send (receive) the

From other end–nodes

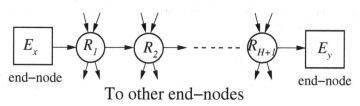

end–node   end–node

To other end–nodes

Fig. 3.14. A route from E_x to E_y in a wavelength-convertible network

signal and that the access fiber from (to) E_x (E_y) does not create a problem regarding the availability of a free channel. New requests for communication may come from other end nodes connected to router nodes $R_1, R_2, \ldots, R_{H+1}$. Any such request involving the router nodes $R_1, R_2, \ldots, R_{H+1}$ is termed an *interfering session*.

Two scenarios will be considered here. In the first scenario, each router node is equipped with a wavelength converter capable of full wavelength conversion. In this case, the request for a lightpath from E_x to E_y will be successful if there is a free channel on each fiber in the route from R_1 to R_{H+1}. Due to the presence of the wavelength converters, the lightpath may very well use one channel on the fiber $R_1 \to R_2$, a different channel on the fiber $R_2 \to R_3$, ..., and a different channel on the fiber $R_H \to R_{H+1}$. The request for a lightpath from E_x to E_y will be blocked if any of the fibers in the route has all n_{ch} channels already assigned to previous requests for lightpaths.

In the second scenario, there is no wavelength converter in the network. This means that the request for a lightpath from E_x to E_y will be successful if there exists some channel c_i, $1 \le i \le n_{ch}$ which has not been assigned to previous requests for lightpaths that use the fibers $R_1 \to R_2$, $R_2 \to R_3$, ..., and $R_H \to R_{H+1}$.

To analyze these two scenarios, a useful metric is the probability that a wavelength is used on a fiber in the route from router node R_1 to R_{H+1}. In the first (second) scenario of the analysis, Q_b (P_b) will denote the probability that the request for a lightpath from E_x to E_y will be blocked, q (p) the probability that a channel is used on any one of the fibers in the route from R_1 to R_{H+1}. Since q $\times n_{ch}$ (p $\times n_{ch}$) is the expected number of allotted channels on any of the fibers in the route from R_1 to R_{H+1}, q (p) is a measure of the *fiber utilization* in the two scenarios.

A simple analysis of the first scenario is given below. In this case, there is a probability q that a channel is used on any one fiber in the route from R_1 to R_{H+1}. The probability that all n_{ch} channels have been used on a given fiber is $q^{n_{ch}}$. The probability that all n_{ch} channels have not been used on a given fiber is $1 - q^{n_{ch}}$. This is the probability that the given fiber does not cause the request for a lightpath from E_x to E_y to be blocked. There are H fibers in the route, so that the probability that none of them will cause the request for a lightpath from E_x to E_y to be blocked is $(1 - q^{n_{ch}})^H$. Thus the probability that the request for a lightpath from E_x to E_y will be blocked is given by the following equation.

$$Q_b = 1 - (1 - q^{n_{ch}})^H \tag{3.1}$$

The above equation may be rewritten to define q for a given Q_b as follows:

$$q = [1 - (1 - Q_b)^{1/H}]^{1/n_{ch}} \approx (\frac{Q_b}{H})^{1/n_{ch}} \tag{3.2}$$

The approximation holds for small $\frac{Q_b}{H}$.

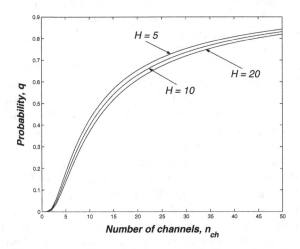

Fig. 3.15. Wavelength utilization increases with the number of wavelengths

In Figure 3.15, for a given blocking probability $Q_b = 10^{-3}$, the utilization q has been plotted against the number of channels n_{ch} for values of H varying from 5 to 20 [32]. As n_{ch} increases, the value of q tends to saturate, becoming 1 as n_{ch} increases to ∞. However, the value of H has a very small effect on q. This means that the length of the route has a small effect and it is a good idea to minimize the number of lightpaths on the edge carrying the maximum number of lightpaths by increasing the number of fibers traversed by a lightpath [32].

A similar analysis of the second scenario is given below. In this case there is a probability p that a channel is used on any one fiber in the route from R_1 to R_{H+1}. The probability that a given channel is available on a single fiber is $1 - p$ and the probability that the channel is available on all the fibers from E_x to E_y is $(1 - p)^H$. The probability that the request for a lightpath from E_x to E_y cannot use that channel is $[1 - (1 - p)^H]$. If the request for a lightpath from E_x to E_y is blocked, none of the n_{ch} channels is available. The probability P_b that the request for a lightpath from E_x to E_y will be blocked is therefore given by the following equation.

$$P_b = [1 - (1 - \mathfrak{p})^H]^{n_{ch}} \tag{3.3}$$

The above equation may be rewritten to define $\mathfrak{p}$ for a given P_b as follows:

$$\mathfrak{p} = 1 - (1 - P_b^{1/n_{ch}})^{1/H} \approx -\frac{1}{H}\ln(1 - P_b^{1/n_{ch}}) \tag{3.4}$$

The approximation holds for large H and $P_b^{1/n_{ch}}$.

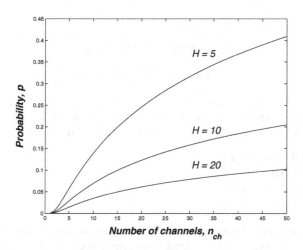

Fig. 3.16. Wavelength utilization is significantly affected by H

If the fiber utilization $\mathfrak{p}$ is plotted against n_{ch} (shown in Figure 3.16) for various values of H, it is clear that the length of the route affects the utilization very significantly [32]. This means that in a network without wavelength converters each lightpath should reduce the number of fibers it traverses [32].

The gain $G = \mathfrak{q}/\mathfrak{p}$ may be used to evaluate the benefit of wavelength converters. For a given blocking probability, (i.e., $P_b = Q_b$), G may be approximated as follows.

$$G = \mathfrak{q}/\mathfrak{p} = \frac{[1 - (1 - P_b)^{1/H}]^{1/n_{ch}}}{1 - (1 - Q_b^{1/n_{ch}})^{1/H}} \approx H^{1-(1/n_{ch})}\frac{P_b^{1/n_{ch}}}{-\ln(1 - P_b^{1/n_{ch}})} \tag{3.5}$$

The plots of G, given in Figure 3.17, against n_{ch} for $P_b = 10^{-3}$ show that the benefits of wavelength converters show a peak for a small value of n_{ch} and then decrease as n_{ch} increases. The improvement when wavelength converters are used is much more significant for large values of H.

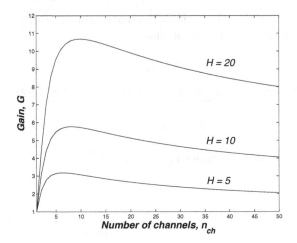

Fig. 3.17. Wavelength changing gain

Exercise 3.2. Consider T_3, another traffic matrix for the same physical topology shown in Figure 3.8. Here, the unit is the capacity of a lightpath.

$$T_3 = \begin{bmatrix} 0.0 & 1.3 & 2.2 & 1.4 \\ 2.1 & 0.0 & 2.5 & 1.1 \\ 1.2 & 2.4 & 0.0 & 1.2 \\ 0.1 & 0.1 & 0.1 & 0.0 \end{bmatrix}$$

Using a minimum number of transmitters and receivers and without using any wavelength converters (so that the wavelength continuity constraint is followed), design the end nodes $E_1, E_2, E_3,$ and E_4. Show the channels you are using for each lightpath. ☐

Exercise 3.3. Consider the network shown in Figure 3.18 with lightpaths as shown. Lightpaths $L_1, L_4,$ and L_5 use channel c_1, lightpath L_2 uses channel c_3, and lightpath L_3 uses channel c_2. Each fiber can support only three channels $c_1, c_2,$ and c_3.

I) Assuming that no router node has a wavelength converter and using dynamic lightpath allocation, show how to set up, if possible, a lightpath from E_3 to E_1 and then a lightpath from E_1 to E_3.

II) Assuming that each router node has a wavelength converter and using dynamic lightpath allocation, show how to set up, if possible, a lightpath from E_1 to E_3, and then another lightpath from E_4 to E_2.

If it was not possible to set up a lightpath in any of these problems, explain why that happened. ☐

Exercise 3.4. Consider again the network shown in Figure 2.23 and suppose that dynamic lightpath allocation is being used. There are a number of existing

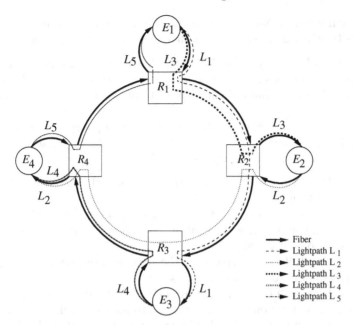

Fig. 3.18. Some lightpaths on the physical topology shown in Figure 2.23

lightpaths, and a request for a new lightpath is being considered. Consider two situations as follows:

Situation 1: Each router node in Figure 2.23 has a WC giving limited wavelength conversion as defined in Figure 2.22.

Situation 2: No router node in Figure 2.23 has a WC.

Create a set of existing lightpaths and a request for the new lightpath such that,

- in Situation 1, the new request will be successful and
- in Situation 2, the new request will not be successful.

□

Exercise 3.5. Consider the same network and the two situations given in Exercise 3.4. Create a set of existing lightpaths and a request for the new lightpath such that in both Situations 1 and 2, the new request will not be successful. □

Exercise 3.6. Consider the physical topology of a ring network with n end nodes $E_1, E_2, \ldots, E_n$ similar to the ring network with end nodes $E_1, E_2, \ldots, E_5$ as shown in Figure 2.23.

Your task is to study, using simulation, the improvement in the blocking probability if each router node is equipped with a full wavelength converter, so that you consider two situations as follows.

Situation 1: No router node has a wavelength converter.
Situation 2: Each router node has a full wavelength converter.

The network uses dynamic lightpath allocation. Each request for a communication specifies a source end node, a destination end node, and a duration of call specified as p units of time. In each unit of time, exactly one call is generated.

Assume that each fiber in the network can support n_{ch} channels. Use $n_{ch} = 8, 16, 32, 64$ and $n = 5, 10, 20, 40$ in your simulation experiments. Start with a network having no lightpaths.

Step 1) Repeat Steps 2 – 5 1,000 times or until the last five requests are not successful.

Step 2) Generate a request for a connection by randomly generating a source end node E_i, a destination end node E_j, $1 \leq i, j \leq n$, and a duration of call $p, 1 \leq p \leq 10$.

Step 3) Check if the lightpaths in the network are such that a lightpath corresponding to the call may be set up.

Step 4) If the call can be set up, update your database to indicate that a new lightpath is now set up in response to the call. Update your statistics that the call was successful. Remember that the call will take p units, so your program has to update the database after p iterations, following the current iteration, to take account of the fact that the lightpath only exists for p iterations.

Step 5) If the call could not be set up, update your statistics that the call was not successful.

After repeating this simulation 100 times, for each n_{ch} and n, and for networks with and without full wavelength converters, draw your conclusions about the effectiveness of wavelength conversion.   □

3.5 Bibliographic Notes

The topics covered in this chapter will be discussed in more detail in later chapters and a review of research in those topics will be included in the corresponding chapters.

The following topics have not been discussed in this book but have received some attention from the research community:

i) Light-trail networks,
ii) Burst switching networks,
iii) Multicasting networks.

This section contains a review of some work on wavelength conversion and these three topics.

3.5.1 Wavelength-Convertible Networks

There has been considerable work on wavelength-convertible networks [32, 60, 230, 310, 314, 322, 373, 374, 375]. Some early studies have shown that if every node can provide complete wavelength conversion, the blocking probability can be significantly reduced [230, 322]. It has been reported that, with fixed routing, wavelength conversion provides about 30 – 40% improvement in blocking probability [314]. A review/survey of the enabling technologies, design methods, how to use the technology in a wavelength routed network, the potential benefits under various network conditions, and some analytical models used in wavelength-convertible networks appears in [310].

The effect of wavelength conversion when the input traffic is non-Poisson has been investigated in [376]. The paper has discussed why a Poisson distribution is not necessarily adequate for characterizing network traffic. A methodology for studying the benefits of wavelength converters for non-Poisson dynamic traffic, using the Pascal approximation, has been presented. For the two topologies considered (mesh-torus and hypercube), the conclusion was that wavelength conversion can only "mildly alleviate" performance degradation for certain scenarios. For other scenarios, wavelength conversion does give significant improvement [376].

As discussed in Section 2.3.6, limited wavelength conversion is a practical and feasible way to give some flexibility with respect to carrier wavelengths (equivalently channel numbers) used in different fibers in the route of a lightpath. The study in [423] gives a simple, approximate probabilistic analysis for single paths in isolation. Another important early work in using limited conversion in rings appears in [169].

Since full wavelength converters (FWCs) are expensive, it is useful to have some nodes (either randomly selected, or selected based on the pattern of lightpaths to be set up) equipped with FWC, with no capability for wavelength conversion for the remaining nodes. This scenario is called *sparse wavelength conversion*, or *partial wavelength conversion*. An early study in this subject appears in [239]. Work summarized in Section 5.5 concludes that, when compared with complete wavelength conversion, sparse wavelength conversion can give nearly the same blocking probability when an "appropriate" number of nodes with complete wavelength converters is used. An important problem in designing sparse wavelength conversion networks is to develop strategies to select which nodes in the network should be equipped with FWCs for maximum benefit. However, [450] reports that the advantages of sparse wavelength conversion is significant only when random channel allocation is used. Benefits similar to that of sparse wavelength conversion may be obtained by the first-fit or the most-used policy.[2] In a network with sparse wavelength conversion, the placement of the wavelength converters is an important issue.

---

[2] The random allocation and the first-fit and the most-used policies will be discussed in Chapter 5.

Since the advantages of wavelength conversion, including sparse wavelength conversion, must be investigated in the context of RWA, the work on wavelength conversion, including placement of wavelength converters, will be reviewed in Section 5.5.

3.5.2 Light-Trail Networks

A lightpath, as explained in this chapter, allows point-to-point communication, from the source to the destination of the lightpath. Light-trail is a somewhat different approach for optical network design. If there is a light-trail from node A_0 to A_p through intermediate nodes $A_1, A_2, \ldots, A_{p-1}$, then this light-trail allows data communication from any node A_i to A_j, $0 \leq i < p$, $i < j \leq p$. This light-trail may be viewed as an optical bus where intermediate nodes $A_1, A_2, \ldots, A_{p-1}$ also have access to the bus [183]. The architecture and the hardware requirements have been explored in [183]. At each intermediate node A_j, $0 < j < p$, the incoming optical signal to the node is split into two parts. One part is used for local processing at node A_j, so that any data intended for node A_j may be extracted and processed. The remaining part of the incoming signal is combined with data originating at node A_j for communication to nodes $A_{j+1}, \ldots, A_p$. This forms the outgoing signal from node A_j that will be communicated to node A_{j+1}. At node A_0 (A_p), there is no incoming (outgoing) signal. Trail routing, wavelength assignment, and fault management of such networks have been discussed in [21]. A testbed for such networks is discussed in [392]. Traffic grooming (discussed in Chapter 9) in light-trail networks has been discussed in [392].

3.5.3 Burst Switching Networks

Another approach for data communication is called *Burst Switching* [307]. If there is a request for data communication from any source node S to a destination node D, the data is assembled at S. This data assembly is called a "burst". In burst switching, a burst is communicated in the following way:

1) source node S determines a route $S = A_0 \rightarrow A_1 \rightarrow A_2 \rightarrow \ldots \rightarrow A_p = D$ through the physical topology,
2) source node S creates a short packet, called a *control packet*, containing the usual header information, including the burst length (denoting the amount of data to be communicated), the route through the physical topology, and the channel number to be used,
3) source node S sends the control packet on a dedicated *control channel*[3] to the next node A_1 from S to D,
4) at node A_1, the control packet is converted to electrical signals and the information in the packet is used to configure the router at A_1 so

---

[3] The same control channel is used to service all requests for communication.

that, when the burst from S arrives at A_1, it will be routed to node A_2,

5) node A_1 sends the control packet, using the control channel, to the next node A_2 in the route $S = A_0 \rightarrow A_1 \rightarrow A_2 \rightarrow \ldots \rightarrow A_p = D$,

6) at nodes A_2, A_3, ..., A_{p-1}, exactly the same operations are carried out as at node A_1, when the control packet arrives at those nodes,

7) at node D, the switches are configured, so that the burst, when it arrives, is sent to an Optical Line Terminal (OLT),

8) after sending the control packet, node A_0 waits for some time, and, without waiting for an acknowledgement for the establishment of the connection, then sends the burst using the channel specified in the control packet,

9) the wavelength on a fiber used by the burst will be released as soon as the burst passes through the fiber.

If the control packet fails, at some intermediate node $A_i, 0 < i < p$, to reserve the bandwidth on the fiber $A_i \rightarrow A_{i+1}$, the burst will have to be dropped. In reliable burst transmissions, a message on the control channel is sent back to A_0 so that node A_0 may retransmit the control packet and the burst later [307]. To be a viable approach the probability of dropping bursts must be acceptably low. Burst switching has been reported in [85, 260, 307, 403, 427, 428]. It has been shown that Optical Burst Switching (OBS) is a viable technology for the next-generation optical networks [85].

3.5.4 Multicasting Networks

In this book the focus is on unicast communication, where each communication has only once source and one destination. *Multicast communication* [334] involves a single source end node S and a set of destination end nodes $\mathcal{D}$. The problem of multicast communication is to optimize, in some sense, network cost and/or performance while determining a scheme for simultaneous communication from source S to all nodes in set $\mathcal{D}$. Multicasting is useful for distributed computing, news-feed, audio/video conference, software and video distribution, and database replication [109, 185, 251, 301, 333, 380]. A significant amount of work has been done in this area and a brief review of some of the work in wavelength routed WDM networks is given below.

The concept of constructing a *multicast tree* is crucial in multicast communication and is informally described below in the context of WDM networks. Given a source S, a set of destinations $\mathcal{D}$, a physical topology, specified as a graph $G = (V, E)$, characteristics of the network and an objective function for optimization, a multicast tree $\mathcal{T}$ is a tree where

i) the nodes of $\mathcal{T}$ are nodes in V,

ii) there is an edge $X \rightarrow Y$ in $\mathcal{T}$ from node X to node Y if node X can communicate directly with node Y,

iii) the root of the tree is node S,

iv) the objective function is optimized with the condition that every node in $\mathcal{D}$ is reachable from S.

Typical objective functions, relevant for WDM networks, include the number of channels and the number of transmitters and receivers used.

In order to support multicasting, it is useful to have a mechanism for communication different from that of lightpaths. The concept of *light-trees*, a generalization of lightpaths, is given in [333]. The idea is that a source end node may communicate, directly in the optical domain, with a number of destinations using one communication. For multicast communication from source S, a transmitter communicates the data using some channel, say c. This optical signal may be split at each intermediate node into two or more independent signals, each using the same channel c to be sent, using different fibers, to other nodes (router nodes or end nodes). Each destination end node has a receiver tuned to receive signals at the same channel c. Architectures to realize this model, and the use of unicasting, broadcasting, and multicasting using light-trees, have been discussed in [333]. Discussions on architectures to achieve multicasting and a critical analysis of such architectures appear in [6]. A survey of architectures for multicasting in optical networks appears in [109].

An analysis of the number of wavelengths needed for multicasting for a given number of destination nodes in an all-optical network appears in [301].

In dynamic multicast communication considered in [251], requests are made and released over time. The network has an arbitrary number of nodes, a fixed number of transmitters and receivers at each node, and a fixed number of channels on each fiber. A multicast connection may be realized by constructing a multicast tree which distributes the message from the source node to all destination nodes such that the channels used on each fiber as well as the receivers and transmitters used at each node are not used by existing communications. To support multicasting, each node (which is capable of acting both as an end node and as a router node) must have a *multicast capable switch* so that an incoming lightpath to the node may be routed, using the same wavelength, to any number of outgoing lightpaths from this node and, optionally, to a receiver. This receiver may retransmit the information carried by the lightpath using a different lightpath. Each retransmission constitutes another hop.

The routing and wavelength assignment (RWA) problem for unicast communication briefly discussed in this chapter and described, in detail, in Chapters 4 and 5 has to be modified somewhat for multicast communication.

Given a source end node S and a set of destination end nodes $\mathcal{D}$, the multicast routing and wavelength assignment (MRWA) problem is that of finding

- a set of edges, each edge representing a fiber in the physical topology, defining the multicast tree, having S as the root and all end nodes in $\mathcal{D}$,
- channels on each fiber in the set to establish the connection from the source to the destination nodes.

Similarly, the multicast wavelength assignment (MWA) problem is that of finding a set of channels to use on a predetermined multicast tree.

In general, it is desirable to find an MRWA or an MWA that is optimal with respect to some cost metric. Typical objectives for optimization include minimizing

- the maximum number of hops from the source to any destination,
- the use of resources, such as the number of channels used,
- some linear combination of the number of receivers and transmitters used.

It may not always be possible to find an MRWA or an MWA for a given multicast request because the needed channels on particular edges in the physical topology, as well as transmitters and receivers at intermediate nodes, may be used by other ongoing multicast communication.

In [251], it was shown that the MRWA problem defined above is, in general, NP-complete, but that the MWA problem can be solved in linear time. An approximation algorithm for MRWA appears in [83]. Efficient solutions to the MRWA problem for simple topologies have been described in [188].

Multicast capable switches are expensive. It therefore makes sense not to use such switches in every node in the network. This problem has been considered in [439]. In [396], techniques for maximizing a different metric called the *Network Capacity* has been studied. For the general mesh topology, the problem is NP-complete while polynomial time algorithms exist for the bidirectional ring. Some greedy algorithms have been developed for this problem.

Algorithms for MRWA taking into account the Quality of Service (such as the maximum allowed delay of messages from the source to any destination) appear in [214].

In [308] there is an analysis of the benefits of limited wavelength conversion in multicasting . Some heuristics for separately determining the multicast tree and the MWA problem that efficiently solve the two problems appear in [399].

4

Route and Wavelength Assignment (RWA) I

In this chapter and in Chapter 5, the important problem of assigning, to each lightpath in a WDM network, a wavelength (equivalently a channel number) and a route through the physical topology will be discussed. In this chapter, networks requiring wavelength continuity constraint (discussed in Section 3.2) are considered so that each lightpath has only one associated channel number. After the RWA is done, it must be guaranteed that whenever two lightpaths L_i and $L_j, i \neq j$, share a fiber, the channel c_i assigned to L_i must be distinct from channel c_j assigned to L_j. Since each fiber in the network can support n_{ch} channels, numbered $1, 2, \ldots, n_{ch}$, an RWA is viable only if the number of channels used is less than n_{ch}.

A considerable amount of research has been done on RWA in networks satisfying the wavelength continuity constraint, using graph-theoretic tools. Major results using this approach are given below. In Chapter 5, RWA using heuristics and integer linear programs will be reviewed.

4.1 RWA as a Graph Coloring Problem

The RWA problem may be viewed as a graph coloring problem in the following way. Let G be the directed graph corresponding to the physical topology of the network. The input to this formulation is G and a set $\mathcal{I}$, called an *instance*, of q source-destination pairs (s_i, d_i) where s_i (d_i) is the source (destination) of the ith communication request, $1 \leq i \leq q$.

Corresponding to any source-destination pair $(s_i, d_i) \in \mathcal{I}$, generally there are many routes in G. Let the routes from s_i to d_i in G be $\rho_i^1, \rho_i^2, \ldots, \rho_i^\aleph$, where $\aleph$ is the number of routes from s_i to d_i. The value of $\aleph$ may be quite large. From these routes, a single route ρ_i has to be selected for the pair (s_i, d_i), for all $i, 1 \leq i \leq q$. This chosen route could, for instance, be a shortest route from s_i to d_i. A *routing* $\Re = \{\rho_1, \rho_2, \ldots, \rho_q\}$ is a set of selected routes, one for each element in $\mathcal{I}$, for which wavelength assignment may be done using graph coloring. The graph to be colored is an undirected graph $G_\Re^\mathcal{I} = (V_\Re, E_\Re)$, called a *conflict graph* [52][1]. The vertex $v_i \in V_\Re$ is the selected route $\rho_i \in \Re$ and an edge $e_{ij} \in E_\Re$ from vertex v_i to vertex v_j exists when routes ρ_i and ρ_j share an edge.

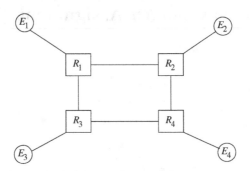

Fig. 4.1. A physical network

Example 4.1. In the physical network shown in Figure 4.1, let there be the following requests for lightpaths:

1. L_1 from E_1 to E_2
2. L_2 from E_1 to E_4
3. L_3 from E_3 to E_2
4. L_4 from E_3 to E_4

The instance $\mathcal{I}$ is therefore $\{(E_1, E_2), (E_1, E_4), (E_3, E_2), (E_3, E_4)\}$. Let the routes chosen for the lightpaths be as follows:

Route number	Route
ρ_1	$E_1 \to R_1 \to R_2 \to E_2$
ρ_2	$E_1 \to R_1 \to R_3 \to R_4 \to E_4$
ρ_3	$E_3 \to R_3 \to R_4 \to R_2 \to E_2$
ρ_4	$E_3 \to R_3 \to R_4 \to E_4$

Figure 4.2 shows these four routes and the set $\Re = \{E_1 \to R_1 \to R_2 \to E_2, \ E_1 \to R_1 \to R_3 \to R_4 \to E_4, \ E_3 \to R_3 \to R_4 \to R_2 \to E_2, \ E_3 \to R_3 \to R_4 \to E_4\}$.

The conflict graph $G_\Re^{\mathcal{I}}$ corresponding to the routing $\Re$ and the instance $\mathcal{I}$ is shown in Figure 4.3.   □

Once the conflict graph is completed, the chromatic number [42] of the conflict graph gives the number of wavelengths (equivalently channels) needed. For instance, the chromatic number of the conflict graph in Figure 4.3 is 3 and the graph may be colored by assigning color 1 to route ρ_1 and ρ_4, color 2

---

[1] also called path interference graph [370]

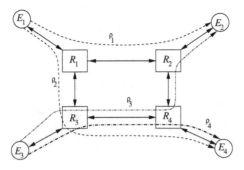

Fig. 4.2. Routes for four lightpaths on the network shown in Figure 4.1

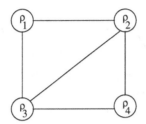

Fig. 4.3. Conflict graph for the lightpaths shown in Figure 4.2

to route ρ_2, and color 3 to route ρ_3. It is easy to see the direct correspondence between the color assigned to a route and the channel number that may be assigned to it. Whenever two routes share a fiber, there is an edge between them. The coloring of the graph ensures that adjacent vertices in the graph are never assigned the same color, exactly the same restriction used when assigning channels to routes. Therefore, once the graph has been colored, the channel assignment can be done simply by associating each color with a channel number.

An important objective is to reduce the chromatic number of the conflict graph for a given instance as much as possible, by properly selecting the route ρ_i from s_i to d_i, for all $i, 1 \leq i \leq q$.

In general, there are many potential routes for each lightpath. In the above example, a specific choice was made for the route for each lightpath. It is quite possible that some other choices for the routes of these lightpaths may reduce the chromatic number of the corresponding conflict graph.

Example 4.2. For the same physical topology shown in the above example, let the selected route for lightpath L_2 be $\rho_2^{new} = E_1 \rightarrow R_1 \rightarrow R_2 \rightarrow R_4 \rightarrow E_4$ (Figure 4.4). Here the routes for the remaining lightpaths L_1, L_3, and L_4 are the same so that the new routing $\Re^{new} = \{E_1 \rightarrow R_1 \rightarrow R_2 \rightarrow E_2, E_1 \rightarrow R_1 \rightarrow R_2 \rightarrow R_4 \rightarrow E_4, E_3 \rightarrow R_3 \rightarrow R_4 \rightarrow R_2 \rightarrow E_2, E_3 \rightarrow R_3 \rightarrow R_4 \rightarrow E_4\}$. The corresponding conflict graph $G_{\Re^{new}}^{\mathcal{I}}$ is shown in Figure 4.5.   □

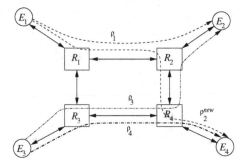

Fig. 4.4. Modified routes for four lightpaths on the network shown in Figure 4.1

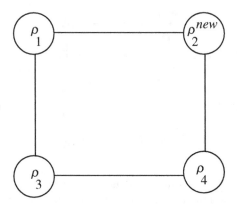

Fig. 4.5. Conflict graph for the lightpaths shown in Figure 4.4

The chromatic number of this graph is 2 and the graph may be colored by assigning color 1 to routes ρ_1 and ρ_4 and color 2 to routes ρ_2^{new} and ρ_3. The route ρ_2^{new} has the same length as the route ρ_2 but gives a graph with a lower chromatic number.

To discuss the RWA problem using graph coloring, $w(G_{\mathfrak{R}}^{\mathcal{I}})$ will denote the chromatic number of graph $G_{\mathfrak{R}}^{\mathcal{I}}$. As illustrated above, for a given $\mathcal{I}$, the value of $w(G_{\mathfrak{R}}^{\mathcal{I}})$ depends heavily on the routing $\mathfrak{R}$. For a given instance $\mathcal{I}$ and the graph G corresponding to the physical topology of a network, $w(G_{\mathfrak{R}_{min}}^{\mathcal{I}})$ will denote the minimum value of $w(G_{\mathfrak{R}}^{\mathcal{I}})$ considering all possible routings $\mathfrak{R}$.

The graph coloring approach to the RWA problem has been studied extensively for a number of different variations of the problem as follows:

1. type of connection,
2. number of lightpaths used in each communication,
3. availability of wavelength converters,
4. special topological properties of the network.

When specifying the set $\mathcal{I}$ of source-destination pairs, if each end node in the network appears exactly once as a source and once as a destination, it is called a *permutation*. If there is only one source and every remaining end node is a destination, it is called *broadcasting*, since data in the source is being sent to all other end nodes in the network. If all possible pairings of end nodes appear in $\mathcal{I}$, it is called *gossiping*.

The notions of single-hop and multi-hop networks have been introduced in Section 1.2.3. Each communication in a single-hop network requires exactly one lightpath. The number of lightpaths for a communication in a multi-hop network is limited by the diameter of the logical topology.

If all nodes (end nodes and router nodes) in a network are equipped with wavelength converters, the problem becomes simpler since the wavelength (and the corresponding channel number) of a lightpath is allowed to change from fiber to fiber. In this case, the problem is that of finding routes for each element in $\mathcal{I}$ such that the maximum number of lightpaths on a fiber is as small as possible.

There are many network topologies which have special properties that may be exploited to make the RWA simpler. For example, if the network is a ring, the RWA problem can be solved easily.

In the following sections, these problems will be reviewed.

4.2 Congestion and Its Relationship to Chromatic Number

Given an instance $\mathcal{I}$ of q source-destination pairs (s_i, d_i) and a routing $\Re$, each lightpath L_i uses route $\rho_i \in \Re$ from source s_i to destination d_i, where the route ρ_i consists of one or more directed edges in the physical topology. For example, the route ρ_1 in Figure 4.2 has directed edges $E_1 \to R_1$, $R_1 \to R_2$, and $R_2 \to E_2$. An alternative way of looking at this is that each directed edge in the physical topology carries zero or more lightpaths. For instance, the directed edge $R_3 \to R_4$ in Figure 4.4 carries lightpaths L_3 and L_4 while the directed edge $R_2 \to R_1$ carries no lightpaths.

Definition 4.1. *The* congestion[2] *of a network G having instance $\mathcal{I}$ and using the routing $\Re$ is the number of lightpaths on the directed edge carrying the maximum number of lightpaths and will be denoted by* $c(G_{\Re}^{\mathcal{I}})$.

Example 4.3. The congestion $c(G_{\Re}^{\mathcal{I}})$ of the network using the routes shown in Figure 4.2 is 3. The congestion of the network using the routes shown in Figure 4.4 is 2.   □

In general, for different routings, the congestion may vary.

---

[2] Also called *load* by some researchers.

Since all dipaths that cross a given link must have different colors, we have the following observation.

Observation 4.1. $\mathtt{w}(G_{\Re}^{\mathcal{I}}) \geq \mathtt{c}(G_{\Re}^{\mathcal{I}})$ *for any instance $\mathcal{I}$ in any digraph G and using routing $\Re$.*

Example 4.4. The chromatic number for the conflict graph shown in Figure 4.3 is 3 while that in Figure 4.5 is 2. Since lightpaths L_1 and L_2 always share the edge $E_1 \rightarrow R_1$ the congestion of any routing must be at least 2. This has been achieved using the routing shown in Figure 4.4. This is an example where the chromatic number and the congestion are the same. □

Exercise 4.1. Show, by giving an example, that the inequality in Observation 4.1 can be strict. □

4.3 Greedy Heuristics for the RWA Problem

It can be shown that the RWA problem is computationally difficult, even when restricted to some simple graphs such as rings and trees (technical details and a proof of this is given in [126]). This implies that the RWA problem cannot be solved optimally in an efficient way, even in some very restricted scenarios. Hence, one has to resort to solving special cases, finding good approximations, or considering heuristic solutions.

4.3.1 A Greedy Heuristic for the RWA Problem

One simple heuristic for the RWA problem is the heuristic that colors the conflict graph in a greedy fashion. Given a graph G, the greedy coloring heuristic for G works as follows:

While there exists a node v in G that is not colored:
Color v with the smallest color that is not used by any of its neighbors.

Example 4.5. For the graph G in Figure 4.6, a greedy coloring of G may proceed by coloring the vertices in the order v_1, v_2, v_3, v_4, v_5. In this case, the coloring would assign color 1 to v_1, color 2 to v_2, color 3 to v_3, color 2 to v_4, and color 3 to v_5. □

Exercise 4.2. Consider the graph in Figure 4.7. Perform a greedy coloring of G, i.e., specify an order of the vertices and then color the vertices greedily according to the specified order. □

The following observation gives a bound on the number of colors used by the greedy coloring heuristic.

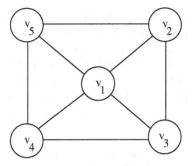

Fig. 4.6. A graph G to be colored

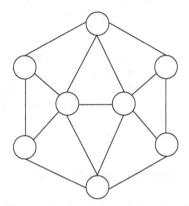

Fig. 4.7. A graph G to be colored

Observation 4.2. *For any graph G, the greedy coloring heuristic achieves a coloring of G that uses at most $\deg(G) + 1$ colors, where $\deg(G)$ is the maximum degree of G.*

Proof. Consider the point in time when a node v in G obtains its color. As the degree of v is at most $\deg(G)$, there is at least one free color to choose from the set $\{1, 2, \ldots, \deg(G) + 1\}$. $\qquad\square$

A simple heuristic for the RWA problem for a directed graph G (corresponding to the physical topology of the network), an instance $\mathcal{I}$ on G, and a routing $\mathfrak{R}$ for the instance $\mathcal{I}$ is as follows:

   Step 1: Construct the conflict graph $G_{\mathfrak{R}}^{\mathcal{I}}$.
   Step 2: Compute a greedy coloring of the conflict graph $G_{\mathfrak{R}}^{\mathcal{I}}$.

Example 4.6. In the physical network in Figure 4.1, let there be the requests for lightpaths as in Example 4.1. Let the routes for the lightpaths be chosen as shown in Figure 4.2. The greedy heuristic for the RWA problem would first construct the conflict graph as shown in Figure 4.3, and then color the conflict graph using the greedy coloring heuristic. Assuming the coloring order $\rho_1, \rho_2, \rho_3, \rho_4$, the vertices of the conflict graph would receive the following colors: color 1 for ρ_1, color 2 for ρ_2, color 3 for ρ_3, and finally color 1 for ρ_4.   □

Exercise 4.3. In the physical network in Figure 4.1, let there be the requests for lightpaths as in Example 4.1. Let the routes for the lightpaths be chosen as shown in Figure 4.4. Apply the greedy heuristic for the RWA problem to this scenario, i.e., construct the corresponding conflict graph along with a greedy coloring.   □

The quality of the obtained solution for the RWA problem can be derived as follows by using Observation 4.2. Let L be the maximum length of a dipath in $\Re$. Then, in the conflict graph $G_\Re^{\mathcal{I}}$, the degree of each node is upper-bounded by $(\mathsf{c}(G_\Re^{\mathcal{I}})-1) \cdot L$. Hence, the greedy coloring of the conflict graph as described above yields

$$\mathsf{w}(G_\Re^{\mathcal{I}}) \le (\mathsf{c}(G_\Re^{\mathcal{I}}) - 1) \cdot L + 1 .$$

4.3.2 An Improved Greedy Heuristic for the RWA Problem

The greedy heuristic for the RWA problem from Section 4.3.1 can be improved in the case where the maximum length L of any dipath in $\Re$ is large. Let a directed graph $G = (V, E)$ correspond to the physical topology of the network. For a given instance $\mathcal{I}$ on G, and a routing $\Re$ for the instance $\mathcal{I}$, the improved greedy heuristic for the RWA problem proceeds as follows:

1. *Construct the set $\Re_1$ of dipaths from $\Re$ of length at least $\sqrt{|E|}$, and the set $\Re_2$ of dipaths from $\Re$ of length less than $\sqrt{|E|}$ (i.e. $\Re_2 = \Re \setminus \Re_1$).*
2. *Give a different color to each of the dipaths in $\Re_1$.*
3. *Apply the greedy coloring heuristic from Section 4.3.1 to the dipaths in $\Re_2$.*

Example 4.7. In the physical network in Figure 4.1, let there be the requests for lightpaths as in Example 4.1. Let the routes for the lightpaths be chosen as shown in Figure 4.2. The improved greedy heuristic for the RWA problem would first color the two "long" lightpaths ρ_2 and ρ_3 by giving color 1 to ρ_2 and color 2 to ρ_3. Then, it would color the two "short" lightpaths ρ_1 and ρ_4 by applying the greedy coloring heuristic from Section 4.3.1, giving color 3 to ρ_1 and color 3 to ρ_4.   □

Exercise 4.4. Consider the physical network in Figure 4.1. Let there be the requests for lightpaths as in Example 4.1. Let the routes for the lightpaths be chosen as shown in Figure 4.4. Apply the improved greedy heuristic for the RWA problem to this scenario, i.e., construct the corresponding coloring of the lightpaths. □

The quality of the obtained solution of the improved greedy heuristic for the RWA problem can be derived as follows. The number of routes in $\Re$ of length at least $\sqrt{|E|}$ is at most $\sqrt{|E|} \cdot c(G_\Re^\mathcal{I})$; otherwise there would be a link load larger than $c(G_\Re^\mathcal{I})$. Each of these dipaths obtains a different color. The remaining dipaths have lengths less than $\sqrt{|E|}$ and conflict with less than $c(G_\Re^\mathcal{I}) \cdot \sqrt{|E|}$ other dipaths; therefore, the greedy coloring ensures that at most $c(G_\Re^\mathcal{I}) \cdot \sqrt{|E|}$ other colors suffice to color them. Overall,

$$\mathtt{w}(G_\Re^\mathcal{I}) \leq 2\sqrt{|E|} \cdot c(G_\Re^\mathcal{I}) .$$

4.4 Specific Networks

Even though it was mentioned in Section 4.3 that the RWA problem is computationally difficult in general, it is shown in this section that it is efficiently solvable or approximable in certain restricted classes of digraphs. [3]

4.4.1 The Bidirectional Path

It is first shown that the RWA problem is efficiently solvable when the underlying graph is a bidirectional path.

The *bidirectional path* of n nodes, denoted by P_n, is the graph whose vertices consist of all integers from 1 to n and whose arcs connect each integer i with $i + 1$ and each integer $i + 1$ with i for $1 \leq i \leq n - 1$.

For any set of requests $\mathcal{I}$ in a bidirectional path P_n, the RWA problem reduces to a dipath coloring problem. In fact, for any request (i, j) there is only one dipath from i to j: the *left-to-right dipath* $(i, i+1, \ldots, j)$ if $i < j$, the *right-to-left* dipath $(i, i-1, \ldots, j)$ if $i > j$. Moreover, left-to-right dipaths and right-to-left dipaths use different links in the underlying bidirectional path P_n and can hence be colored independently.

In the proof of the following theorem, an efficient, recursive, optimal coloring of a set of left-to-right (or right-to-left) dipaths on a bidirectional path is given.

Theorem 4.1. *Let P be a set of left-to-right (or right-to-left) dipaths on a bidirectional path. Then P can be efficiently colored with an optimal number of colors.*

---

[3] For any graph-theoretic notion that may not be defined in this book, the reader is referred to the introductory books [42, 199, 240].

Proof. The coloring procedure for left-to-right dipaths is given here only. The procedure for right-to-left dipaths is obtained analogously.

The coloring is described recursively. If $n = 2$, then all the requests are colored with different colors. As all the requests go across the same link, they must be colored with different colors in any coloring. Hence, the number of colors used in this coloring is optimal.

Let the theorem be true for any bidirectional path on $n - 1$ nodes. For a set of left-to-right dipaths P on a bidirectional path with n nodes $1, \ldots, n$, it is necessary to

1) transform the set of dipaths P on the path with nodes $1, \ldots, n$ into a set of dipaths P' on the path with nodes $2, \ldots, n$,
2) then color P' recursively,
3) and then construct a coloring of P from the solution for P'.

In order to obtain 1), all the dipaths starting in node 1 are transformed as though they were starting in node 2. This transformation is shown in Figure 4.8.

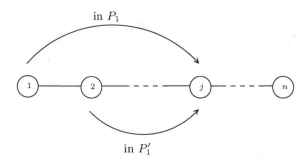

Fig. 4.8. Transformation of the dipaths in P

More precisely, P' is constructed as follows:

$$P_1 = \{(1, j) \in P \mid 2 \leq j \leq n\},$$
$$P_1' = \{(2, j) \mid (1, j) \in P_1, 2 < j \leq n\},$$
$$P' = (P \setminus P_1) \cup P_1'.$$

Now, P' is a set of dipaths on a path with $n - 1$ nodes, i.e., the nodes $2, \ldots, n$. P' can now be colored recursively with an optimal number of colors. In order to color the dipaths in P, the coloring of P' is now used as follows:

a) Give to the dipath (i, j) with $i > 1$ the same color assigned to it in the coloring of P'.
b) Give to $(1, j)$ the color assigned to $(2, j)$ in the coloring P'.
c) Color dipaths $(1, 2)$.

It should be noticed that step c) needs extra colors only if the load of link $(1,2)$ is strictly larger than the load of links $(2,3), (3,4), \ldots, (n-1,n)$. Therefore, a valid color assignment for P is obtained using an optimal number of colors.   □

Example 4.8. In a bidirectional path of seven nodes, let P be the following set of left-to-right dipaths: $P = \{(1,3), (2,4), (1,5), (3,6), (5,7)\}$. The coloring procedure in the proof of Theorem 4.1 proceeds as follows:

First, the following sets P' are constructed:

Recursion level 1: $P' = \{(2,3), (2,4), (2,5), (3,6), (5,7)\}$,
Recursion level 2: $P' = \{(3,4), (3,5), (3,6), (5,7)\}$,
Recursion level 3: $P' = \{(4,5), (4,6), (4,7)\}$,
Recursion level 4: $P' = \{(5,6), (5,7)\}$,
Recursion level 5: $P' = \{(6,7)\}$.

Then, coming back from the recursion, the following colors are assigned to the dipaths:

Recursion level 5: Color(5,7)=1,
Recursion level 4: Color(3,6)=2,
Recursion level 3: Color(1,5)=1,
Recursion level 2: Color(2,4)=3,
Recursion level 1: Color(1,3)=2.   □

Exercise 4.5. Consider a bidirectional path of seven nodes. Let P be the following set of (left-to-right) dipaths: $P = \{(1,3), (2,7), (3,5), (4,6)\}$. Color the dipaths in P by applying the procedure from the proof of Theorem 4.1. □

Exercise 4.6. Consider a bidirectional path of seven nodes. Let P be the following set of (left-to-right and right-to-left) dipaths: $P = \{(1,3), (2,7), (3,5), (4,6), (7,5), (6,3), (4,1), (2,1)\}$. Color the dipaths in P by applying the procedure from the proof of Theorem 4.1.   □

4.4.2 The Bidirectional Ring

In this section, it is shown that the RWA problem is efficiently approximable when the underlying graph is a bidirectional ring.

The *bidirectional ring* of n nodes, denoted by R_n, is the graph whose vertices consist of all integers from 1 to n and whose arcs connect integer n with 1, integer 1 with n, and each integer i with $i+1$ and each integer $i+1$ with i for $1 \leq i \leq n-1$.

Let $\mathcal{I}$ be a set of requests on a bidirectional ring. For any request (i,j) there are two dipaths from i to j: the clockwise and the anticlockwise dipath, as shown in Figure 4.9. Moreover, clockwise and anticlockwise dipaths use different links of the ring and can be colored independently.

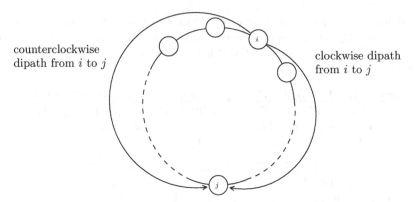

Fig. 4.9. Routing on a ring

In the proof of the following theorem, an efficient approximate coloring of a set of clockwise (or anticlockwise) dipaths on a bidirectional ring is given.

Theorem 4.2. *For any set of clockwise (anticlockwise) dipaths P on a ring, P can be efficiently colored such that the number of colors used is at most twice the load of P.*

Proof. The coloring procedure for clockwise dipaths is given here only. The procedure for anticlockwise dipaths is obtained analogously.

Considering a set of clockwise dipaths P on a ring of n nodes, the goal is to

1) transform the set of dipaths P on the ring of n nodes into a set of dipaths P' on the path with $n+1$ nodes,
2) then color P' using the coloring for the bidirectional path from Section 4.4.1,
3) and then construct a coloring of P from the solution for P'.

In order to obtain 1), each dipath (i, j) containing the link $(n, 1)$ is split into two dipaths $(i, 1)$ and $(1, j)$ as shown in Figure 4.10. Namely, let

$$P_1 = \{(i, j) \in P \mid 1 \le i \le j \le n\}$$

be the set of dipaths that do not contain the link $(n, 1)$ and define:

$$P_2 = \{(i, 1) \mid \text{there is } (i, j) \in P \setminus P_1\},$$
$$P_3 = \{(1, j) \mid \text{there is } (i, j) \in P \setminus P_1\}.$$

The set $P' = P_1 \cup P_2 \cup P_3$ is a set of dipaths on a path with $n+1$ nodes (i.e., the nodes $1, 2, \ldots, n, 1'$, where the node $1'$ is a copy of node 1) of the same load as the load of P in the original ring.

P' can now be colored with an optimal number of colors by using the algorithm from Section 4.4.1. In order to obtain the desired coloring of the set P, the following is done:

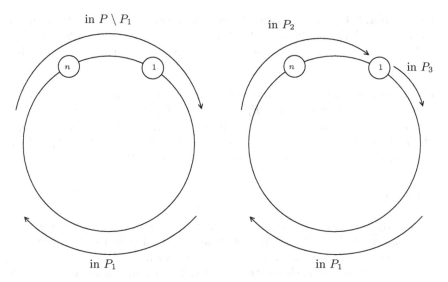

Fig. 4.10. Illustration of dipaths splitting on a ring

a) Assign to each $(i, j) \in P_1$ the same color as in P'.
b) For each $(i, j) \in P \setminus P_1$ if, in the coloring of P', the color assigned to $(i, 1)$ is equal to the color assigned to $(1, j)$ then assign to (i, j) this color; otherwise, if the color assigned to $(i, 1)$ differs from the color assigned to $(1, j)$ then assign to (i, j) an extra color not used for any other dipath.

The number of extra colors is upper-bounded by the load of the link $(n, 1)$. Hence, the total number of colors used is at most twice the load of P.   □

Example 4.9. In a bidirectional ring of seven nodes, let P be the following set of clockwise dipaths: $P = \{(1, 3), (2, 4), (1, 5), (3, 6), (5, 7), (5, 1), (6, 2), (7, 4)\}$. The coloring procedure in the proof of Theorem 4.2 proceeds as follows.

First, the following sets P_1, P_2, P_3, P' are constructed:

$P_1 = \{(1, 3), (2, 4), (1, 5), (3, 6), (5, 7)\}$,
$P_2 = \{(5, 1), (6, 1), (7, 1)\}$,
$P_3 = \{(1, 2), (1, 4)\}$,
$P' = \{(1, 3), (2, 4), (1, 5), (3, 6), (5, 7), (5, 1), (6, 1), (7, 1), (1, 2), (1, 4)\}$.

Then, the dipaths in P' are colored using the algorithm from Section 4.4.1:

Color(7,1)=1, Color(6,1)=2, Color(5,1)=3, Color(5,7)=1, Color(3,6)=2, Color(1,5)=1, Color(2,4)=3, Color(1,3)=2, Color(1,2)=3, Color(1,4)=4.

Finally, the desired coloring of the dipaths in P is obtained:

Color(1,3)=2, Color(2,4)=3, Color(1,5)=1, Color(3,6)=2, Color(5,7)=1, Color(5,1)=3, Color(6,2)=4, Color(7,4)=5.   □

Exercise 4.7. Consider a bidirectional ring of seven nodes. Let P be the following set of clockwise dipaths: $P = \{(1,3),(2,7),(3,5),(4,6),(5,2),(6,1),(7,4)\}$. Color the dipaths in P by applying the procedure from the proof of Theorem 4.2. □

Exercise 4.8. Consider a bidirectional ring of seven nodes. Let P_1 be the following set of clockwise dipaths: $P_1 = \{(1,3),(2,7),(3,5),(4,6),(5,2),(6,1),(7,4)\}$. Let P_2 be the following set of anticlockwise dipaths: $P = \{(7,5),(6,3), (4,1),(2,1),(3,6),(2,7),(1,5),(4,6)\}$. Color the dipaths in $P = P_1 \cup P_2$ by applying the procedure from the proof of Theorem 4.2. □

4.4.3 Trees

As already mentioned in Section 4.3, even in the case of trees the RWA problem is computationally difficult. Moreover, as the routing in a tree is fixed, the RWA problem reduces to a dipath coloring problem.

In this section, it is shown that the RWA problem is solvable exactly for some class of trees. The star of $n+1$ nodes, denoted by S_n, is the graph whose vertices consist of all integers from 0 to n and whose arcs connect integer 0 with each integer i and each integer i with 0 for $1 \leq i \leq n$. The star of $n+1$ nodes is depicted in Figure 4.11.

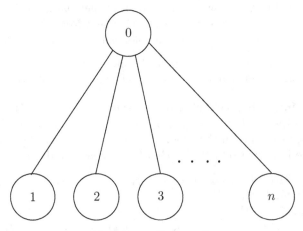

Fig. 4.11. The star S_n

In the proof of the following theorem, an efficient optimal coloring of a set of dipaths on a star is given.

Theorem 4.3. *For any n and any set of dipaths P on S_n, P can be efficiently colored with an optimal number of colors.*

Proof. Given a set of dipaths P on S_n, define a bipartite multigraph $H = (V(H), E(H))$ where $V(H) = \{a_1, \ldots, a_n\} \cup \{b_1, \ldots, b_n\}$ and $E(H)$ contains the edge (a_i, b_j) if and only if P contains a dipath from i to j, $1 \le i, j \le n$. The maximum degree Δ_H of a node in H is at most the load of P. An example of this construction is shown in Figure 4.12.

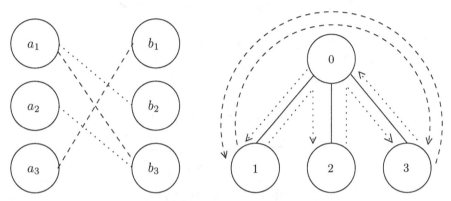

Fig. 4.12. Given the set of dipaths $P = \{(0,1), (1,2), (1,3), (2,3), 3,0), (3,1)\}$ on S_3, the corresponding graph H and the coloring of P

To color the set of dipaths (i, j), $i, j \ge 1$, of length 2 in P is equivalent to edge coloring H, that is, to assigning colors to the edges of H such that no two edges of the same color have a common endpoint. By Hall's theorem [42], the edges of H can be efficiently colored with Δ_H colors.[4]

Once the edge coloring of H has been done, all the dipaths to be still considered are of type $(i, 0)$ or $(0, i)$ and can be now colored using an optimal number of colors.   □

Exercise 4.9. Consider the set of dipaths $P = \{(0,2), (2,1), (1,3), (2,3), (1,0), (3,1)\}$ on S_3. Construct the corresponding graph H, an edge coloring of H, and the corresponding coloring of P. For coloring the edges of H, you may use any (potentially nonoptimal) coloring algorithm, e.g. greedy coloring.   □

A *spider* is a tree with at most one node of degree larger than 2. An example of a spider is shown in Figure 4.13.

---

[4] The exact details of the edge coloring of H are beyond the scope of this book.

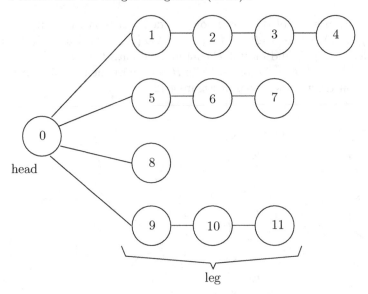

Fig. 4.13. A spider graph

Theorem 4.4. *If G is a spider then, for each set of dipaths P on G, P can be efficiently colored with an optimal number of colors.*

Proof. If G is a path or a star, then G can be colored optimally by using the procedures from the proofs of Theorem 4.1 and 4.3. Otherwise G is a spider with at least three legs. In such a case one can first color the dipaths going through the head as a star. One can then complete the coloring of the dipaths on each leg in a path.   □

Exercise 4.10. Consider the spider from Figure 4.13. Consider the set of dipaths $P = \{(2, 6), (1, 3), (4, 1), (8, 7), (6, 7), (5, 10), (9, 11)\}$. Construct a coloring of the dipaths in P.   □

4.5 Specific Instances

In this section, some special instances of the RWA problem in a digraph $G = (V, E)$ will be considered.

4.5.1 One-to-All Communication

In a *one-to-all* (broadcast) instance, one node v, called the source, has to be connected with each other node in the graph. In this case, the set of connection requests is
$$I = \{(v, w) \mid w \in V\}.$$

Given a digraph $G = (V, E)$ and a one-to-all instance I_{one} on G with source v, the problem is to set up $|V| - 1$ dipaths from the source of the one-to-all Communication process to any other node in V. Let $d(v)$ denote the out-degree of $v \in V$ and $d_{\min}(G) = \min_{v \in V(G)} d(v)$. When v is the source of the process there must exist at least $(|V| - 1)/d(v)$ dipaths out of the $|V| - 1$ dipaths originated at v that share the same edge incident to v. Therefore,

$$\mathtt{w}(G_{\Re_{min}}^{I_{one}}) \geq \left\lceil \frac{|V| - 1}{d_{\min}(G)} \right\rceil . \tag{4.1}$$

On the other hand, if G is k-arc-connected, the following upper bound holds. (A digraph G is k-arc-connected if it is necessary to remove at least k arcs in order to disconnect G.)

Theorem 4.5. *For any k-arc-connected digraph $G = (V, E)$,*

$$\mathtt{w}(G_{\Re_{min}}^{I_{one}}) \leq \left\lceil \frac{|V| - 1}{k} \right\rceil . \tag{4.2}$$

Proof. Let v be the source of the process. Let the node set $V \setminus \{v\}$ be partitioned, in an arbitrary way, into $s = \lceil (|V| - 1)/k \rceil$ subsets, say $V_1, \ldots, V_s$, of size at most k each. Since G is k-arc-connected, for each $i = 1, \ldots, s$, it is possible to choose k arc-disjoint dipaths to connect v to the k nodes in V_i (see [42]);[5] the same color can be assigned to these dipaths. Hence, the information from v to each other node in G can be routed in one round using a total of at most $s = \lceil (|V| - 1)/k \rceil$ colors.   □

A digraph G is *maximally arc-connected* if its minimum degree is equal to its arc-connectivity. From (4.1) and (4.2), the following theorem is obtained

Theorem 4.6. *For any maximally k-arc-connected digraph G,*

$$\mathtt{w}(G_{\Re_{min}}^{I_{one}}) = \left\lceil \frac{|V| - 1}{d_{\min}(G)} \right\rceil .$$

The above theorem gives the exact value of the number of colors necessary to perform one-to-all communication in various classes of important networks.

Exercise 4.11. Consider the graph G in Figure 4.14. Assume $v = 000$ as the source of a one-to-all communication instance. Give a routing of the requests along with a coloring of the routing dipaths.   □

---

[5] The details of this construction are beyond the scope of this book.

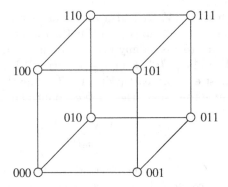

Fig. 4.14. A graph G for a one-to-all communication instance

4.5.2 All-to-All Communication

In an *all-to-all* (gossip) instance, each node in the network has to be connected with each other node in the graph. In this case, the set of connection requests is

$$I = \{(u, v) \mid u, v \in V, u \neq v\}.$$

All-to-all communication on a bidirectional ring will be considered in Section 5.2.

Exercise 4.12. Consider the graph G in Figure 4.14. Consider an all-to-all communication instance on G. Try to construct (by hand) a routing of the requests along with a coloring of the routing dipaths.   □

4.6 Bibliographic Notes

Surveys on algorithmic and graph-theoretic results for wavelength routing in all-optical networks can be found in [38, 165, 226].

The hardness of the RWA problem (even when restricted to rings and trees) is proved in [126]. The hardness of the congestion problem is considered in [38]. The greedy coloring heuristic from Section 4.3.1 is presented and analyzed in, for example, [42]. The greedy and the improved greedy heuristics from Sections 4.3.1 and 4.3.2 for the RWA problem are presented and investigated in [2]. In this paper, it is also shown that the bounds in Sections 4.3.1 and 4.3.2 cannot be improved in general by presenting a digraph G and an instance $\mathcal{I}$ that can be routed with a set of dipaths $\Re$ of maximum length L and load π when $\mathbf{w}(G_{\Re_{min}}^{\mathcal{I}}) = \Omega(\min\{L, \sqrt{|E|}\} \cdot \pi)$.

Theorem 4.1 follows from the fact that coloring left-to-right (or right-to-left) dipaths on a line is equivalent to color interval graphs [42]. Theorem 4.1 gives a simple proof of this result. Theorem 4.2 is due to Tucker [389]. As for any set $\mathcal{I}$ of requests on a ring, there exists an efficient algorithm that determines a set of dipaths $\Re$ of minimum load [154, 406], using Theorem 4.2, for any set $\mathcal{I}$ of requests on a ring, and one can efficiently construct a set of dipaths $\Re$ and a coloring of $\Re$ such that the number of colors used is at most twice the load of $\Re$. The bound of Theorem 4.2 is improved in [219], where it is proved that for any set of clockwise (anticlockwise) dipaths P on a ring, P can be efficiently colored such that the number of colors used is at most $\frac{3}{2}$ times the load of P. It is an open problem whether the constant $\frac{3}{2}$ in the previous statement can be improved.

A survey on wavelength routing in all-optical tree networks can be found in [76]. The best known lower bound for dipath coloring on a tree is described in [192, 233], where it is shown that there exists a tree T and a set of dipaths P on T such that $\mathrm{w}(T^P_{\Re_{min}})$ is at least $\frac{5}{4}$ times the load of P. The best known upper bounds for dipath coloring on a tree is described in [218, 105], where it is shown that, given a tree T and a set of dipaths P on T, it is possible to efficiently find a valid color assignment to the dipaths P such that the number of colors used is at most $\frac{5}{3}$ times the load of P. It is an open problem whether the gap between the upper and the lower bound on trees can be closed further. The proof of Theorems 4.3 and 4.4 can be found in [161, 162].

The proof of Theorems 4.5 and 4.6 can be found in [52, 53]. By Mader's theorem [42], Theorem 4.6 gives the exact value of $\mathrm{w}(G^{I_{one}}_{\Re_{min}})$ for the wide class of vertex-transitive graphs, e.g., the d-dimensional hypercube H_d, the $r \times s$ mesh $M_{r,s}$, the d-dimensional torus C^d_m, and any Cayley digraph G of degree d. Whereas for an arbitrary set of communication requests the RWA problem is computationally difficult, for the one-to-all communication the computation of $\mathrm{w}(G^{I_{one}}_{\Re_{min}})$ can be done in polynomial time also in general graphs based on network flows [52, 53].

4.6.1 One-to-Many and All-to-All Communication

For a general network G and an arbitrary instance I, the problem of determining $\mathrm{w}(G^{\mathcal{I}}_{\Re_{min}})$ has been proved to be NP-hard in [126]. In particular, it has been proved that determining $\mathrm{w}(G^{\mathcal{I}}_{\Re_{min}})$ is NP-hard for trees and cycles. In [161], these results have been extended to binary trees and meshes.

The problem of determining $\mathrm{c}(G^{\mathcal{I}}_{\Re_{min}})$ for a given digraph G and instance I has been extensively investigated, especially in the context of arc-forwarding index and edge-forwarding index [193, 267, 360], as well as the brief overview on forwarding indices in [38].

From Observation 4.1, it is known that

$$\mathrm{w}(G^{\mathcal{I}}_{\Re_{min}}) \geq \mathrm{c}(G^{\mathcal{I}}_{\Re_{min}}) . \tag{4.3}$$

Hence, it is natural to ask more precisely about the general relation between $\mathrm{w}(G^{\mathcal{I}}_{\Re_{min}})$ and $\mathrm{c}(G^{\mathcal{I}}_{\Re_{min}})$, especially the question in which cases equality can be achieved in (4.3), or in which cases one can at least come close to equality in (4.3).

In order to obtain equality in (4.3), a routing $\Re$ must be found such that $c(G_{\Re}^{\mathcal{I}}) = c(G_{\Re_{min}}^{\mathcal{I}})$, and the associated conflict graph $Confl(G, \mathcal{I}, \Re)$ is $c(G_{\Re_{min}}^{\mathcal{I}})$-vertex colorable. On the other hand, the following general relation was derived between the parameters $w(G_{\Re_{min}}^{\mathcal{I}})$ and $c(G_{\Re_{min}}^{\mathcal{I}})$ in [2]:

Theorem 4.7 ([2]).

1. Let I be an instance in a network G. Let $L = \max_{(x,y) \in I} \text{dist}(x, y)$. Then, $w(G_{\Re_{min}}^{\mathcal{I}}) = O(L \cdot c(G_{\Re_{min}}^{\mathcal{I}}))$.
2. For every L and c, there exists a directed graph G and an instance $\mathcal{I}$ such that $L = \max_{(x,y) \in \mathcal{I}} \text{dist}(x, y)$, $c(G_{\Re_{min}}^{\mathcal{I}}) = c$, and $w(G_{\Re_{min}}^{\mathcal{I}}) = \Omega(L \cdot c)$.

Theorem 4.7 shows that the gap between $w(G_{\Re_{min}}^{\mathcal{I}})$ and $c(G_{\Re_{min}}^{\mathcal{I}})$ can be quite substantial in general. As a consequence, it becomes of high interest to investigate the relation between $w(G_{\Re_{min}}^{\mathcal{I}})$ and $c(G_{\Re_{min}}^{\mathcal{I}})$ for specific important graph classes G and/or instances $\mathcal{I}$, and there has been an ongoing effort in this direction. Especially, one-to-many instances and the all-to-all instance have been thoroughly investigated in the literature.

One-to-many instances have been investigated in [39]. Using a network flow approach, equality was shown for (4.3) in this case:

Theorem 4.8 ([39]). $w(G_{\Re_{min}}^{\mathcal{I}}) = c(G_{\Re_{min}}^{\mathcal{I}})$ for any one-to-many instance $\mathcal{I}$ in any digraph G.

Results about the all-to-all instance I_A can be found in [7, 37, 38, 161, 336], and especially on rings in [52, 53, 237, 405]. For the all-to-all instance, equality in (4.3) has been proved for specific networks like paths, cycles, hypercube [52, 53], toroidal mesh of even side [37], and trees [161], as well as for more general graph classes like clique-compound graphs [7], cartesian product of cycles and paths [336], and cartesian product of complete graphs [37]. In [383], bounds on the number of wavelengths for inflated networks are given for the all-to-all instance. In [291, 384], bounds on the optical index of (families of) circulant 4-regular graphs are given in terms of the arc-forwarding index. There is an ongoing effort to settle the question whether the optical index is equal to the arc-forwarding index for general graphs.

Question 4.1. Does the equality $w(G_{\Re_{min}}^{\mathcal{I}_A}) = c(G_{\Re_{min}}^{\mathcal{I}_A})$ hold for the all-to-all instance I_A in any symmetric digraph G?

Instances different from one-to-many instances and from the all-to-all instance are of importance in the context of graph embeddings and will be discussed later on in Section 5.

4.6.2 Permutations

Some lower bounds on the numbers of wavelengths required for permutation instances in any network were given in [302]. From the algorithmic point of

view, determining $w(C_n{}^{\mathcal{I}_1}_{\Re_{min}})$ is NP-hard, which can be shown by a modifica-
tion of the NP-hardness proof of the wavelength problem in a ring for general
instances, given in [126]. However, polynomial 2-approximation algorithms
exist for both problems [126, 309, 354]. The problem $c(C_n{}^{\mathcal{I}_1}_{\Re_{min}})$ is solvable
in polynomial time, which follows from the more general results of [406] and
[153], respectively. This motivates study of the best possible upper bounds on
the above two measures. In [304], the following is shown.

Theorem 4.9 ([304]). *The following inequalities hold, for* $w(C_n{}^{\mathcal{I}_1}_{\Re_{min}})$ *and for*
$c(C_n{}^{\mathcal{I}_1}_{\Re_{min}})$, *for an* $\mathcal{I}_1$ *permutation request on the* n-*vertex ring* C_n:

$$w(C_n{}^{\mathcal{I}_1}_{\Re_{min}}) \leq \left\lceil \frac{n}{3} \right\rceil, \quad c(C_n{}^{\mathcal{I}_1}_{\Re_{min}}) \leq \left\lceil \frac{n}{4} \right\rceil$$

Both bounds are the best possible for worst-case instances.

4.6.3 Miscellaneous

Approximate multicommodity flow for WDM networks design. Motivated by
the quest for efficient algorithms for the Routing and Wavelength Assignment
problem (RWA), [62] addresses approximations of the fractional multicommodity flow problem which is the central part of a complex randomized rounding algorithm for the integral problem. Through the use of dynamic shortest
path computations and other combinatorial approaches, [62] improves on the
best known algorithm for approximations of the fractional multicommodity
flow. This algorithm is of practical interest for the work on multi-fiber networks with wavelength translators.

Design of Multifiber WDM Networks. [144, 143] addresses the design of multi-
fiber optical networks. It is shown that the wavelength assignment constraints
change from peer conflicts in single fiber networks to group conflicts in multi-
fiber networks. A new model is developed for the wavelength assignment problem based on conflict hypergraphs which structurally capture the group conflicts. This model allows for adapting hypergraph coloring approximation algorithms to the wavelength assignment problem, and for validating them on
real-world networks.

*Lightpath Assignment for Multifiber WDM Optical Networks with Wavelength
Translators.* [99] considers the problem of finding a lightpath assignment for
a given set of communication requests on a multifiber WDM optical network
with wavelength translators. Given such a network, the number w of wavelengths available on each fiber, the number k of fibers per link, and c the
number of partial wavelength translations available on each node, the problem is to decide whether it is possible to find a w-lightpath for each request
in the set such that there is no link carrying more than k lightpaths using the
same wavelength nor node where more than c wavelength translations take

place. The main theoretical result is the writing of this problem as a particular instance of integral multicommodity flow, hence integrating routing and wavelength assignment in the same model. Three heuristics are provided, mainly based upon randomized rounding of fractional multicommodity flow and enhancements that are three different answers to the trade-off between efficiency and tightness of approximation, and their practical performances are discussed on both theoretical and real-world instances.

Wavelength Assignment in Trees and 'Rings. In the context of the standard wavelength assignment problem, two important network topologies have been considered: bidirected trees and undirected rings.

For trees, one direction of research [18, 19] was to define the class of greedy wavelength assignment algorithms that use randomization, extending in this way the class of deterministic greedy algorithms which have been studied extensively in recent years. The limitations of these algorithms were studied proving several lower bounds while a wavelength assignment algorithm was developed that assigns at most $7L/5 + o(L)$ wavelengths on any set of paths of load L on a binary tree network of depth $o(L^{1/3})$, with high probability. For the analysis of the upper and lower bounds, new tail inequalities for random variables following hyper-geometrical probability distributions were developed.

The randomized wavelength assignment algorithm can be thought of as applying some kind of randomized rounding on fractional path colorings. These are equivalent to solutions to the natural relaxation of the integer linear program corresponding to the wavelength assignment instance. [79] shows that each symmetric set of paths of load L on a binary tree has a fractional path coloring of cost $1.367L$. By applying similar techniques with those in [18, 19], one obtains an $1.367 + o(1)$ approximation algorithm for wavelength assignment of symmetric sets of paths on binary trees (again, with some restrictions on the depth of the tree). Symmetric sets of paths are important since many services that will be supported by all-optical networks in the future will require bidirectional reservation of optical bandwidth.

Another direction of research was to approximate the optimal solution of the wavelength assignment problem using almost optimal solutions to the corresponding fractional path coloring problem as a guide. In [71], by simplifying and extending previous work, polynomial time algorithms are presented that compute almost optimal fractional path colorings in bounded-degree trees and in rings. The methods cover the case of multiple fibers as well. By applying a novel randomized rounding technique and using known wavelength assignment algorithms as subroutines, approximation algorithms are obtained with improved approximation ratios in bounded-degree trees and in rings with one or multiple fibers. The randomized rounding technique can be applied to any network. The approach gives improved existential upper bounds for the cost of the optimal wavelength assignment as a function of the cost of the corresponding optimal fractional path coloring solution. For the analysis, new tail inequalities are used for generalizations of occupancy problems.

Traffic Grooming. In a WDM network, routing a request consists in assigning it a route in the physical network and a wavelength. If each request uses at most $1/C$ of the bandwidth of the wavelength, we say that the grooming factor is C. This means that on a given edge of the network one can groom (group) at most C requests on the same wavelength. With this constraint, the objective can be either to minimize the number of wavelengths (related to the transmission cost) or minimize the number of Add Drop Multiplexers (ADM) used in the network (related to the cost of the nodes).

The problem of traffic grooming in WDM rings or paths with all-to-all uniform unitary traffic has been addressed in [48, 50]. The goal is to minimize the total number of SONET add-drop multiplexers (ADMs) required. It has been shown that this problem corresponds to a partition of the edges of the complete graph into subgraphs, where each subgraph has at most C edges (where C is the grooming ratio) and where the total number of vertices has to be minimized. Using tools of graph and design theory, the problem is solved optimally for rings for practical values and infinite congruence classes of values for a given C. In [46] the problem is solved for rings and $C = 6$. In [43, 44] the problem is studied for paths.

In [47] an optimal solution to the Maximum All Request Path Grooming (MARPG) problem is given motivated by a traffic grooming application. The MARPG problem consists in finding the maximum number of connections which can be established in a path of size N, where each arc has a capacity or bandwidth C (grooming factor). A greedy algorithm is presented to solve the problem and an explicit formula for the maximum number of requests that can be groomed.

Shared Risk Resource Groups. Multiple link failure models, in the form of Shared Risk Link Groups (SRLGs) and Shared Risk Node Groups (SRNGs), are becoming critical in survivable optical network design. [96, 98, 119] classify both of these forms of failures under a common scenario of shared risk resource groups (SRRG) failures. Techniques are developed for the minimum color path problem to tolerate multiple failures arising from a shared resource group failure. The minimum color s-t-cut problem is studied and proved NP-complete and hard to approximate. Efficient MILP formulation and heuristic algorithms are also provided.

Miscellaneous. A state-of-the-art survey on efficient access to optical bandwidth, routing and grooming in WDM networks is contained in [40]. General bounds on the number of wavelengths and switches in all-optical networks are derived in [31]. The problem of efficient routing and wavelength assignment in trees and rings under different scenarios, with and without wavelength conversion, is considered in [14, 16, 17, 18, 19, 68, 70, 69, 72, 73, 74, 75, 76, 77, 78, 79, 128, 127, 129, 130, 131, 132, 133, 134, 158, 163, 212, 213, 217, 275, 369].

Wavelength assignment on trees of rings has been considered in [57, 58]. Problems related to wavelength routing in shortest-path all-optical networks, with and without wavelength conversion, are considered in [136, 137, 138]. Resource allocation problems in multifiber WDM tree networks have been studied in [135]. Multicasting in optical networks has been considered in [160, 164, 257, 258]. Multihop all-to-all broadcast on WDM optical networks has been investigated in [182]. On-line permutation routing on WDM all-optical networks is considered in [179]. Efficient protocols and wavelengths requirements for permutation routing on all-optical multistage interconnection networks are presented in [180, 181].

A survey of traffic grooming using combinatorial design is contained in [49]. Approximation algorithms for the traffic grooming problem are presented in [145, 146]. The traffic grooming problem in bidirectional WDM ring networks is considered in [51, 97]. The problem of minimizing the number of ADMs in optical networks is considered in [45, 55, 65, 67, 66, 150, 151, 123, 124, 125, 210, 211, 256, 347, 348, 400, 430, 441]. Graph decompositions with application to wavelength add-drop multiplexing for minimizing SONET ADMs are studied in [95]. The design of low-cost survivable wavelength-division-multiplexing (WDM) networks is studied in [120]. To achieve survivability, lightpaths are arranged as a set of rings. [147, 148] considers routing of wavebands for all-to-all communications in all-optical paths and cycles. Nash equilibria in non-cooperative all-optical networks are investigated in [59].

On-line routing in all-optical networks where requests can dynamically change and are given at different times have been covered in [20, 34, 242]. Probabilistic algorithms for on-line routing in all-optical networks have been studied in [33, 149, 243, 244]. Studies of the routing model, where wavelength converters are allowed, appear in [15, 17, 159]. Specific communication problems using the *linear cost* model appear in [36].

5

Route and Wavelength Assignment (RWA) II

As mentioned in Section 3.3, RWA may be done for static (also called off-line) or for dynamic lightpath allocation [60, 433][1]. The following situations have been investigated by researchers:

Case I) Networks where all router nodes have a full wavelength converter at each output port.

Case II) Networks where no router node has a full wavelength converter.

Case III) Networks where some router nodes have a full wavelength converter at each output port.

Case IV) Networks where some or all router nodes have a limited wavelength converter at each output port.

In this chapter, the first two situations will be discussed in detail. Research on the remaining two situations will be summarized in the bibliographic notes section. As discussed in Section 3.4, if a network has a full wavelength converter at each output port of every router node, a route, such that every fiber in the route has n_{ch} or less lightpaths, is a valid route, where n_{ch} is the number of channels that can be accommodated by each fiber in the network. The fact that the wavelength continuity constraint is not needed allows a relatively simple formulation for RWA in such networks. The assumption that every node has a full wavelength converter, however, is not realistic in practical networks. In Section 3.2, it was mentioned that, if there is no wavelength converter in the network, the wavelength continuity constraint has to be satisfied so that each lightpath will have associated channel number which is used by the lightpath on every fiber in its route. The solutions for off-line RWA using mathematical programming are computationally intractable even for medium-sized networks. Heuristics are useful for solving, within a reasonable amount of time, the off-line RWA problem in practical networks [440].

---

[1] Some papers [433] also discuss a third category "incremental" where new connection requests come; but once a connection is set up, it continues indefinitely.

For dynamic RWA, a quick solution is very important and it is not practical to search for all possibilities of RWA. The use of a designated end node to act as a central repository of a lightpath database is one possibility in dynamic RWA. In the central repository approach, every request for a lightpath has to be communicated to this designated end node which determines whether or not the request can be granted. The problem with the approach is that the designated end node could become a bottleneck. A better solution is to use a distributed algorithm which will run efficiently on a number of end nodes. Since different requests for lightpaths may potentially use the same channel on a given fiber, it is important to view channels as resources that may be used by many lightpaths. Once a lightpath is set up, for every fiber in its path, the lightpath must have exclusive access to the channel it uses on the fiber. To ensure this, some locking protocol is needed in distributed algorithms.

In this chapter schemes for off-line as well as dynamic RWA using mathematical programming or heuristics will be covered.

5.1 Off-line Route and Wavelength Assignment

The off-line solution for the RWA problem takes, as input, the physical topology of the network, the set of lightpaths[2] to be established, and the number of channels n_{ch} that may be carried by a fiber. This problem is complicated due to the fact that there are many choices, in general, for the route of each lightpath in the physical topology of a nontrivial network. Since the number of channels n_{ch} in a WDM network is approximately 100 and is increasing with time, the solution space containing all possible combinations of the routes through the physical topology and all possible channel numbers of each of the lightpaths is, generally, enormous.

The objectives of the two formulations given below are the same — minimize the *congestion*,[3] the maximum number of lightpaths that share the same fiber. Minimizing the value of congestion is one way to reduce the cost of a network since a lower value of congestion means that the design requires fewer channels on each fiber and is therefore more economical. If, after finding the minimum value of congestion, it turns out that the value of congestion is more than the number of channels n_{ch} on a fiber, it means that RWA is not feasible for this set of lightpaths on this network.

5.1.1 Exact Solution of the RWA Problem in Networks with Full Wavelength Converters

In the network, there are $\mathcal{N}_E$ end nodes and $\mathcal{N}_R$ router notes. Each node (either an end node or a route-node) is assigned a unique number $n, 1 \leq n \leq$

---

[2] Each lightpath is defined by the source and the destination for the lightpath.

[3] The notion of congestion was introduced in Section 4.2.

$\mathcal{N}_E + \mathcal{N}_R$. To specify the problem mathematically using an Integer Linear Program[4] (ILP), the following formulation, henceforth called $FORM1$, may be used.

Formulation $FORM\ 1$ is based on the idea of network flow programming and treats each lightpath as a commodity[5].

Notation used in $FORM\ 1$

m : the number of lightpaths to be set up.

$\mathcal{N}_E\ (\mathcal{N}_R)$: the number of end nodes (router nodes) in the network.

E : the set of all pairs of node (end node or router node) numbers (i, j), such that there is a directed edge from node N_i to node N_j in the physical topology. Each directed edge represents a fiber in the network.

source(k) : the node number (which must be an end node) corresponding to the source of the kth lightpath, $1 \le k \le m$.

destination(k) : the node number (which must be an end node) corresponding to the destination of the kth lightpath, $1 \le k \le m$.

Ω_{max} : the congestion in the network.

X_{ij}^k : an integer variable having a value of 0 or 1 where

$$X_{ij}^k = \begin{cases} 1 \text{ if the } k^{\text{th}} \text{ lightpath is routed through the edge } (i, j) \in E \\ \quad \text{in the physical topology,} \\ 0 \text{ otherwise.} \end{cases}$$

Once $FORM\ 1$ is solved, for the kth lightpath, the value of X_{ij}^k is known for each edge $(i, j) \in E$ in the physical topology. Since the edge (i, j) appears in the route of the kth lightpath if and only if $X_{ij}^k = 1$, the solution completely defines the route of the kth lightpath, for all $k, 1 \le k \le m$.

The formulation for $FORM\ 1$

Objective function:

$$\text{minimize}\ \ \Omega_{max} \tag{5.1}$$

---

[4] As mentioned in Section 1.3, a formulation involving both binary variables and some continuous variables is called a mixed integer linear program (MILP). A formulation that involves only integer variables is called an integer linear program (ILP).

[5] Fundamental concepts of flows of commodities on networks and MILP formulations for problem solving using network flow programming are briefly reviewed in Appendix 3.

subject to

1. Ω_{max} is the largest number of lightpaths over any fiber.

$$\sum_{k=1}^{m} X_{ij}^k \leq \Omega_{max}, \forall(i,j) \in E \tag{5.2}$$

2. Each lightpath starts from its source and ends at its destination.

$$\sum_{j:(i,j)\in E} X_{ij}^k - \sum_{j:(j,i)\in E} X_{ji}^k = \begin{cases} 1 & \text{if source}(k) = i, \\ -1 & \text{if destination}(k) = i, \\ 0 & \text{otherwise.} \end{cases} \tag{5.3}$$

(5.3) has to be repeated for all values of $k, 1 \leq k \leq m$ and for all $i, 1 \leq i \leq \mathcal{N}_E + \mathcal{N}_R$.

Justification of *FORM* 1

In (5.2), X_{ij}^k is the flow for the kth lightpath on edge $(i,j) \in E$. If the variable X_{ij}^k has a value 1, it means that the kth lightpath passes through the edge (i,j). Thus $\sum_{k=1}^{m} X_{ij}^k$ is the total number of lightpaths on edge (i,j). Ω_{max} must be greater than or equal to the largest possible value of $\sum_{k=1}^{m} X_{ij}^k$, considering all edges in E. Since the objective is to minimize Ω_{max}, (5.2) ensures that Ω_{max} is the congestion — the maximum, over all edges, of the sum of all flows on an edge.

The sum of flows, for commodity k, on all incoming edges for node N_i, is $\sum_{j:(j,i)\in E} X_{ji}^k$. Similarly, the sum of flows for commodity k on all outgoing edges for node i is $\sum_{j:(i,j)\in E} X_{ij}^k$. The intent of (5.3) is to specify that the difference between these two sums of flows is

- 0, if node N_i is an intermediate node in a route from the source of the kth lightpath to its destination,
- 1, if node N_i is the source of the kth lightpath; this ensures that each lightpath starts from its source,
- −1, if node N_i is the destination of the kth lightpath; this ensures that each lightpath terminates at its destination.

These are called the flow balance equations [4].

The complexity of *FORM* 1

For all $(i, j) \in E$ and for all $k, 1 \leq k \leq m$, binary $(0/1)$ variables X_{ij}^k are needed. The variable Ω_{max} can be a continuous variable. Therefore, the formulation has $|E|m$ binary variables and one continuous variable. The number of constraints is $|E|$ for (5.2) and $(\mathcal{N}_E + \mathcal{N}_R)m$ for (5.3). The total number of constraints is $|E| + (\mathcal{N}_E + \mathcal{N}_R)m$. In this ILP, the complexity is determined by the number, $|E|m$, of binary variables [294]. In a 14-node network with 40 edges and a total of 50 lightpaths, a total of 2,000 $(0/1)$ variables are needed. In other words, even for a relatively small network, such a formulation is computationally intractable. Techniques for reducing the search space in this type of problem have been discussed in [286].

Exercise 5.1. Change the objective function for the formulation to find routes for the lightpaths such that the total sum of the number of edges (where each edge represents a fiber) used in all the lightpaths is minimum.

Is this a good objective function? Justify your answer.   □

Exercise 5.2. Consider the physical topology of the network given in Figure 5.1. The value of $n_{ch} = 4$ so that each fiber can carry four channels. The network has four end nodes $E_1, \ldots, E_4$ and the following lightpaths:
$$\{(E_1, E_2), (E_1, E_3), (E_3, E_2), (E_2, E_1), (E_4, E_1)\}.$$

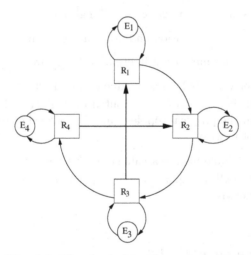

Fig. 5.1. The physical topology of a network

Show an RWA using a minimum number of channels assuming that each router node is equipped with a full wavelength converter. Justify that your solution requires the minimum number of channels.   □

5.1.2 Exact Solution of the RWA Problem in Networks Without Wavelength Converters

In order to consider networks without wavelength converters, it is necessary to ensure that the wavelength continuity constraint is satisfied. This means that, in a WDM network with m logical edges, for all $k, 1 \leq k \leq m$, the kth lightpath must have some channel number $c_k, 1 \leq c_k \leq n_{ch}$, associated with it. The kth lightpath must use the channel c_k on all fibers in its route through the physical topology. Formulation $FORM\ 2$ given below is a simpler version of a formulation in [370].

Notation used in $FORM\ 2$

$\mathcal{N}_E$: the number of end nodes in the network.

$\mathcal{N}_R$: the number of router nodes in the network.

m : the number of lightpaths.

E : the set of all pairs of node (end node or router node) numbers (i, j), such that there is a directed edge from node N_i to node N_j in the physical topology.

source(k) : the source end node of the kth lightpath.

destination(k) : the destination end node of the kth lightpath.

Ω_{max} : the maximum number of lightpaths that share the same fiber.

X_{ij}^{wk} : an integer variable having a value of 0 or 1 where
$$X_{ij}^{wk} = \begin{cases} 1 \text{ if the } k^{\text{th}} \text{ lightpath is routed through the edge } (i,j) \in E \\ \quad \text{in the physical topology and is assigned channel } w, \\ 0 \text{ otherwise.} \end{cases}$$

Y_j^k : an integer variable having a value of 0 or 1 where
$$Y_j^k = \begin{cases} 1 \text{ if the } k^{\text{th}} \text{ lightpath is assigned channel } j, \\ 0 \text{ otherwise.} \end{cases}$$

The formulation for $FORM\ 2$

Objective function :

$$\text{minimize} \quad \Omega_{max} \tag{5.4}$$

subject to

1. Ω_{max} must be the number of channels used on the edge with the maximum number of lightpaths.

$$\sum_{w=1}^{n_{ch}} \sum_{k=1}^{m} X_{ij}^{wk} \leq \Omega_{max}, \forall (i,j) \in E \qquad (5.5)$$

2. Satisfy the flow equations for the lightpath.

$$\sum_{j:(i,j)\in E} X_{ij}^{wk} - \sum_{j:(j,i)\in E} X_{ji}^{wk} = \begin{cases} Y_w^k & \text{if source}(k) = i, \\ -Y_w^k & \text{if destination}(k) = i, \\ 0 & \text{otherwise.} \end{cases} \qquad (5.6)$$

(5.6) has to be repeated for all $k, w, i, 1 \leq k \leq m, 1 \leq w \leq n_{ch}, 1 \leq i \leq \mathcal{N}_E + \mathcal{N}_R$.

3. Each lightpath must have exactly one wavelength associated with it.

$$\sum_{w} Y_w^k = 1, \forall k, 1 \leq k \leq m \qquad (5.7)$$

4. At most one lightpath on any edge $(i,j) \in E$ may use channel w.

$$\sum_{k=1}^{m} X_{ij}^{wk} \leq 1, \forall (i,j) \in E, \forall w, 1 \leq w \leq n_{ch} \qquad (5.8)$$

Justification for *FORM 2*

1. In (5.5), $\sum_{k=1}^{m} X_{ij}^{wk}$ is the number of lightpaths using channel w on the edge (i,j). $\sum_{w=1}^{n_{ch}} \sum_{k=1}^{m} X_{ij}^{wk}$ gives the total number of lightpaths on the edge (i,j) using any channel. This must be less than or equal to Ω_{max}. Since Ω_{max} is being minimized, it ensures that Ω_{max} is the maximum number of lightpaths on any edge.

2. To understand (5.6), it is convenient to view the network as having n_{ch} layers, the wth layer for channel $w, 1 \leq w \leq n_{ch}$. The variable X_{ij}^{wk} is applicable for the wth layer corresponding to the edge (i,j). If $Y_w^k = 0$, the kth lightpath is not assigned channel w. If $Y_w^k = 1$, the kth lightpath is assigned channel w and the flow equation in (5.6) is exactly like (5.3) in Section 5.1.1. The intent is to find, for the kth lightpath, exactly one layer w so that (5.6) may be used to define a route for the lightpath.

3. The purpose of (5.7) is to ensure that exactly one channel is assigned to the kth lightpath by specifying that Y_w^k is 1 for exactly one channel w.

4. The purpose of (5.8) is to specify that the same channel w is never used for more than one lightpath that passes through the edge (i,j).

The complexity of _FORM_ 2

There are two classes of binary (0/1) variables — X_{ij}^{wk} and Y_w^k. There is one variable X_{ij}^{wk} for each edge $(i,j) \in E$, each value of $w, 1 \le w \le n_{ch}$, and each value of $k, 1 \le k \le m$. There is one variable Y_w^k for each value of $w, 1 \le w \le n_{ch}$, and for each value of $k, 1 \le k \le m$. Therefore, the formulation has $mn_{ch}(|E|+1)$ binary variables. The number of constraints is $|E| + mn_{ch}(\mathcal{N}_E + \mathcal{N}_R) + m + |E|n_{ch} = |E|(n_{ch}+1) + m(n_{ch}(\mathcal{N}_E + \mathcal{N}_R) + 1)$.

Exercise 5.3. Consider the physical topology of the network given in Figure 5.2. Each of the routers is equipped with full wavelength converters. Suppose that the RWA problem is to set up three lightpaths — one from E_1 to E_3, one from E_2 to E_4, and one from E_3 to E_2.

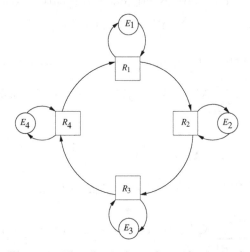

Fig. 5.2. The physical topology of a network

Show how the problem may be specified as an ILP using the formulation described in Section 5.1.1.   □

Exercise 5.4. Solve the problem in Exercise 5.3 using the formulation described in Section 5.1.2, using a mathematical programming package (e.g., CPLEX [206]).   □

Exercise 5.5. The formulation described in Section 5.1.1 does not attempt to take into account the length of the route used by a lightpath. Change the formulation so that the formulation also minimizes, whenever possible, the length of the route of the lightpaths, after minimizing the congestion. (Hint: the objective function should be a composite linear function of the number of channels and the sum of the lengths of routes used by the lightpaths.)   □

Exercise 5.6. The formulations described in Sections 5.1.1 and 5.1.2 are for static RWA. Using the same ideas, develop formulations for dynamic RWA with and without a full wavelength converter at each router node. Some suggestions are as follows:

When a new request for a lightpath from source s to destination d arrives, such a formulation has to be solved. A number of lightpaths are already in operation. The details (including the route and channel(s) used by each existing lightpath) about these lightpaths are already known. The RWA for the existing lightpaths cannot be changed. If the wavelength continuity constraint has to be followed, the availability of channels on each fiber must be saved in a database. In a network with full wavelength converters at each router node, the database only needs to store how many channels are available on each fiber. The objective function is up to you — it could be the congestion of the network, the length of the route for the new lightpath from s to d, or some linear function of these two. $\square$

Exercise 5.7. The problem described in Exercise 5.6 solves dynamic RWA using ILP. Another option is to use a search for a shortest path or a heuristic search for such a route. Develop an algorithm for such a breadth-first search and for a heuristic search. In your algorithm, do not change the RWA for the existing lightpaths. $\square$

Exercise 5.8. A major problem with the formulations described in Sections 5.1.1 and 5.1.2 is the fact that the number of 0/1 variables is enormous. One reason is that all possibilities for the route of each lightpath have to be considered. Significant savings can be achieved by limiting the number of routes that are considered for each lightpath in the following way.

For each source-destination pair, $\mathcal{R}$ routes may be precomputed and stored. The value of $\mathcal{R}$ should be small and has to be fixed ahead of time. It is a good idea to make these routes edge-disjoint, to the extent possible. If $\mathcal{R} = 3$, this means finding three edge-disjoint routes between every ordered pair of end nodes.

Let the precomputed routes from s to d be $\rho_{sd}^1, \rho_{sd}^2, \ldots, \rho_{sd}^{\mathcal{R}}$. When carrying out RWA for a lightpath from s to d, the search for a route could only consider the routes $\rho_{sd}^1, \rho_{sd}^2, \ldots, \rho_{sd}^{\mathcal{R}}$.

Modify formulations described in Sections 5.1.1 and 5.1.2 to incorporate this idea. $\square$

5.2 Route and Wavelength Assignment in a Bidirectional Ring

It is possible to develop RWA algorithms for topologies whose connectivity can be easily defined using some simple rules. Examples of such topologies include the ring, the torus, and the de Bruijn graph. An analysis for RWA to allow complete connectivity in a bidirectional ring following the algorithm in [121] is given below. Algorithms for RWA in some other topologies are available in [268].

Theorem 5.1. *The number of channels necessary and sufficient for complete connectivity is* $\mathfrak{W}$ *where*

$$
\mathfrak{W} = \begin{cases}
\frac{\mathcal{N}_E^2 - 1}{8} & \text{, if } \mathcal{N}_E \text{ is odd,} \\[2ex]
\frac{\mathcal{N}_E^2 - 1}{8} & \text{, if } \mathcal{N}_E/2 \text{ is odd,} \\[2ex]
\frac{\mathcal{N}_E^2}{8} & \text{, if } \mathcal{N}_E/2 \text{ is even}
\end{cases}
\tag{5.9}
$$

Proof. The case where the number of nodes $\mathcal{N}_E$ in the network is odd is given below. The case of even $\mathcal{N}_E$ has been dealt in [121] in a similar way.

In a bidirectional ring with $\mathcal{N}_E$ end nodes, for each source-destination pair (s, d), there are only two choices for the route — either go clockwise in the ring or go anticlockwise. The shortest-path routing has been used here.

The proof below is for the pairs of end nodes for which the routing involves the ring that permits routing in the clockwise direction. The proof for the pairs of end nodes for which the routing involves routing in the anticlockwise direction is identical.

The channel assignment is done using a recursive process.

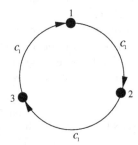

Fig. 5.3. A ring with three nodes needs only one channel

The base case of the process is a network having $\mathcal{N}_E = 3$. As shown in Figure 5.3, this requires only one channel, say c_1. The base case satisfies the claim for $\mathcal{N}_E = 3$.

Let the theorem be true for all $\mathcal{N}_E, \mathcal{N}_E \leq K$, so that, for the network with K end nodes, numbered $1, 2, \ldots, K$, shown in Figure 5.4, a total of $(K^2 - 1)/8$ channels are sufficient for complete connectivity. It should be noted that there are exactly $(K + 1)/2$ nodes between point A and point B (going in the clockwise direction) and $(K - 1)/2$ nodes between point B and point A.

The recursive process is to add two nodes $(K+1)$ and $(K+2)$ at insertion points A and B as shown in Figure 5.4 and assign channels. All the $(K^2-1)/8$ channels already assigned for communication between nodes $1, 2, \ldots, K$ will continue to be used for communication as before. New channels are needed for communication from each of the existing nodes in the ring shown in Figure 5.4 to the new nodes $(K+1)$ and $(K+2)$ as well as for communication from nodes $(K+1)$ and $(K+2)$ to each of the already existing nodes in the ring. Only the channels required for communication using the clockwise ring need to be considered since the case of the channels required for communication using the counter-clockwise ring is identical. For assignment, it is convenient to view the ring as having two regions. Let an imaginary line join points A and B as shown by the dotted line in Figure 5.4. The part of the ring to the right (left) of the dotted line will be called *region* 1 (2) in the discussions below. The channel assignments can be done as follows:

- (Region 1) Assign channels $c_1, c_2, \ldots, c_{(K+1)/2}$ for communication from node $(K + 1)$ to nodes $1, 2, \ldots, (K + 1)/2$. Assign channels $c_1, c_2, \ldots, c_{(K+1)/2}$ for communication from nodes $1, 2, \ldots, (K+1)/2$ to node $(K + 2)$.
- (Region 2) Assign channels $c_1, c_2, \ldots, c_{(K-1)/2}$ for communication from node $(K+2)$ to nodes $(K+3)/2, (K+5)/2, \ldots, K$. Assign channels $c_1, c_2, \ldots, c_{(K-1)/2}$ for communication from nodes $(K+3)/2, (K+5)/2, \ldots, K$ to node $(K + 1)$.
- (Region 2) Assign channel $c_{(K+1)/2}$ for communication from node $(K + 2)$ to node $(K + 1)$.

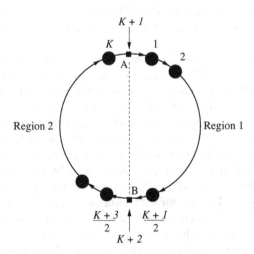

Fig. 5.4. A ring with K nodes

In the K node network shown in Figure 5.4, $(K^2 - 1)/8$ channels were needed for complete connectivity. After inserting nodes $K+1$ and $K+2$ into the network shown in Figure 5.4, the number of additional channels needed is $(K+1)/2$. Therefore the total number of channels needed for communication in a ring network with $(K+2)$ nodes is $(K^2 - 1)/8 + (K+1)/2 = (K^2 - 1 + 4K + 4)/8 = ((K+2)^2 - 1)/8$. The claim made above is therefore true for any network with $\mathcal{N}_E$ nodes, $\mathcal{N}_E$ odd. $\square$

Exercise 5.9. Show how you will define full connectivity for a ring network with seven end nodes. $\square$

5.3 A Heuristic for Route and Wavelength Assignment

The algorithms described in Section 5.1 give optimal RWA but are not feasible for practical networks from a computational perspective. This section describes a simple heuristic for RWA [27] in a network where there is no wavelength converter. The input to this heuristic, henceforth called *heu-RWA*, is a set of m source-destination pairs (s_i, d_i) where s_i (d_i) is the source (destination) of the ith lightpath. The output of *heu-RWA* is a set of pairs (ρ_i, c_i) where ρ_i and c_i represent the route and the channel assigned to the lightpath from s_i to d_i.

In *heu-RWA*, for each lightpath, a shortest route from the source to the destination of the lightpath is chosen. A shortest route has been described as a *Minimum Number of Hops* (MNH) route in [27]. In general, for a given pair (s_i, d_i), there may be a number of MNH routes. There are two phases in *heu-RWA* — finding a route for each lightpath and assigning a channel number to each lightpath.

In the first phase, for each lightpath, a "good" route through the physical topology is determined. Given a list of source-destination pairs, one for each lightpath, a set of good routes is a list of MNH routes that reduces the congestion as much as possible. The minimum number of channels needed cannot be less than the congestion of the network. This is the motivation for selecting a set of "good" routes. There is no guarantee, of course, that the use of a set of good routes necessarily leads to the use of a minimum number of channels. It is also important to note that the routes found by the heuristic below does not necessarily give the minimum congestion, but tries to reduce it as much as possible. In the second phase, for each lightpath L_i and its selected route ρ_i, a channel c_i is assigned in such a way that c_i does not conflict with the channel c_j assigned to any other lightpath L_j that shares an edge with lightpath L_i.

The steps involved in allocating the routes in Phase I are as follows:

Phase I

Step 1) Repeat Step 2, for all $i, 1 \leq i \leq m$.

Step 2) Create a list $[\rho_i^1, \rho_i^2, \ldots, \rho_i^\zeta]$, where there are ζ MNH routes from s_i to d_i and ρ_i^j is the jth MNH route from s_i to d_i, $1 \leq j \leq \zeta$.

Step 3) Select a list of routes $\Re = [\rho_1, \rho_2, \ldots, \rho_m]$ by selecting route ρ_1^p as the MNH route ρ_1 from s_1 to d_1 for some value of p, ρ_2^s as the MNH route ρ_2 from s_2 to d_2 for some value of s, $\ldots$, ρ_m^r as the MNH route ρ_m from s_m to d_m for some value of r.

Step 4) Determine the congestion corresponding to the m MNH routes in $\Re$.

Step 5) If the congestion may be reduced by assigning another MNH route ρ_i^j to ρ_i, for any $i, 1 \leq i \leq m, 1 \leq j \leq \zeta$, do the replacement and repeat Step 5 until the congestion cannot be reduced any further.

When Steps 1–5 are over, for each source-destination pair (s_i, d_i), there is an MNH route ρ_i in $\Re$. This is the "good route" that the above greedy heuristic finds for each source-destination pair.

In Phase II, a channel c_i is assigned to each route ρ_i as follows:

Phase II

Step 1) Sort the routes in the list $\Re$ in descending order according to the number of edges in each route (i.e., the longest route first and the shortest route last).

Step 2) Repeat Steps 3–5 for the longest route ρ_j in the list $\Re$ that has not been allocated a channel number yet.

Step 3) Form a list χ of channel numbers of all routes ρ_i in the list $\Re$ which has been already assigned a channel number and routes ρ_i and ρ_j share one or more edge(s).

Step 4) Assign to lightpath j the lowest channel number that does not appear in χ.

Step 5) If all routes in the list $\Re$ have been assigned a channel, stop. Otherwise go back to Step 2.

Example 5.1. The problem is to establish the following lightpaths on the physical topology shown in Figure 5.5:

    I) L_1 from E_4 to E_3,
    II) L_2 from E_4 to E_1,
    III) L_3 from E_6 to E_3, and
    IV) L_4 from E_5 to E_3.

In Figure 5.5, the links between nodes are shown as bidirectional so that if there is a bidirectional link $X \leftrightarrow Y$, it means that a lightpath can go from X to Y or from Y to X.

After Steps 1 and 2 of Phase I are over, Table 4.1 shows lightpaths L_1, L_2, L_3, and L_4 and the corresponding MNH routes.

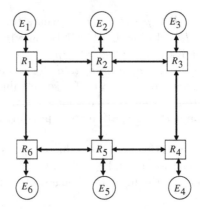

Fig. 5.5. The physical topology of a network

Table 5.1. Lightpaths and their MNH routes

Lightpath	MNH routes
L_1	$E_4 \to R_4 \to R_3 \to E_3$
L_2	$E_4 \to R_4 \to R_5 \to R_2 \to R_1 \to E_1$
	$E_4 \to R_4 \to R_3 \to R_2 \to R_1 \to E_1$
	$E_4 \to R_4 \to R_5 \to R_6 \to R_1 \to E_1$
L_3	$E_6 \to R_6 \to R_5 \to R_2 \to R_3 \to E_3$
	$E_6 \to R_6 \to R_5 \to R_4 \to R_3 \to E_3$
	$E_6 \to R_6 \to R_1 \to R_2 \to R_3 \to E_3$
L_4	$E_5 \to R_5 \to R_2 \to R_3 \to E_3$
	$E_5 \to R_5 \to R_4 \to R_3 \to E_3$

After Step 3 is over, the selected list of MNH routes is $\Re = [E_4 \to R_4 \to R_3 \to E_3, E_4 \to R_4 \to R_5 \to R_2 \to R_1 \to E_1, E_6 \to R_6 \to R_5 \to R_2 \to R_3 \to E_3, E_5 \to R_5 \to R_2 \to R_3 \to E_3]$. Here, for each lightpath, the first MNH route in Table 4.1 has been selected for inclusion in $\Re$.

In Step 4, the number of lightpaths on each edge is determined. The congestion is 3 on the edge $R_5 \to R_2$, due to three lightpaths on this edge.

In Step 5, it may be verified that by replacing the route $E_4 \to R_4 \to R_5 \to R_2 \to R_1 \to E_1$ by the route $E_4 \to R_4 \to R_5 \to R_6 \to R_1 \to E_1$, the congestion may be reduced to 2. It may also be verified that the congestion cannot be reduced any further by any other replacement of routes.

Therefore, Phase I ends with list $\Re = [E_4 \to R_4 \to R_3 \to E_3, E_4 \to R_4 \to R_5 \to R_6 \to R_1 \to E_1, E_6 \to R_6 \to R_5 \to R_2 \to R_3 \to E_3, E_5 \to R_5 \to R_2 \to R_3 \to E_3]$.

In Phase II Step 1, the routes in the list $\Re$ have to be sorted using the number of edges as the sorting key, in decreasing order. This gives $\Re = [E_4 \to R_4 \to R_5 \to R_6 \to R_1 \to E_1, E_6 \to R_6 \to R_5 \to R_2 \to R_3 \to E_3, E_5 \to R_5 \to R_2 \to R_3 \to E_3, E_4 \to R_4 \to R_3 \to E_3]$.

There will be four iterations of Steps 2–5, one for each entry in list $\mathfrak{R}$.

Iteration 1) In Step 2, the lightpath L_2 having the maximum number of edges and the route $E_4 \rightarrow R_4 \rightarrow R_5 \rightarrow R_6 \rightarrow R_1 \rightarrow E_1$ is considered. In Step 4, the lightpath L_2 is assigned channel number 1.

Iteration 2) In Step 2, the lightpath L_3 having the route $E_6 \rightarrow R_6 \rightarrow R_5 \rightarrow R_2 \rightarrow R_3 \rightarrow E_3$ is considered.

In Step 3, it is observed that lightpaths L_2 and L_3 have no edge in common. Therefore, in Step 4, the lightpath L_3 is also assigned channel 1.

Iteration 3) In Step 2, the lightpath L_4 having the route $E_5 \rightarrow R_5 \rightarrow R_2 \rightarrow R_3 \rightarrow E_3$ is considered.

In Step 3, it is observed that lightpaths L_3 and L_4 have edges $R_5 \rightarrow R_2$, $R_2 \rightarrow R_3$, and $R_3 \rightarrow E_3$ in common. Therefore, the list $\chi = [1]$. In Step 4, the lightpath L_3 is assigned channel 2.

Iteration 4) In Step 2, the lightpath L_1 having the route $E_4 \rightarrow R_4 \rightarrow R_3 \rightarrow E_3$ is considered.

In Step 3, it is observed that lightpaths L_1 and L_2 have an edge in common. Therefore, the list $\chi = [1]$. In Step 4, the lightpath L_1 is assigned channel 2.

$\square$

Exercise 5.10. Consider the physical topology given in Figure 5.5. Show how to use the heuristic given above to set up the following lightpaths:

    I) Lightpath L_1 from E_1 to E_5,
    II) Lightpath L_2 from E_3 to E_2,
    III) Lightpath L_3 from E_2 to E_5.

$\square$

5.4 Dynamic Route and Wavelength Assignment

In dynamic lightpath allocation [166], processing the request for a lightpath involves finding a possible route from the source to the destination of the requested lightpath and the assignment of a free channel to every fiber in the route, without disturbing any of the lightpaths already in operation. As mentioned in Section 3.3.1, when using dynamic lightpath allocation, the request for a lightpath may not always succeed. In a case where the request for a lightpath does not succeed, the request will be blocked. After a request for a lightpath is made, it takes some time to determine whether or not the request may be successfully handled. Two metrics are useful in evaluating any scheme for dynamic RWA — the *blocking probability* and the *setup time*.

The blocking probability is measured by the ratio of the number of blocked requests for lightpaths to the total number of attempts to set up lightpaths using a large number of trials [343]. This gives an estimate of the probability that a request for a lightpath might get blocked. The setup time is the time a source node has to wait until it knows what channel number has been allocated and can be used [343]. A good algorithm for dynamic RWA would be expected to have both a low setup time and a low blocking probability.

Since RWA is known to be a difficult problem, a common approach is to decouple the issue of routing from that of channel assignment. Three approaches have been investigated for routing as follows:

1) *Fixed routing* where the route ρ_{sd} for each source destination pair (E_s, E_d) of end nodes is predetermined. If a request for communication from E_s to E_d arrives, the route ρ_{sd} has to be examined to see if a lightpath may be established from E_s to E_d using route ρ_{sd}. If a lightpath may not be set up, the request has to be blocked. An obvious choice for ρ_{sd} is the shortest route from any end node E_s to end node E_d, determined using, for example, Dijkstra's algorithm [4]. The advantage of the fixed routing approach is the simplicity and speed of determining the route. The disadvantage is that more requests for connections are likely to fail in this approach. This is due to the fact that there is only one choice for the route to be used for each request for connection and it is quite possible that routes for many existing connections are already using the same fiber, so that a new request for connection from E_s to E_d may be blocked even though other routes from E_s to E_d may have available channels.

2) *fixed-alternate routing* where a number κ of routes $\rho_{sd}^1, \rho_{sd}^1, \ldots, \rho_{sd}^\kappa$ for each source destination pair (E_s, E_d) is predetermined. To find a usable route, from source E_s to destination E_d, route ρ_{sd}^1 is examined first to see if a lightpath may be established from E_s to E_d using route ρ_{sd}^1. If this fails, routes $\rho_{sd}^2, \rho_{sd}^2, \ldots, \rho_{sd}^\kappa$ are successively examined to see if a lightpath may be established from E_s to E_d using any of these routes. The first route, say $\rho_{sd}^i, 1 \leq i \leq \kappa$, which may be used to establish a lightpath from E_s to E_d, is actually used to set up the connection. This is also a simple scheme and has improved performance compared to the fixed routing scheme since κ routes from E_s to E_d, rather than a single route, are examined before a decision to block the request is taken. As in the former strategy, a new request for connection from E_s to E_d may still be blocked in this strategy even though other routes from E_s to E_d may have available channels.

3) *Adaptive routing* where the route is chosen by searching a database to find an "optimum" route. One way to implement adaptive routing is to search a database storing the network state, defined by the nodes of the network, the edges in the network, each edge representing a fiber, and the channels currently in use on each edge. The optimum route

could be one having the shortest length, having the same channel available on every edge on the route, or one using the least congested route [433]. An advantage of this approach is that the blocking probability is, in general, lower compared to the other two approaches given above. A disadvantage is the time to determine a route, so that the setup time may be relatively long. Further, it is necessary to maintain a global table. Keeping the database up to date when processing the request for a new connection request may create difficulties.

In the above routing strategies, the channel for the lightpath on each fiber in the route being considered has to be determined. If the network has a full wavelength converter on every output port in every router node, the channel for the lightpath on each fiber in the route can be determined easily. In this situation, the only requirement for successful assignment of channels for establishing a lightpath, once a route for the connection is determined, is that the number of existing lightpaths on each fiber, in the route under consideration, must be less than n_{ch}, the number of channels on a fiber.

Since wavelength converters are expensive, networks where the wavelength continuity constraint is satisfied have been investigated widely. The discussions below assume that every lightpath satisfies the wavelength continuity constraint. In other words, the lightpath uses the same channel number on every fiber in its route.

A number of heuristics [433] have been proposed to allot, if possible, a channel number c_{sd} for a given route ρ_{sd} from end node E_s to E_d, such that no lightpath using any edge in ρ_{sd} uses channel number c_{sd}. The first step is to determine the set, $\mathcal{C}$, of channel numbers that have not been allotted to any existing lightpath that uses any edge in route ρ_{sd}. Each element in set $\mathcal{C}$ has a value from 1 to n_{ch} (the number of channels on a fiber) and is a viable candidate for the new lightpath from E_s to E_d. The next step is to select a channel number c_{sd} from the set of channels $\mathcal{C}$.

A summary of some widely used strategies to select a channel number c_{sd} is given below:

1) *Random channel assignment* selects, at random, one channel number (often with uniform probability) from set $\mathcal{C}$.

2) *First-fit channel assignment* selects the channel having the lowest channel number. This favors the selection of lower channel numbers so that connection requests with longer routes have a higher chance of getting an available channel number.

3) *Least-used channel assignment* selects the channel which is used least often in the network. The idea is to balance the load on all channels. However, this makes it difficult to find an available channel number if the route is relatively long.

4) *Most-used channel assignment* selects the channel which is used most often in the network. The idea is to pack all the channel numbers

so that the difficulty of the least-used channel assignment strategy is avoided.

Two approaches exist for a dynamic RWA: centralized and distributed. In the first approach, some designated end node in the network acts as the *central agent* that keeps track of all the allocated channels in the current ongoing communication in the network. To establish a lightpath, from a source end node E_s to a destination end node E_d, the source node E_s sends a request to the central agent. The central agent either responds positively, with a channel number and a route from E_s to E_d, or negatively, indicating failure. The communication gets blocked if there is a negative response from the central agent.

The problem with the first approach is that all requests have to be processed by the central agent so that the central agent could become a bottleneck. This is particularly true if there is a large number of requests for lightpaths, each request for communicating a relatively small amount of data. In the second approach, a search for a route and a channel for a lightpath from any source to any destination is done using a *distributed search algorithm* involving the nodes that are potential intermediate nodes in setting up the lightpath. Here again, either the distributed search is successful, resulting in a route and a channel number, or a failure. In the latter case, the communication has to be blocked. The algorithms given below are taken from [343]. A distributed algorithm similar to that in [343] also appears in [274].

The discussions below are taken from [343] and assume that the network has no wavelength converter so that the wavelength continuity constraint is applicable.

5.4.1 Dynamic Routing Using a Central Agent

In the first approach, there is a central agent maintaining a database of lightpaths currently active in the network. This database contains the route and the channel used for every lightpath currently in operation. Whenever there is a request for a lightpath from a source E_s to a destination E_d, the request is sent to this central agent to establish the lightpath. The central agent carries out a limited search of the database of lightpaths, and either returns a route and a channel number for the lightpath, or returns a message that the search was unsuccessful. The search cannot be exhaustive since a low setup time is very important in a dynamic RWA.

Two strategies are considered below. Since existing lightpaths cannot be disturbed, each of these strategies has to take into account the channels currently being used along the edges of the route(s) being considered and determines a channel number that can be used. If no such channel number can be found, possibly because every channel is being used in some edge on the route(s), the connection gets blocked.

In Strategy 1, a single fixed route, typically a shortest route from E_s to E_d, is used as the route for a lightpath. For each source-destination pair (E_s, E_d), the fixed route is precomputed and saved in the central agent. Precomputing the routes is useful since it is very important to reduce the setup time. If there is no channel that is not used in any one of the fibers on the route, the request for a lightpath from E_s to E_d is blocked; otherwise a lightpath from E_s to E_d may be established and the database in the central agent is updated with the information pertaining to the new lightpath.

In Strategy 2, a fixed number of alternate routes between the source E_s and the destination E_d are considered. This is the fixed alternate route approach mentioned earlier. For each source-destination pair (E_s, E_d), a number of routes, that should preferably be edge-disjoint, are precomputed and saved in the central agent. These routes are examined, one at a time, until either all the routes have been examined without success or a route has been found with a channel number that is not used in any one of the fibers in the route. In the former case, the request for a lightpath from E_s to E_d is blocked; otherwise a lightpath from E_s to E_d may be established and the database in the central agent is updated with the information pertaining to the new lightpath.

The centralized approach is simple and a connection can never get blocked if some channel is available on every edge in the route(s) that are examined. The major problem with the centralized approach is that the central agent could be a bottleneck and has to be reliable.

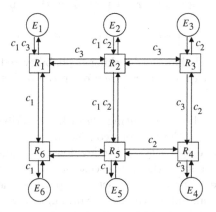

Fig. 5.6. The physical topology of a network

Example 5.2. The physical topology of a network is shown in Figure 5.6. The figure also shows zero or more channel numbers next to each edge. This means that the specified channel(s) has (have) been already allocated to establish lightpaths in response to earlier requests. This information about the availability of channels on each edge is stored in the database of the central agent. In this example, each fiber can support three channels c_1, c_2, and c_3.

Let there be two independent requests for lightpaths, one from E_1 to E_5 and one from E_2 to E_5.

If Strategy 1 is used, let the fixed route from E_1 to E_5 (E_2 to E_5) be $E_1 \rightarrow R_1 \rightarrow R_2 \rightarrow R_5 \rightarrow E_5$ ($E_2 \rightarrow R_2 \rightarrow R_5 \rightarrow E_5$). It may be readily verified that the request for a lightpath from E_1 to E_5 cannot be granted since no channel is available on all edges in the route $E_1 \rightarrow R_1 \rightarrow R_2 \rightarrow R_5 \rightarrow E_5$. The request for a lightpath from E_2 to E_5 can be granted since channel c_3 is available on all the edges in the route $E_2 \rightarrow R_2 \rightarrow R_5 \rightarrow E_5$.

If Strategy 2 is used, let there be two routes for each pair of end nodes in the network. For the two requests for lightpaths given above, let the two routes from E_1 to E_5 (E_2 to E_5) be $E_1 \rightarrow R_1 \rightarrow R_2 \rightarrow R_5 \rightarrow E_5$ and $E_1 \rightarrow R_1 \rightarrow R_6 \rightarrow R_5 \rightarrow E_5$ ($E_2 \rightarrow R_2 \rightarrow R_5 \rightarrow E_5$ and $E_2 \rightarrow R_2 \rightarrow R_3 \rightarrow R_4 \rightarrow R_5 \rightarrow E_5$). It may be readily verified that the request for a lightpath from E_1 to E_5 can be granted using the route $E_1 \rightarrow R_1 \rightarrow R_6 \rightarrow R_5 \rightarrow E_5$ and channel c_2. The request for a lightpath from E_2 to E_5 can be granted as in the previous case.   □

The performance of the strategy using a central agent has been studied via Monte-Carlo simulation to determine the variation of blocking probability with the number of wavelengths used in the network [343].

Figure 5.7 shows the general pattern[6] of the variation of blocking probability with the number of wavelengths used in the network for two cases:

1) when the route is a shortest path from the source to the destination and
2) when two alternate routes are explored to find a usable route.

It should be noted that the blocking probability reduces significantly as alternate routes are explored in routing; often the reduction is one or several order of magnitude.

Exercise 5.11. The dynamic routing algorithm described in this section assumes that each link in the network consists of exactly one fiber. In reality, when laying down fibers, it is typical to lay down a bundle of many fibers, not just one. This is done since the actual cost of the fiber itself is very low and the major component of the cost of setting up a fiber network is the cost of digging the trenches and filling them up. Normally, only a few of the fibers will be used and the rest kept for future use. Later on, if a fiber, currently in use, becomes faulty, it is easy to switch over to another unused fiber. Wavelength assignment in multi-fiber networks has been considered in [247].

When setting up lightpaths, if it is decided that a lightpath will use the link from node N_i to node N_j, there is a choice of any one of the fibers from N_i to N_j.

---

[6] This figure is taken from [343] and is relevant to a specific topology. More cases are available in [343].

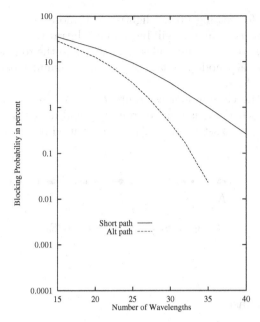

Fig. 5.7. Plot showing the variation of blocking probability with the number of wavelengths used

Consider a WDM network using the topology shown in Figure 5.6 where each link consists of four fibers, each carrying $n_{ch} = 4$ channels. Modify the strategy discussed in this Section to take advantage of the availability of four fibers on each link. Study the blocking probability of this network as done in Figure 5.7 and compare it to a network where each link consists of exactly one fiber carrying $n_{ch} = 4 \times 4$ channels.

Explain why the network with 4 fibers/link, each with 8 channels/fiber has better performance than a network with 1 fiber/link with 32 channels/fiber. For discussions on this type of situation, see [24].   □

5.4.2 Dynamic Routing Using a Distributed Algorithm

In the distributed approach given below, the usable wavelength is determined through two phases: a *request phase* and a *response phase*. It is assumed that the physical topology is bidirectional, so that if there is a fiber from node N_i to node N_j, there is also a fiber from node N_j to node N_i. In practical networks this is the case. Here, ρ_{sd} will stand for a shortest route from E_s to E_d. In this discussion,[7] it is assumed that only the shortest route ρ_{sd} will

---

[7] Another possibility, not considered here, is to try a limited number of routes to establish the lightpath before concluding that the lightpath cannot be set up.

be investigated for setting up the lightpath. If more than one shortest route exists from E_s to E_d, any one will be chosen arbitrarily. In each end node E_s, ρ_{sd} is precomputed for all end nodes $E_d, s \neq d$, and the routes stored in E_s. It is assumed that each node is aware of the set of channels being used on each of its outgoing edges.

Let ρ_{sd} be the chosen shortest route $E_s = N_0 \rightarrow N_1 \rightarrow \ldots \rightarrow N_k = d$ from E_s to E_d, where N_i is a node (router or end node). The symbol $\mathcal{C}^i$ will be used to denote the set of channels currently used over the link $N_i \rightarrow N_{i+1}$ for other communications.

$$N_0 = s \quad N_1 \qquad N_2 \qquad\qquad N_{i-1} \quad N_i \quad N_{i+1} \qquad N_k = d$$

Fig. 5.8. Route from E_s to E_d

The request phase

In the request phase, a request message flows from the source E_s towards the destination E_d using the route ρ_{sd}. The request message received by a node $N_i, 0 \leq i \leq k$, in the route from E_s to E_d includes a nonempty set C_i of possible channel numbers, each of which might be used to form a lightpath from E_s to N_i using the route $E_s = N_0 \rightarrow N_1 \rightarrow \ldots \rightarrow N_i$, a part of the route ρ_{sd} from E_s to E_d (Figure 5.8). The algorithm executed at node N_i, is given below. This algorithm is initiated at node $N_i, i \neq 0$, when it receives a request message from node N_{i-1}. This algorithm is initiated at node $N_0 = s$ when E_s is processing a request for a lightpath from E_s to E_d.

Node N_i computes a list C_{i+1} by excluding, from the set C_i, the set of channels $\mathcal{C}^i$ currently being used over the link $N_i \rightarrow N_{i+1}$ for other communications (i.e., $C_{i+1} = C_i - \mathcal{C}^i$). Node N_i communicates the set C_{i+1} to node N_{i+1}. The request phase is over when the request reaches the destination E_d. At this point, if the set of possible channels C_k received by E_d is empty, it is not possible to set up a lightpath from E_s to E_d, using the route ρ_{sd}. Otherwise, the destination E_d picks a channel number arbitrarily from the set C_k and enters the response phase. A formal description is given below.

Algorithm 1 : Processing during the request phase at node N_i.

---

begin

    if $i = 0$ **then**

        $C_1 \longleftarrow \{1, 2, \ldots, n_{ch}\} - C^0$

        if $C_1 = \emptyset$ **then**

            send message that the request for a lightpath cannot be set up using route ρ_{sd}

        end

        else

            send a request message, including set C_1, to node N_1

        end

        stop

    end

    if $i = k$ **then**

        randomly pick an element e, representing the channel number to be used for the lightpath, from C_k

        send a response message, including element e, back to N_{k-1} and stop

    end

    else

        $C_{i+1} \longleftarrow C_i - C^i$

        if $C_{i+1} \neq \emptyset$ **then**

            send a request message, including set C_{i+1}, to node N_{i+1} and stop

        end

        else

            send a response message, indicating that the request cannot be serviced, back to N_{i-1} and stop

        end

    end

end

The response phase

In the response phase, a response message flows back from a node in the path ρ_{sd} towards the source node E_s using the route ρ_{sd} backwards. The response message sent by a node in the route indicates whether the request for a lightpath using the route ρ_{sd} is possible and, if so, what channel should be used for the lightpath. If a lightpath could not be established, the communication might be blocked or the source node might attempt to establish a connection subsequently. The algorithm executed at node N_i is given below. This algorithm is initiated

- at node $N_i, 0 < i < k$ when it receives a response message from node N_{i+1},
- at node N_k, after it receives a request message from node N_{k-1},
- at node N_i, after it receives a request message from node N_{i-1}, and determines that the set $C_{i+1} = \emptyset$.

Algorithm 2 : Processing during the response phase at node N_i.

---

begin

    if $i = 0$ **then**

        if *the response message indicates that the lightpath can be set up using some channel* c_k **then**

            | start communication using channel c_k on the fiber from $N_0 \rightarrow N_1$

        end

        else

            | send message that the request for a lightpath cannot be set up using route ρ_{sd}

            | stop

        end

    end

    else

        if *the response message indicates that the lightpath can be set up using some channel* c_k **then**

            | set the router at N_i, so that signal, using channel c_k on the fiber $N_{i-1} \rightarrow N_i$, is routed to the fiber $N_i \rightarrow N_{i+1}$

        end

        send the response message to node N_{i-1}

        stop

    end

end

Example 5.3. In this example the same physical topology of a network shown in Figure 5.6 will be used. One difference is that, in this example, each fiber is capable of supporting five channels so that $C_0 = \{1, 2, 3, 4, 5\}$. Let there be a request for a lightpath from E_1 to E_5 and the selected route be $E_1 \rightarrow R_1 \rightarrow R_2 \rightarrow R_5 \rightarrow E_5$.

Request phase:

Step 1) At node E_1, the set C_1 is formed by excluding from C_0 the channel(s) already in use on the edge $E_1 \rightarrow R_1$. Thus $C_1 = \{1, 2, 3, 4, 5\} - \{1, 3\} = \{2, 4, 5\}$. The set C_1, being nonempty, is sent to node R_1.

Step 2) At node R_1, the set $C_2 = \{2, 4, 5\} - \{3\} = \{2, 4, 5\}$ is computed. The set C_2, being nonempty, is sent to node R_2.

Step 3) At node R_2, the set $C_3 = \{2, 4, 5\} - \{1, 2\} = \{4, 5\}$ is computed. The set C_3, being nonempty, is sent to node R_5.

Step 4) At node R_5, the set $C_4 = \{4, 5\} - \{1\} = \{4, 5\}$ is computed. The set C_4, being nonempty, is sent to node E_5.

Step 5) At node E_5 (the destination of this lightpath) the decision is made that the request for a lightpath will be successful and a channel for the lightpath using the route ρ_{sd} is selected. Any channel from the set $\{4, 5\}$ is a valid candidate. The node E_5 arbitrarily may pick 4, for instance, as the chosen channel number for the lightpath.

Response phase:

The response phase sends the message to the source E_s that channel number 4 may be used to establish a lightpath. This message uses the route $E_1 \rightarrow R_1 \rightarrow R_2 \rightarrow R_5 \rightarrow E_5$ backwards, i.e., $E_5 \rightarrow R_5 \rightarrow R_2 \rightarrow R_1 \rightarrow E_1$, in four steps.   □

This approach gives the same result as the approach involving the central agent (Section 5.4.1) so long as requests are arising at infrequent intervals where there is a guarantee that a request for a lightpath cannot occur until the previous request for a lightpath has been processed completely. This is not a realistic assumption, and the algorithm must take care of requests for lightpaths that have any possible temporal relationship to each other.

To illustrate this problem informally, let there be two requests for lightpaths, the first from E_1 to E_5 and the second from E_2 to E_5. The second request for a lightpath occurs before the request phase for the first lightpath is over. The steps for the first lightpath will be exactly the same as the Steps 1–5 outlined above for a lightpath from E_1 to E_5. The steps for the second lightpath will be similar to the Steps 1–5. For the second lightpath, the route will be $E_2 \rightarrow R_2 \rightarrow R_5 \rightarrow E_5$ and will be carried out independently of the Steps 1–5 for the first lightpath.

When processing the request for the second lightpath, at node R_2, the channels in the set $\{3, 4, 5\}$ are still available for the second lightpath. This means that two independent searches are working with the information that channels 4 and 5 are available to them. After the two searches are over, the same channel on the edge $R_2 \rightarrow R_5$ may be allocated to both the requests. A similar situation may happen at node R_5 as well.

The problem is that the channels on an edge are resources that may be used by any lightpath using that edge. However, more than one lightpath cannot use the same channel on a given edge. To ensure that channels are not allotted to more than one lightpath request, some scheme is needed for reserving, at node N_i, the channels in C_{i+1} before forwarding the set C_{i+1} to N_{i+1}.

A brief and informal description of a distributed algorithm for RWA using the shortest route ρ_{sd} for the lightpath from E_s to E_d is given below. This may be easily extended to consider alternative routes as done for the central agent approach (Section 5.4.1). More details of this and other algorithms for distributed RWA are available in [344].

It is clear from the above discussions that a period of time is needed from the time the request for a lightpath from E_s to E_d is received to the time a decision can be made on whether the lightpath can be set up or not. This time is called a *setup time* for the lightpath. For a given link $N_i \rightarrow N_{i+1}$ on a route ρ_{sd}, a channel c_p might be unavailable due to two reasons:

- some other communication is using the channel c_p using a different lightpath sharing the same link $N_i \rightarrow N_{i+1}$, or

- node N_i has reserved the channel c_p for some other lightpath sharing the same link but has not yet received the corresponding response message.

The duration of the unavailability due to the first reason depends on the call duration, the time for data communication after the lightpath is set up, while the unavailability due to the second reason depends on the expected setup time for the other lightpath that is using the link $N_i \rightarrow N_{i+1}$.

Using the standard convention of representing unavailability of shared resources by locks, two types of locks are needed: C-lock and T-lock. A channel is assigned a C-lock on a link if some ongoing communication is using the channel on that link. A channel included in C_{i+1} is assigned a T-lock on a link $N_i \rightarrow N_{i+1}$ during the setup time for processing that request for a lightpath. A C-lock is released when the corresponding communication is over. A T-lock on a link $N_i \rightarrow N_{i+1}$ is released when N_i receives a response message from N_{i+1}. The distributed algorithm is informally discussed using the same example used above.

Example 5.4. In the case of the network shown in Figure 5.6, each node will maintain, on each of its outgoing edges, a C-lock on the channels already in use. For instance, the node E_2 will maintain a C-lock on the channels 1 and 2 for edge $E_2 \rightarrow R_2$. The case of the request for a lightpath from E_1 to E_5 is discussed below.

Request phase:

Step 1) At node E_1, the set of channels $C_1 = \{2, 4, 5\}$ is formed. For the edge $E_1 \rightarrow R_1$, a T-lock is put on each of these channels, indicating that these channels are now being considered for allocation in response to a request for a lightpath. The set C_1 is sent to node R_1.

Steps 2–5 are identical to Steps 2–5 of the request phase described in example 5.3, except that, in each case, a T-lock is put on each of the channels in $\{2, 4, 5\}$ before the set of channels is forwarded to the next node in the route.

Response phase:

The response phase sends the message using the route $E_1 \rightarrow R_1 \rightarrow R_2 \rightarrow R_5 \rightarrow E_5$ backwards in four steps as before. One important difference is that each node receiving a response message releases the T-locks on the channels it was maintaining for this request and puts a C-lock on the selected channel, channel number 4 in this example.

The problem of allocating the same channel on an edge to two or more lightpaths clearly disappears when T-locks and C-locks are used. More details of the distributed algorithm are available in [343].   □

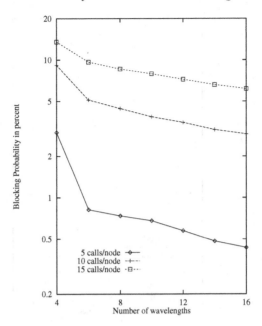

Fig. 5.9. Plot showing the variation of blocking probability with the number of channels used

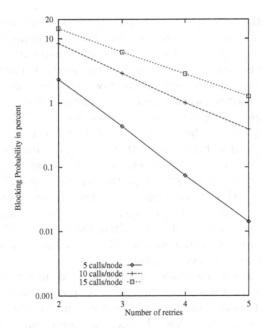

Fig. 5.10. Plot showing the variation of blocking probability with the retries

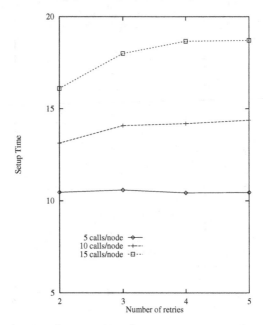

Fig. 5.11. Plot showing the variation of setup time against the number of retries

In studying the performance of the strategy for a distributed algorithm, the blocking probability and the setup time to establish a communication between a source-destination pair are the most important parameters. The blocking probability may be estimated, as before, by the ratio of occurrences of blocked connections to the total number of connections and the setup time by the time interval between the time when a request for communication was made and the time when the processing of the request was completed. When the processing of a request is complete, the source node either finds the channel number to be used for the lightpath or it finds that a lightpath cannot be established. The setup time is measured using unit time such that a node executes all the instructions in the distributed algorithm in one time unit. The maximum number of calls (each call representing a request for communication) that a node can handle at a time is determined by the number of transmitters in each end node, since the number of transmitters limits the number of lightpaths that can start from an end node. In this study the maximum number of calls/node was chosen to be 5, 10, and 15.

If a request for communication did not succeed during the first attempt, after some time the request may be "retried" one or more times, with the expectation that some valid RWA will be found in the second or subsequent attempts. Figure 5.9 shows how the blocking probability changes with the number of channels available in the network and the number of "retries". For each request for a lightpath arriving at a node i and attempting to use the edge $i \rightarrow j$, a T-lock was assigned for all channels not used in the route up to

i and not used in the edge $i \to j$. As a result, even with the availability of a larger number of channels in the network, the blocking probability improved only slightly (Figure 5.9). Increasing the number of retries proves to be quite effective (Figure 5.10) in reducing the blocking probability since every additional retry reduces the blocking probability by an order of magnitude. One would expect a large setup time as more retries are attempted. Interestingly enough, the setup time increases at a very slow rate (Figure 5.11); the increase of setup time is negligibly small with the increase of retries. It is interesting to note that, for the specific situation considered in the simulation experiment using a smaller number of channels and up to five retries gives an acceptably small blocking probability. More details are available in [343].

Exercise 5.12. Why is it that the availability of a large number of channels does not significantly improve the blocking probability? Can you think of a scenario when this would not be true?   □

Exercise 5.13. Formally describe the algorithms during the request phase and the response phase when T-locks and C-locks are used. When a request for a lightpath reaches a node N_i, and all channels on the next edge in the path have T-locks, your algorithm should simply block the request.   □

Exercise 5.14. Develop another distributed algorithm where such a request waits for some time t_{wait}, with the expectation that the lock of some of the channels with a T-lock will be released before time t_{wait} elapses. Study, using simulation, the behavior of this system and compare it with the algorithm described above.   □

Exercise 5.15. Develop a distributed algorithm that processes a request for a lightpath from E_s to E_d by exploring k alternate paths from E_s to E_d rather than a single path.

Do you expect this to have a better performance? Explain.   □

5.5 Bibliographic Notes

Extensive work on RWA has been reported in the last 15 years. This section is a brief review of some of the literature on RWAs. This includes RWA in networks where

   i) the wavelength continuity constraint must be satisfied,
   ii) each (or some) node has a Full Wavelength Converter (FWC),
   iii) each (or some) node has a Limited Wavelength Converter (LWC). [8]

---

[8] Described in Section 2.3.6.

A critical survey of RWA research up to 1997 has been given in [426], including the performance of analytical methods and heuristics, control and management of all-optical networks, and prototype systems and commercial networks. Another survey of RWA, including heuristics for wavelength assignment, appears in [433].

Fixed-point approximation models for networks using fixed-alternate routing was studied in [277]. Blocking probability models for a wavelength-continuous path were proposed in [60], which also examined least-used routing. A method to find an optimal logical path (consisting of one or more lightpaths) from a given source to a destination, based on a simple graph model of the physical topology, is given in [87]. An analytical model for fixed-alternate routing has been proposed [186] to consider the effect of blocking probability of paths with different numbers of hops and different wavelength-assignment policies. The paper also considered dynamic routing and compared the performance of alternate routing with that of dynamic routing. Dynamic routing schemes such as least-loaded routing achieve significantly better blocking performance, when compared with fixed shortest-path routing, in wavelength-continuous and wavelength-convertible networks [220]. Adaptive wavelength routing has been examined in [283], and it was shown that adaptive routing outperforms constrained routing schemes such as alternate routing. An integer linear program (ILP) formulation is proposed in [28] for the exact solution of the routing and wavelength allocation (RWA) problem. Lower bounds are discussed and new lightpath allocation heuristic algorithms based on the heu-RWA heuristic described in Section 5.3. The problem of wavelength assignment in Manhattan Street Networks has been addressed in [225], using a two-stage process involving simulated annealing.

The approach in [320] provides a reliable, distributed protocol for lightpath setup and teardown in an underlying optical layer which provides links between electronic ATM switches. Different wavelength reservation policies, as well as the effectiveness of wavelength conversion for distributed control in optical networks, have been considered in [431]. A review of dynamic RWA, followed by a critical comparison of two distributed control and management protocols, appears in [432]. Each protocol has some advantages and disadvantages with respect to setup time and blocking probabilities under different conditions. In distributed wavelength provisioning, the information on the lightpaths on every fiber in the network is updated only at intervals. This delay is unavoidable in distributed systems, so that the "current" global information about wavelength availability cannot be guaranteed at any particular place and time. The performance of the protocol has been compared to a link-state-based approach [321].

The problem of maximizing the number of lightpaths, given a set of source-destination pairs in a network with no wavelength converters, has been initially formulated in [231] as an integer linear program. LP-relaxation of the ILP formulation makes the approach viable for practical-sized networks. In [298], it is shown how, with the adoption of a cost function (sum of the amount of flow on all the links), the resulting formulation may be solved using Linear Programming rather than Integer Linear Programs for some special topologies, such as the ring networks under some assumptions. For arbitrary networks with full wavelength conversion, a rounding method gives fast, efficient solutions that are near-optimal solutions.

The concept of a *tunnel*, defined as a group of consecutive channel numbers that are switched together, has been introduced in [195]. The paper discusses the architecture of such a system, and its advantages and disadvantages. The RWA problem in this approach is to determine if an existing tunnel may be used for a new request for connection or whether a new tunnel has to be established.

In [227] it was shown that better network usage is possible if routing at the IP layer and RWA at the optical layer are considered together. The idea is to decide whether the existing topology can support the new request and, if so, which route is "good", and whether it is better to establish a new lightpath and if so, what is a good RWA.

If, in dynamic lightpath allocation, some existing lightpaths are allowed to be "rearranged" in order to support a new request for connection, the corresponding RWA algorithm is termed in [331] as *rearrangeably nonblocking*. The goal of [331] is to design, for a bidirectional ring topology and a torus topology, an online RWA algorithm which is rearrangeably nonblocking using the minimum number of wavelengths, with few rearrangements of existing lightpaths. A similar study for tree networks appears in [330].

The problem of routing a set of lightpath demands for which the start and end dates are known is discussed in [234]. The objective in [234] is to minimize the number of channels in the network. A tabu search meta-heuristic has been formulated to find approximate solutions.

Future networks will support diverse applications with different characteristics (e.g., arrival rates and call holding times). In [325] the algorithm in [451] has been modified to include the cases of multiple classes of calls and multicast traffic.

5.5.1 Sparse Wavelength Conversion and RWA

Analytical models in [373] evaluate the blocking performance of networks under dynamic Poisson traffic for a given conversion density (the fraction of nodes with full wavelength conversion capability). Three regular network topologies — the ring, the mesh-torus, and the hypercube — have been analyzed. The results show that either a relatively small number of converters is sufficient for a certain level of performance or that conversion does not offer a significant advantage. The benefits of wavelength conversion are largely dependent on the network load, the number of available wavelengths, and the connectivity of the network. It was shown in [373] that the advantages of wavelength conversion are much higher in a larger network and as the network size increases, performance increases dramatically initially with conversion density. Simulation-based optimization approaches have been outlined in [410]. The utilization statistics of FWC are determined using simulation and these statistics is used to allocate the FWCs in an optimal way. Using simulations on regular and irregular networks under both uniform and nonuniform traffic, it was shown that this approach can significantly reduce the number of FWCs. An approximate analytical model for fixed-alternate routing using sparse wavelength conversion appears in [314]. This paper includes simulation studies of the relationships between alternate routing and wavelength conversion on three representative network topologies. A new dynamic RWA algorithm, called the *weighted least-congestion routing and first-fit wavelength assignment (WLCR-FF)*, that considers the distribution of free wavelengths and the hop length of each route jointly, has been proposed in [92]. In networks with sparse or full wavelength conversion, the proposed WLCR-FF algorithm is shown to be better than strategies such as the fixed routing or the fixed-alternate routing. In order to reduce the blocking probability, the need to consider, simultaneously, the RWA and the wavelength conversion was pointed out in [92]. Networks with sparse conversion also have been considered in [451]. In this work, a network with point-to-point calls is decomposed into a number of path subsystems, which are analyzed separately to give approximate solutions. These solutions are combined to form a solution for the overall network.

The following two different reasons for connection blocking have been considered in [261]:

- a free channel is not available,
- the global information is outdated.

Distributed RWA with sparse wavelength conversion capabilities, including an analysis of the blocking probability, has been considered in [261]. These analytical models are shown to be highly accurate, when compared to simulation results.

5.5.2 Limited Wavelength Conversion and RWA

Ring networks, tree networks, and (with certain restrictions) networks of arbitrary topology, with fixed wavelength conversion capabilities, have been investigated in [319]. It was shown that significant improvements in traffic-carrying capacity can be obtained in such networks by providing very limited wavelength conversion capability within the network. For instance, it was shown that there are ring and star networks that only require wavelengths to be shifted to their nearest wavelengths. The problem of wavelength assignment in bidirectional ring networks with wavelength converters was considered in [84]. A tight lower bound for the number of wavelengths needed to support all possible logical topologies using as few wavelength converters as possible was developed in [84]. An approximate analytical method for RWA for networks with limited wavelength conversion, validated using simulation of some regular topologies as well as the NSFNET and the ARPANET, appears in [290]. An analysis of static and dynamic RWA for regular, all-optical networks with limited wavelength conversion is given in [238]. For these topologies it was shown that having one alternate choice for the next hop gives as much improvement as having twice as many wavelengths. A dynamic RWA algorithm in all-optical networks with limited range wavelength conversions has also been outlined in [238]. The objective was to accommodate as many requests as possible with the least number of wavelength conversions. For each new request, a wavelength weight gives the correlation of different wavelengths on different fibers in all possible routes. In [238] the sum of the weights on each candidate route is used to choose the best RWA. The benefits of limited wavelength conversion in some regular all-optical WDM networks (torus, hypercube) was explored in [350]. It was shown that, for these regular networks, most of the performance advantages of full conversion may be obtained using limited wavelength conversion. A switch architecture to achieve limited wavelength conversion has been proposed in [350]. The model in [60] has been extended in [388] to calculate the average blocking probability in all-optical networks when limited range wavelength conversion is available. It was shown that the performance improvement obtained by full wavelength conversion over no wavelength conversion can be almost achieved by using limited wavelength conversion with the degree of conversion being only 1 or 2. A heuristic for wavelength and converter assignment in a WDM network with limited wavelength conversion capabilities, using fixed shortest-path routing, has been proposed in [352].

5.5.3 Placement of Wavelength Converters

A heuristic to minimize the call-blocking probability by properly placing a given number of wavelength-convertible nodes and selecting the routes of the lightpaths has been described in [187]. The heuristic first determines the location of wavelength-convertible nodes assuming the use of the shortest-path route for every lightpath. The actual route of the lightpaths is then decided upon to further reduce the blocking probability. A heuristic algorithm for optimal converter placement in networks with sparse wavelength conversion has been proposed in [13] using a dynamic traffic model and fixed routing. The goal is to place a given number of wavelength converters such that an objective function (either the average blocking probability or the maximum blocking probability over all paths) is minimized. Given a network topology, the number of full wavelength converters available, and the traffic between the end nodes, the optimal placement of the wavelength converters in the network has been studied in [375]. This paper gives a dynamic programming approach to placing the full wavelength converters to minimize the average blocking probability. The approach has been extended to obtain optimal placements for the bus and ring topologies. In [412] it was concluded that the wavelength converter placement and RWA algorithms are "closely related" so that a wavelength converter placement mechanism that works well for a particular

RWA algorithm might not be appropriate with a different RWA algorithm. This paper looks at converter placement problems using the fixed-alternate routing (FAR) algorithm [186] and least-loaded routing (LLR) algorithm [249] and has proposed heuristics for placing wavelength converters to minimize the overall blocking probability. An efficient algorithm for the optimal placement of a given number of wavelength converters in all-optical networks with arbitrary topologies has been proposed in [381]. It was established that this approach works quite well in networks with high connectivity when the number of converters to be placed is about half that of the number of nodes in the network. A heuristic for placing limited wavelength converters to minimize the average blocking probability in mesh topologies appears in [394]. In [450] it was reported that the advantages of sparse wavelength conversion are significant only when random channel allocation is used. Benefits similar to that of sparse wavelength conversion may usually be obtained by the first-fit or the most-used allocation policy. A number of routing and channel assignment strategies have been studied in [245] to conclude that they "do not work well in the presence of wavelength conversion since they usually only take into consideration the distribution of available wavelengths, and do not explicitly consider the lengths of routes". In this paper a strategy to carry out RWA and converter placement together has been presented.

5.5.4 Multi-fiber Networks

An important class of WDM networks is one where each pair of nodes may have any number of fibers linking them. A network where each link between nodes consists of a bundle of fibers is called a *multi-fiber* network[24]. In such a network, even if there is no wavelength converter, when setting up a lightpath, there is flexibility in selecting the appropriate fiber in the bundle constituting each link in the network. This is called *virtual wavelength translation*[24] since the effect is similar to wavelength translation. RWA for multi-fiber networks, along with the problem of allocation of wavelength converters, has been studied in [433]. In addition to the heuristics already discussed in Section 5.4, the paper reviews and analyzes RWA for multi-fiber networks *Min-Product* (MP), *Least-loaded* (LL), *Max-sum* (M$\sum$), and *Relative-Capacity-Loss* (RCL) and proposes a distributed version of RCL. It was shown that the most-used and the first-fit allocation have similar blocking probabilities except that the network state has to be known globally in the most-used allocation policy so that the first-fit allocation is easier to implement, with lower overhead. The random allocation gives the lower bound on the call blocking probability. If all nodes in the network have full wavelength converters, this gives the upper bound on the call blocking probability. The performance of the first-fit policy is similar to that of random allocation with sparse wavelength conversion. It was also observed that the first-fit policy works poorly if the blocking probability is high. A model for the multi-fiber network called the *multi-fiber link-load correlation* (MLLC) model has been proposed in [250] to evaluate the blocking performance of multi-fiber networks. This model has not made the independent wavelength load assumption [32]. The MLLC model is shown to be accurate and it was observed that a small number of fibers in the bundle constituting each link is sufficient to guarantee high network performance [250]. The problem of RWA to maximize the number of requests honored using dynamic lightpath allocation for multi-fiber networks without wavelength converters has been considered in [276, 418]. In [419] a new wavelength assignment algorithm has been proposed, called the Relative Least Influence (RLI). This is a dynamic centralized algorithm that allows fixed as well as fixed-alternate routing for use in single-fiber networks as well as multi-fiber networks. The algorithm takes into account the effect of establishing a call on all the potential paths. Simulation results have been used to determine the viability of the approach. In [362], it was shown that the advantages of virtual wavelength translation in multi-fiber are substantial — comparable, in some cases, to that of having full wavelength translators in each node without the associated cost.

Logical Topology Design I

Topologies for data communication may be broadly categorized as *regular* or *irregular*. The topologies with fixed and simple structural properties are called regular topologies. The topologies which have no such structural properties are characterized as irregular. In this chapter, the application of regular and *almost regular* topologies for broadcast-and-select networks is discussed. The logical topologies for wavelength-routed networks, derived using MILP or heuristics, are discussed in Chapter 7. Such topologies belong to the category of irregular topologies.

Regular topologies have useful graph-theoretical properties that have been studied in detail. Regular topologies are very attractive in designing multiprocessor systems [205, 371, 356], with potentially thousands of processors. An important requirement in a multiprocessor system is that every processor must be able to communicate with any other processor very quickly. For inter-processor communication, many schemes for connections between processors, often called *interconnection architectures*, have been proposed. Some examples of such interconnection architectures are the *shufflenet* [194], the *de Bruijn graph* [305, 345, 364, 359], the *2-dimensional mesh* [205], the *hypercube* [56, 111], the *Kautz graph* [300], and the *multimesh* [102].

As discussed in Section 1.1.2, the logical topology of a WDM network is determined by the lightpaths in the network. Even though the route and wavelength assignment (RWA), discussed in Chapters 4 and 5, imposes restrictions on logical topologies, establishing complex logical topologies has become increasingly feasible with the large number of channels allowed on a single fiber.

For multi-hop broadcast-and-select networks, discussed in Section 3.1, defining any logical topology is possible, subject to RWA restrictions, simply by tuning the transmitters and receivers in the end nodes appropriately. The complexity of the logical topology is limited only by the number of transmitters and receivers at each end node and the number of channels n_{ch} permitted on a fiber. Interconnection architectures are attractive, as possible logical topologies for multi-hop broadcast-and-select networks, due to the following reasons:

- opto-electronic and electro-optical conversions in a multi-hop network are expensive so that a network with a low diameter is preferable to a network with a larger diameter,
- designing a logical topology with a very large number of end nodes is not a problem since interconnection architectures have well-defined rules for connectivity,
- routing algorithms in regular topologies are easy to define and do not involve any routing table.

A brief review of the de Bruijn graph and some related graph theoretic terms are given in Appendix 2. As explained in Appendix 2, a de Bruijn graph with d^k vertices, for some d and k, is called a $B(d, k)$. The rules for defining the directed edges between the d^k vertices have also been reviewed in Appendix 2. The de Bruijn graph is a possible candidate as a logical topology in a broadcast-and-select network. Each vertex u in a $B(d, k)$ will correspond to an end node E_u in a broadcast-and-select network with d^k end nodes. As explained in Appendix 2, a vertex u in a $B(d, k)$ is assigned a k-digit address $u_k u_{k-1} \ldots u_2 u_1$, where $u_i \in \{0, 1, 2, \ldots, d-1\}$, for all $i, 1 \leq i \leq k$. It is convenient to associate the same address $u_k u_{k-1} \ldots u_2 u_1$ with end node E_u in a broadcast-and-select network, using $B(d, k)$ as its logical topology.

Let there exist an edge $u \to v$ in a $B(d, k)$ and let end node E_u (E_v) in the broadcast-and-select network correspond to vertex u (v) in $B(d, k)$. Then a logical edge $E_u \Rightarrow E_v$ must exist in the logical topology for the broadcast-and-select network (so that there is a lightpath from E_u to E_v). In a $B(d, k)$ the vertices with addresses in the form $jj \ldots j$, $j \in \{0, 1, 2, \ldots, d-1\}$, have a self-loop and exactly $d-1$ incoming and $d-1$ outgoing edges to other vertices. All other vertices have exactly d incoming and d outgoing edges and no self-loops. In a broadcast-and-select network, each end node with an address in the form $jj \ldots j$, $j \in \{0, 1, 2, \ldots, d-1\}$, must have $d-1$ transmitters and $d-1$ receivers. The remaining end nodes must have d transmitters and d receivers.

The low diameters and the straightforward routing algorithms are attractive features of interconnection architectures (e.g., the de Bruijn graph). However, in general, there are the following major problems in directly using a regular topology as the logical topology of a multi-hop broadcast-and-select network:

- The number of vertices in an interconnection architecture must be given by some well-defined formula. For example, if a de Bruijn graph $B(d, k)$ is chosen as the model for the logical topology of a broadcast-and-select network, the number of end nodes $\mathcal{N}_E$ in the network must be given by the formula $\mathcal{N}_E = d^k$. In other words, if d is fixed, the number of end nodes can only be d, d^2, d^3, etc. It is not realistic to require that the number of end nodes in a broadcast-and-select network satisfy such a relationship.

- The regular topology does not take into account the traffic in the network. It is therefore quite possible that the amount of traffic on some logical edges is high while others have relatively very low traffic so that they are underutilized.

Optical network researchers have proposed [286, 300, 337] generalizations of interconnection architectures to take care of the first problem. In such generalizations, there is considerable flexibility in the number of end nodes in the network, retaining the advantages of low diameter and a simple routing algorithm of interconnection architectures. It is also desirable to make such topologies scalable so that additional end nodes may be added without modifying a large number of existing lightpaths. One generalization, based on the idea of the de Bruijn graph, is given below [209].

6.1 A Scalable Topology Based on the de Bruijn Graph

Given two integers N_S and d, this topology is described as a digraph $G_S = (V_S, E_S)$ where V_S is the set of all vertices in the graph, $|V_S| = N_S$ and E_S the set of all directed edges in the graph. The graph G_S is based on the de Bruijn graph $B(d, k)$. Some important features of this topology are as follows:

- it allows any arbitrary number N_S of vertices, $d^k \leq N_S \leq (d+1)^k$,
- it is scalable so that it is possible to add new vertices to the graph with relatively little perturbation of the graph,
- it has a low diameter,
- the routing strategy is simple and does not involve any routing table.

If the number of vertices $N_S = d^k$ (or $N_S = (d+1)^k$), the topology is the same as the de Bruijn graph $B(d, k)$ (or $B(d+1, k)$). These cases need not be discussed any further and it is assumed, in the following discussions, that $d^k < N_S < (d+1)^k$.

One interesting aspect of the approach is that the regularity condition is followed somewhat loosely. The topology can be characterized as *almost regular* since the number of incoming and outgoing edges are both almost equal to d. This is in contrast to existing low diameter topologies which are based on regular graphs where each vertex has the same number of incoming and outgoing edges.

Let Z_{d+1} be the set of all digits $\{0, 1, \ldots, d\}$ and let Z_{d+1}^k be the set of all k-digit strings $x_k \ldots x_2 x_1$, where digit $x_i \in Z_{d+1}$, for all $i, 1 \leq i \leq k$. The set Z_{d+1}^k may be partitioned into $k+1$ subsets $S_0, S_1, \ldots, S_k$ such that all strings with no occurrence of digit d are included in S_0 and all strings in Z_{d+1}^k having x_j as the leftmost occurrence of digit d are included in $S_j, 1 \leq j \leq k$. It is easy to see that $|S_0| = d^k$ and $|S_j| = d^{k-j}(d+1)^j, 1 \leq j \leq k$.

Each vertex x of G_S will be assigned an address, a unique element of the set Z_{d+1}^k, a string $x_k \ldots x_2 x_1$, $x_i \in Z_{d+1}, 1 \leq i \leq k$. Since the number of vertices, $\mathcal{N}_S$, is greater than d^k, some end nodes will have addresses containing the digit d. In order to describe the rule for assigning addresses to vertices, an ordering relation between any pair of strings $x_k \ldots x_2 x_1$ and $y_k \ldots y_2 y_1$ is defined as follows.

Definition 6.1. *If* $x_k \ldots x_2 x_1 \in S_i$, $y_k \ldots y_2 y_1 \in S_j$, *and* $i < j$ *then* $x_k \ldots x_2 x_1 < y_k \ldots y_2 y_1$.

Definition 6.2. *For two strings* $x_k \ldots x_2 x_1$, $y_k \ldots y_2 y_1 \in S_j, 0 \leq j \leq k$, *if* t *is the largest integer such that* $x_t \neq y_t$ *then* $x_k \ldots x_2 x_1 < y_k \ldots y_2 y_1$ *if* $x_t < y_t$.

Definition 6.3. *For any string* $x_k \ldots x_2 x_1 \in S_j, j > 0$, *the term* i-*conjugate of* $x_k \ldots x_2 x_1$ *will depict the string* $x_k \ldots x_{j+1} i x_{j-1} \ldots x_2 x_1$ *obtained by replacing the leftmost occurrence of digit* d *in* $x_k \ldots x_2 x_1$ *by* i *for any* $i, 0 \leq i \leq d-1$.

If the string $x_k \ldots x_2 x_1 \in S_0$, it does not have any i-conjugate.

Example 6.1. If $d = 2$ and $k = 2$, $Z_3^2 = \{00, 01, 02, 10, 11, 12, 20, 21, 22\}$, the sets $S_0 = \{00, 01, 10, 11\}$, $S_1 = \{02, 12\}$, and $S_2 = \{20, 21, 22\}$. The ordering between the elements in Z_3^2 is $00 < 01 < 10 < 11 < 02 < 12 < 20 < 21 < 22$. The 0-conjugate (1-conjugate) of the string 12 is 10 (11). □

Since $\mathcal{N}_S < (d + 1)^k$, all strings of Z_{d+1}^k will not be used as addresses for the vertices in G_S. The $\mathcal{N}_S$ smallest strings from Z_{d+1}^k will be used as addresses for the $\mathcal{N}_S$ vertices of G_S.

Definition 6.4. *A string, already assigned as the address of some vertex in* G_S, *will be called a used string; otherwise the string will be called an unused string.*

Example 6.2. If $d = 2$ and $k = 2$, the vertices in a graph with $\mathcal{N}_S = 6$ will be assigned addresses 00, 01, 10, 11, 02, and 12. All the strings 00, 01, 10, 11, 02, and 12 are used strings. The strings 20, 21, and 22 are unused strings. □

The following observations are direct consequences of the above discussions.

Property 6.1. All strings of S_0 are used strings.

Property 6.2. If $x_k \ldots x_2 x_1 \in S_j$ is a used string, then for all $i, 0 \leq i < d$, the i-conjugate of $x_k \ldots x_2 x_1$ is a used string.

Property 6.3. If $x_k \ldots x_2 x_1 \in S_j$ is a used string, then all strings of $S_0, S_1, \ldots,$ S_{j-1} are used strings.

6.1.1 The Topology of the Network

To define the topology of the network, the set of edges E_S of the digraph G_S is defined as follows. Let two vertices u and v in G_S have addresses $u_k \dots u_2 u_1$ and $v_k \dots v_2 v_1$. The edge $u \to v$ will be in G_S if

> **Rule 1:** $u_{k-1} u_{k-2} \dots u_2 u_1 j$ is a used string for some $j \in Z_{d+1}$ and $u_{k-1} u_{k-2} \dots u_1 j = v_k \dots v_2 v_1$.
>
> **Rule 2:** $u_{k-1} u_{k-2} \dots u_2 u_1 j$ is an unused string for some $j \in \{0, 1, \dots, d-1\}$ and $v_k v_{k-1} \dots v_2 v_1$ is the j-conjugate of u.
>
> **Rule 3:** $u_{k-1} u_{k-2} \dots u_2 u_1 d$ is an unused string, $u_{k-1} u_{k-2} \dots u_2 u_1 0$ is a used string, and $v_k \dots v_2 v_1$ is the 0-conjugate of $u_k \dots u_2 u_1$.

Rule 1 is the same as the rule for defining edges in a de Bruijn graph. If $u_{k-1} u_{k-2} \dots u_2 u_1 j, j \in Z_{d+1}$, is an unused string, it means that the vertex with the address $u_{k-1} u_{k-2} \dots u_2 u_1 j$ does not exist. Rule 2 is to take care of the situation when the vertex $u_{k-1} u_{k-2} \dots u_2 u_1 j$ does not exist. In this case the jth edge from u is to the j-conjugate of u. Rule 3 is for a special situation where there is an edge from $u_k \dots u_2 u_1$ to $u_{k-1} u_{k-2} \dots u_2 u_1 0$ but the vertex with the address $u_{k-1} u_{k-2} \dots u_2 u_1 d$ does not exist so that the dth edge cannot go from $u_k \dots u_2 u_1$ to $u_{k-1} u_{k-2} \dots u_2 u_1 d$. Only in this special case, the dth edge goes to the 0-conjugate of $u_k \dots u_2 u_1$.

As an example, a graph with six vertices and a graph with seven vertices, both for $d = 2$, are shown in Figures 6.1 and 6.2. In these graphs, a dotted line from vertex u to vertex v represents an edge when v is the j-conjugate of u, for some j. Otherwise the edges are shown by solid lines. It is important to note that, when the j-conjugate of u is computed, $0 \le j \le d - 1$.

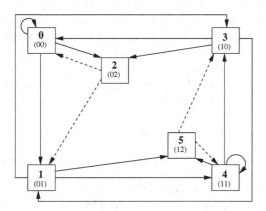

Fig. 6.1. A graph with $\mathcal{N}_S = 6$ vertices

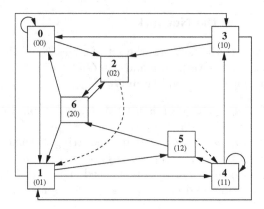

Fig. 6.2. A graph with $\mathcal{N}_S = 7$ vertices

Theorem 6.1. *In G_S, each vertex has an out-degree between d and $d+1$.*

Proof. Let a vertex $u \in V_S$ be assigned the address $u_k \dots u_2 u_1 \in S_j$.

i) If $j = 0$, for every vertex v having an address $u_{k-1} \dots u_2 u_1 t \in S_0, 0 \leq t < d$, the edge $u \to v$ exists. If $u_{k-1} \dots u_2 u_1 d$ is a used string then u has a directed edge to the vertex having an address $u_{k-1} \dots u_2 u_1 d$ and hence u has out-degree $d + 1$; otherwise u has out-degree d.

ii) If $j > 0$, the string $u_{k-1} \dots u_2 u_1 t \in S_{j+1}$. For any $t, 0 \leq t < d$, if $u_{k-1} \dots u_2 u_1 t$ is a used string, then u has an outgoing edge to the vertex having address $u_{k-1} \dots u_2 u_1 t$. Otherwise, the t-conjugate of u is a used string (Property 6.2) and u has a directed edge to the vertex having the t-conjugate of u as its address for $0 \leq t < d$. Therefore, u has an out-degree of at least d.

If the address $u_{k-1} \dots u_2 u_1 d$ is a used string there will be a $(d+1)^{\text{th}}$ edge from $u_k \dots u_2 u_1$ to $u_{k-1} \dots u_2 u_1 d$ so that u has an out-degree of $d + 1$.

If the address $u_{k-1} \dots u_2 u_1 d$ is an unused string, and the vertex with the address $u_{k-1} \dots u_2 u_1 0$ exists, the 0th edge from u is from $u_k \dots u_2 u_1$ to $u_{k-1} \dots u_2 u_1 0$. In this case it is guaranteed that the 0-conjugate of u is a used string (Property 6.2) and the 0th edge from u is not an edge to the 0-conjugate of u. In this case, the $(d+1)^{\text{th}}$ edge is from u to the 0-conjugate of u, so that u has an out-degree of $d + 1$.

In summary, u has out-degree d or $d + 1$.   □

Theorem 6.2. *In G_S, each end node has an in-degree between d and $d+2$.*

Proof. Let a vertex $u \in V_S$ be assigned the address $u_k \dots u_2 u_1 \in S_j$. Then, for all $t, 1 \leq t < d$, the string $t u_k \dots u_3 u_2 \in S_{j-1}$. The string $t u_k \dots u_3 u_2$ must be a used string (Property 6.3) and is the address of some vertex $v \in V_S$. Thus the directed edge $v \to u$ exists and u has an in-degree at least d.

There may be edges of the form $w \to u$ because u is a j-conjugate of w, for some j, and there can be one edge from $d u_k \dots u_3 u_2$ to $u_k \dots u_2 u_1 \in S_j$.

Let the largest used string from Z_{d+1}^k be an element of $S_{\hat{M}}$, for some $\hat{M}, 1 \leq \hat{M} \leq k$. Clearly u cannot be a j-conjugate of w if $w \in S_p$, for any $p > \hat{M}$, because all strings having d in a digit position higher than $\hat{M}$ are unused strings. Also, u cannot be j-conjugate of w if $p < \hat{M} - 1$, because all strings in $S_{\hat{M}-1}$ are used strings and therefore there is never a need to connect the outgoing edge of a vertex having an address in S_p, $p < \hat{M} - 1$, to one of its j-conjugates. Hence if u is a j-conjugate of w, $w \in S_{\hat{M}-1}$ or $w \in S_{\hat{M}}$.

There are two cases to consider, depending on whether $\hat{M} < k$ or not.

Case 1) If $\hat{M} < k$, there is no possibility of a vertex having an address $du_k \ldots u_3 u_2$, so there cannot be an edge $du_k \ldots u_3 u_2 \rightarrow u_k \ldots u_2 u_1 \in S_j$. Since there can be at most two vertices, such that there can be an edge from each of these vertices to u because u is their j-conjugate, the in-degree of u can be at most $d + 2$.

Case 2) If $\hat{M} = k$, there can be an edge $du_k \ldots u_3 u_2 \rightarrow u_k \ldots u_3 u_2 u_1$.

In addition there may be at most two edges $w^1 \rightarrow u, w^1 \in S_k$ and $w^2 \rightarrow u, w^2 \in S_{k-1}$ because u is a j-conjugate of w^1 and w^2. Let the address of w^1 (w^2) be $dw_{k-1}^1 \ldots w_2^1 w_1^1$ ($w_k^2 dw_{k-2}^2 \ldots w_2^2 w_1^2$). If $w^1 \rightarrow u$ exists, $w_{k-1}^1 = d$; otherwise $w_{k-1}^1 \ldots w_2^1 w_1^1 t$ is a used string. In this situation, $u_{k-1} = d$. If $u_{k-1} = d$, u cannot be a j-conjugate of any vertex $w^2 \in S_{k-1}, 1 \leq j < d$. In other words, both $w^1 \rightarrow u, w^1 \in S_k$, and $w^2 \rightarrow u, w^2 \in S_{k-1}$, cannot exist. Thus the total number of incoming edges cannot exceed $d + 2$.   □

In conclusion, if G_S is used as the logical topology of a broadcast-and-select network, taking into account the self-loops, each end node in the network will have between $d-1$ and $d+1$ transmitters and between $d-1$ and $d+2$ receivers.

6.2 Addition of an End node to an Existing Network

Adding a new end node E_w to an existing broadcast-and-select network using G_S as its logical topology is the same as adding a new vertex w to a graph G_S giving a new graph G_S^{new}. The logical topology of the broadcast-and-select network after adding the end node E_w will be that defined by the new graph G_S^{new}. The graph G_S^{new} should be such that

i) the address of the new vertex w must be assigned the smallest unused string in Z_{d+1}^k and
ii) the edges in the new graph still satisfy the Rules 1–3 described in Section 6.1.1.

Let the unused string allotted to w be $w_k \ldots w_2 w_1 \in S_j$, for some $j, 1 \leq j \leq k$. To insert w, the following three-phase process may be used:

Phase 1) redefine some existing directed edges to create the edges to w,

Phase 2) define requisite edges from $\mathfrak{w}$, and

Phase 3) if $\mathfrak{w}_1 = 0$, define an additional edge in the network.

In Phase 1, any existing vertex x with an address $t\mathfrak{w}_k \ldots \mathfrak{w}_3\mathfrak{w}_2$, such that $0 \le t < d$, must satisfy the condition $t\mathfrak{w}_k \ldots \mathfrak{w}_3\mathfrak{w}_2 \in S_{j-1}$. Before the addition of the vertex $\mathfrak{w}$, $\mathfrak{w}_k \ldots \mathfrak{w}_2\mathfrak{w}_1$ was an unused string so that the vertex x had an edge to the $\mathfrak{w}_1$-conjugate of x, that is, to the vertex with the address $t\mathfrak{w}_k \ldots \mathfrak{w}_j\mathfrak{w}_1\mathfrak{w}_{j-2} \ldots \mathfrak{w}_2$. For every $t, 0 \le t < d$, the edge $t\mathfrak{w}_k \ldots \mathfrak{w}_3\mathfrak{w}_2 \to t\mathfrak{w}_k \ldots \mathfrak{w}_j\mathfrak{w}_1\mathfrak{w}_{j-2} \ldots \mathfrak{w}_2$ must be replaced by the edge $t\mathfrak{w}_k \ldots \mathfrak{w}_3\mathfrak{w}_2 \to \mathfrak{w}_k \ldots \mathfrak{w}_1$. Each such replacement consists of deleting one edge and adding the new edge. This phase involves deletion of d edges and addition of d edges.

It may be readily verified that the situation that an existing vertex x has an address $d\mathfrak{w}_k\mathfrak{w}_{k-1} \ldots \mathfrak{w}_3\mathfrak{w}_2$ where $\mathfrak{w}_k \ldots \mathfrak{w}_2\mathfrak{w}_1$ is an unused string before the addition of vertex $\mathfrak{w}$ cannot happen. Therefore this situation need not be considered.

In Phase 2, there are two cases to consider when defining the outgoing edges from $\mathfrak{w}$. Case 1 applies when the string $\mathfrak{w}_{k-1} \ldots \mathfrak{w}_1 t$ is a used string. This can happen only if $j = k$. Case 2 is for all remaining situations where the string $\mathfrak{w}_{k-1} \ldots \mathfrak{w}_1 t$ is an unused string so that there is no vertex with this address. This will always happen when $j \ne k$.

Case 1: In this case, an edge $\mathfrak{w}_k \ldots \mathfrak{w}_2\mathfrak{w}_1 \to \mathfrak{w}_{k-1} \ldots \mathfrak{w}_2\mathfrak{w}_1 t$ must be created. This follows from Rule 1 in Section 6.1.1 and involves addition of up to $d + 1$ edges.

Case 2: The tth outgoing edge from $\mathfrak{w}$ will be to the t-conjugate of $\mathfrak{w}$ for all $t, 0 \le t < d$, where $\mathfrak{w}_{k-1} \ldots \mathfrak{w}_1 t$ is an unused string. In other words, the tth outgoing edge from $\mathfrak{w}$ will be $\mathfrak{w}_k \ldots \mathfrak{w}_3\mathfrak{w}_2\mathfrak{w}_1 \to \mathfrak{w}_k \ldots \mathfrak{w}_{j+1}t\mathfrak{w}_{j-1} \ldots \mathfrak{w}_1$. This phase follows from Rule 2 in Section 6.1.1 and involves up to d edges.

Phase 3 is needed only if $\mathfrak{w}_1 = 0$. In this case an edge $t\mathfrak{w}_k \ldots \mathfrak{w}_3\mathfrak{w}_2 \to t\mathfrak{w}_k \ldots \mathfrak{w}_{j+1}0\mathfrak{w}_{j-1} \ldots \mathfrak{w}_2$ has to be added. This phase follows from Rule 3 in Section 6.1.1 and involves one edge.

When the process is over, the resulting topology follows the rules given in Rules 1–3 in Section 6.1.1 with the vertex $\mathfrak{w}$ included in the graph. From the above discussion, it follows that the addition of a vertex to an existing graph G_S requires $O(d)$ edge removals and $O(d)$ edge insertions to create the new graph G_S^{new}.

Example 6.3. To add a vertex to the graph $G_S = (V_S, E_S)$ with $\mathcal{N}_S = 6, d = 2$, and $k = 2$ shown in Figure 6.1, the address of the new vertex must be the smallest unused string, 20, for the graph G_S. As the new vertex with address 20 is inserted, the edges $02 \to 00$ and $12 \to 10$ (edges from vertices with addresses 02 and 12 to their respective 0-conjugates) are replaced by the edges $02 \to 20$ and $12 \to 20$. As discussed above, the edges $20 \to 00$, $20 \to 01$, and $20 \rightarrow 02$ have to be inserted. These changes produce the new topology as shown in Figure 6.2. □

To add a new end node to an existing broadcast-and-select network using the topology shown in Figure 6.1, the lightpaths $02 \Rightarrow 00$ and $12 \Rightarrow 10$ are to be taken down and new ones $02 \Rightarrow 20$ and $12 \Rightarrow 20$ have to be set up. Also, lightpaths $20 \Rightarrow 00$, $20 \Rightarrow 01$, and $20 \Rightarrow 02$ are to be defined.

6.3 A Routing Scheme for This Topology

The routing problem is to find a "good" route in graph G_S from any vertex x, having an address $x_k \ldots x_2 x_1$, to any vertex y, having an address $y_k \ldots y_2 y_1$. A recursive algorithm for this problem is given below. In the algorithm, ℓ_{xy} will denote the length of the longest suffix of the string $x_k \ldots x_2 x_1$ that is also a prefix of $y_k \ldots y_2 y_1$ and σ_{xy} will denote the string of length $2k - \ell_{xy}$ given by $x_k \ldots x_2 x_1 y_{k-\ell_{xy}} \ldots y_2 y_1$. The string σ_{xy} has $k - \ell_{xy} + 1$ substrings, each of length k. The ith substring of length k is the substring that starts from digit number i in σ_{xy} and will be denoted by $\sigma_{xy}^i, 1 \le i \le k - \ell_{xy} + 1$. It is easy to see that $\sigma_{xy}^1 = x$ and $\sigma_{xy}^{k-\ell_{xy}+1} = y$.

Property 6.4. If $\sigma_{xy}^i, 1 < i \le k - \ell_{xy}$ are *all* used strings, then the following route (where the vertices in the route are identified by their addresses) from x to y exists:

$$
\begin{aligned}
&x_k \ldots x_2 x_1 \rightarrow \\
&x_{k-1} \ldots x_2 x_1 y_{k-\ell_{xy}} \rightarrow \\
&x_{k-2} \ldots x_2 x_1 y_{k-\ell_{xy}} y_{k-\ell_{xy}-1} \rightarrow \\
&\ldots \\
&x_{\ell_{xy}+1} x_{\ell_{xy}} \ldots x_2 x_1 y_{k-\ell_{xy}} \ldots y_3 y_2 \rightarrow \\
&x_{\ell_{xy}} \ldots x_2 x_1 y_{k-\ell_{xy}} \ldots y_3 y_2 y_1.
\end{aligned}
\tag{6.1}
$$

Clearly $y_k \ldots y_3 y_2 y_1 = x_{\ell_{xy}} \ldots x_2 x_1 y_{k-\ell_{xy}} \ldots y_3 y_2 y_1$, and this gives the shortest path from x to y and has a length $k - \ell_{xy}$.

Property 6.5. If $x \in S_0$, σ_{xy}^i is a used string, for all $i, 1 < i \le k - \ell_{xy}$.

Proof. Let $y \in S_j$. Clearly $j \le k - \ell_{xy}$. Let $j = k - \ell_{xy} - p$, for some $p \ge 0$. The vertex v with address $x_{k-t} \ldots x_2 x_1 y_{k-\ell_{xy}} y_{k-\ell_{xy}-1} \ldots y_{k-\ell_{xy}-t+1}$ satisfies the following conditions:

Condition 1) if $t > p$, the address of v must be in set S_{t-p},
Condition 2) if $t \le p$, the address of v must be in set S_0,
Condition 3) v is a vertex in the path shown in (6.1).

If the address of v is in S_0, then this address is a used string. If the address of v is not in S_0, then $t \le k - \ell_{xy} - 1$, so that $t - p = t + j - (k - \ell_{xy}) \le j - 1$, and hence every string in S_{t-p} is a used string by Property 6.3 since $y \in S_j$. □

The following properties are useful only when σ_{xy}^j is an unused string, for some $j, 1 < j < k - \ell_{xy}$, and σ_{xy}^i is a used string, for all $i, 1 \le i < j$. Also, let the largest used string be in group $S_{\hat{M}}$, for some $\hat{M}, 1 \le \hat{M} \le k$. In the following discussions, α (β) will denote the string σ_{xy}^{j-1} (σ_{xy}^j).

Property 6.6. Either $\alpha \in S_{\hat{M}}$ or $\alpha \in S_{\hat{M}-1}$.

The edge $\alpha \to \beta$ does not exist since there is no vertex with an address β. The construction indicates that there must be an edge $\alpha \to \alpha_1$ where α_1 is a z-conjugate of α, for some $z, 1 \le z < d$.

Property 6.7. $\alpha_1 \in S_{\hat{M}-1}$ $(S_{\hat{M}-2})$ if $\alpha \in S_{\hat{M}}$ $(S_{\hat{M}-1})$.

Property 6.8. The number of occurrences of digit d in $\sigma_{\alpha_1 y}$ is 1 less than the number of occurrences of digit d in $\sigma_{\alpha_1 d}$.

The algorithm is described in the form of a function FindPath that takes the string σ_{xy} as its argument and returns a path from x to y. This function returns a path in the form of a sequence of vertices from x to y. In the algorithm describing the function FindPath, the symbol $\mathbb{P}_1 \| \mathbb{P}_2$ has been used to denote concatenation of two paths $\mathbb{P}_1$ and $\mathbb{P}_2$ so that if P_1 is a path from vertex u to vertex v and P_2 is a path from vertex v to vertex w, then $\mathbb{P}_1 \| \mathbb{P}_2$ is a path from u to w by following the edges of $\mathbb{P}_1$ to go from u to v and then by following the edges of path $\mathbb{P}_2$ to go from v to w.

Algorithm FindPath(σ_{xy})

Step 1) If all substrings of σ_{xy} are used strings then
$$\{$$

$$\text{Let path } \mathbb{P} = x_k \ldots x_2 x_1 \to$$
$$x_{k-1} \ldots x_1 y_{k-\ell_{sd}} \to$$
$$\ldots \to$$
$$x_{\ell_{xy}+1} \ldots x_1 y_{k-\ell_{xy}} \ldots y_3 y_2 \to$$
$$x_{\ell_{xy}} \ldots x_1 y_{k-\ell_{xy}} \ldots y_2 y_1$$

Return $\mathbb{P}$.

$$\}$$
Step 2) Let $\beta = \sigma_{xy}^j$ be the unused substring of σ_{xy} with the lowest value of j and let $\alpha = \sigma_{xy}^{j-1}$.
Step 3) Let $\mathbb{P}_1 = \text{FindPath}(\sigma_{x\alpha})$, the path from x to α.
Step 4) Let $\mathbb{P}_2 =$ the edge $\alpha \to \alpha_1$ where α_1 is the z-conjugate of α for any z.
Step 5) Let $\mathbb{P}_3 = \text{FindPath}(\sigma_{\alpha_1 y})$, the path from α_1 to y.
Step 6) Let $\mathbb{P} = \mathbb{P}_1 \| \mathbb{P}_2 \| \mathbb{P}_3$.
Step 7) Return $\mathbb{P}$.

Theorem 6.3. *The distance from any vertex x to any vertex y cannot exceed $2k$.*

Proof. Using the algorithm given above, the length of the path from x to y is $k - \sigma_{xy} + \tilde{p}$, where $\tilde{p}$ is the number of extra edges added because of using Step 2 in the algorithm. Each time Steps 2–6 are used, one extra edge is needed in the path. In the routing from x to y given above, the intent is to get rid of $k - \ell_{xy}$ digits from the address of x where ℓ_{xy} is the number of digits in the longest suffix of x. When Step 4 is used, the number of occurrences of the digit d in the address of the intermediate node α_1 is 1 less than the number of occurrences of the digit d in the address of the intermediate node α. The maximum number of times Steps 2–6 may need to be used is r, where r is the number of occurrences of digit d in the address of the source x. Since $r \leq k$, $\tilde{p} \leq r \leq k$. This immediately leads to the theorem.   □

Example 6.4. To find a path from 01 to 00 in the graph with six vertices shown in Figure 6.1, the process is as follows.

Here, $s = 01$ and $d = 00$ so that $\sigma_{sd} = 0100$, $\sigma_{sd}^1 = 01$, $\sigma_{sd}^2 = 10$, and $\sigma_{sd}^3 = 00$. Since σ_{sd}^i is a used string for all $i, 1 \leq i \leq 3$, the path is $01 \to 10 \to 00$.   □

Example 6.5. To find a path from 02 to 11 in the graph with six vertices shown in Figure 6.1, the process is as follows.

In this case, $\sigma_{sd} = 0211$. Now, the substring $\sigma_{sd}^2 = 21$ is not a used string. From vertex 02, the edge to vertex 00 (the 0-conjugate of 02) must be used. Since 01 and 11 are used strings, the path from 00 is $00 \to 01 \to 11$. Therefore, the path from 02 to 11 is $02 \to 00 \to 01 \to 11$. The path length has increased by 1 since the edge from 02 to its 0-conjugate was used.   □

In [286], the principle of shuffle interconnection between end nodes in a shuffle net is generalized (the generalized version can have any number of end nodes in each column) to obtain a scalable network topology called GEMNET [286]. A similar idea of generalizing the Kautz graph has been studied in [300], showing a better diameter and network throughput than GEMNET. Both these scalable topologies are given by regular digraphs. In [337] the multimesh architecture [102] has been generalized for use as a logical topology.

Exercise 6.1. Define a graph G_S with $d = 2, k = 2$, and $\mathcal{N}_S = 8$.   □

Exercise 6.2. Add a vertex to the graph G_S you defined in Exercise 6.1.   □

Exercise 6.3. Consider a graph G_S with $d = 2, k = 4$, and $\mathcal{N}_S = 20$. Find a route from the vertex with address 1120 to the vertex with address 1121.   □

Exercise 6.4. Write a program to find the average distance between any source to any destination in the above topology for $\mathcal{N}_S$ vertices, using the routing scheme given above. Use different values of $\mathcal{N}_S$ for a given value of d and k and plot the average distance against $\mathcal{N}_S$.   □

Exercise 6.5. Write a program to study the link loading characteristics of graph G_S as follows:

Suppose the task is to send one unit of information from E_s to E_d for all $s, d, 1 \leq s, d \leq \mathcal{N}_S, s \neq d$. Using the routing scheme given above, find how many units, u_{max}, are flowing on the edge carrying the maximum number of units and find how many units, u_{min}, are flowing on the edge carrying the minimum number of units. Also find the average number of units, u_{avg}, flowing on the edges. If the ratio u_{min}/u_{max} or the ratio u_{avg}/u_{max} is much less than 1, what conclusions to you draw?

In this experiment, use all possible values of $\mathcal{N}_S$, for each value of d and k, and try $2 \leq d, k \leq 4$.   □

In Exercise 6.5, the specific routing given in Section 6.3 was used to study the link loading characteristics. It is possible that a better routing exists. The following exercise addresses this possibility.

Exercise 6.6. Using a mathematical programming package (such as the CPLEX [206]), study the optimum link loading characteristics of graph G_S as follows:

Suppose the task is to send one unit of information from E_s to E_d for all $s, d, 1 \leq s, d \leq \mathcal{N}_S, s \neq d$. Specify linear constraints to find the "best" routing that minimizes the number of units that flow on the edge carrying the maximum number of units.

In this experiment use all possible values of $\mathcal{N}_S$, for each value of d and k, and try $2 \leq d, k \leq 4$.

Compare your results in this experiment with those you obtained in Exercise 6.5.   □

6.4 Bibliographic Notes

A number of regular topologies have been studied for WDM networks. These topologies include the Manhattan Street Network [270], the Perfect Shuffle [194], the Hypercube [111], the de Bruijn graph [359], and the Cayley graph [379]. The k-dimensional bidirectional square lattice, the twin shuffle, and the de Bruijn graph networks were compared in [268] for the maximum and average distance between source and destination. It was shown in [359] that a network topology based on the de Bruijn graph performs better than shufflenet [194] as a physical topology. In [300], bounds on the average node distance for the Kautz graph were compared to those for de Bruijn graphs in [359] and it was shown that the average queueing delay in a network topology based on the Kautz digraph network gives better results than a network topology based on the de Bruijn digraph. There are severe restrictions on these regular topologies with respect to the number of nodes in the network.

In [207], a new scalable topology has been defined using a generalized version of the shufflenet. A scalable version of the topology based on the

Kautz digraph has been studied and the results reported in [393]. In [337] a new topology called the *generalized multimesh* has been proposed for WDM networks. This topology is based on the multimesh architecture proposed in [102] for multi-processor systems. The generalized multimesh has been characterized as a *semiregular* structure, where the simplicity of interconnection and routing is comparable to the torus network. However, the multimesh and the generalized multimesh have significantly superior topological properties. The generalized multimesh allows an arbitrary number of end nodes in the network.

There has been significant research on multicast communication[1] in broadcast-and-select networks. Reviews on this topic appear in [109, 185, 380]. Heuristics to obtain multicasting schedules that result in low average packet delays under changing multicast traffic conditions have been proposed in [327]. Multicasting using channel sharing and time division multiplexing has been applied in [387] to the GEMNET architecture [207]. The problem of designing minimum length schedules for packet transmissions has been considered in [328]. Heuristics to generate "nearly optimal" schedules have been proposed in [328]. Algorithms for scheduling multicast traffic have been proposed in [278] and have been analyzed using discrete-time queueing systems. A hybrid multicast scheduling algorithm that takes into account the average utilizations of the data channels and the receivers, with good performance for wide ranges of traffic conditions and channel resources, has been proposed in [254].

---

[1] Multicasting has been briefly reviewed in Section 3.5.4.

7

Logical Topology Design II

This chapter discusses techniques for determining the set of lightpaths defining the logical topology of the network and for developing an optimal strategy for data communication to meet the user requirements. Here static (or off-line) lightpath allocation, introduced in Section 3.3.1, is considered. The notion of a traffic matrix was introduced in Section 1.4. In this discussion, the traffic requirements of the user will be denoted by a traffic matrix $T = [t(s,d)]$, where $t(s,d)$ is the traffic from end node E_s to end node E_d.

The formulation for the logical topology design problem must optimize network performance by determining

1) the set of lightpaths (i.e., the set of logical edges E_L in graph G_L) to be created,
2) for each lightpath in E_L,
   - its route through the physical topology,
   - its channel number on each fiber in its route,
3) the strategy for routing the traffic T over the logical topology.

Determining an optimal logical topology, the RWA for each lightpath in the logical topology, and a routing strategy on the logical topology to optimize the network performance are not independent problems. Some early investigators have looked at the problem of optimizing all three problems together in one formulation, using Mixed Integer Linear formulation (MILP) [116, 241, 322, 323, 324]. These formulations turn out to have extremely large numbers of integer variables, and are intractable for practical networks. This fact has motivated the development of heuristic approaches for finding good solutions efficiently.

One popular heuristic [116] is to split the problem into three *independent* subproblems as follows:

Subproblem 1) (**Find Logical Topology**) Determine the logical topology to be mapped on the physical topology, by choosing the lightpaths in terms of their source and destination nodes.

Subproblem 2) (**Carry out RWA for each lightpath**) Determine the route of each lightpath over the physical topology as well the channel number[1] assigned to each lightpath.

Subproblem 3) (**Find an optimal traffic routing strategy**) Route the traffic, as specified by the traffic matrix T, over the logical topology obtained.

This decomposition may be suboptimal, since solving the subproblems in sequence and combining the solutions may not result in the optimal solution for the fully integrated problem. The decomposition may even be infeasible since some later subproblem may have no solution, given the solution obtained for an earlier subproblem. However, since the original problem is intractable for practical networks, this heuristic is used widely [25, 231, 289, 323].

As pointed out in Section 1.4, once the logical topology is defined, there are, in general, a number of different logical paths from E_s to E_d. The logical paths from E_s to E_d will be denoted by $\wp_{sd}^1, \wp_{sd}^2, \ldots, \wp_{sd}^\aleph$, where there are $\aleph$ logical paths from E_s to E_d. The user requirement is that the network must communicate all traffic specified in the traffic matrix $T = [t(s, d)]$. This means that traffic $t(s, d)$ must be communicated from E_s to E_d, for all end node pairs (E_s, E_d) having $t(s, d) \neq 0$. In general, traffic $t(s, d)$ may be communicated from E_s to E_d using any one or more of the logical paths $\wp_{sd}^1, \wp_{sd}^2, \ldots, \wp_{sd}^\aleph$.

Once the logical topology is determined, the problem is to determine how much of the traffic $t(s, d)$ is to be sent using the logical path $\wp^i(s, d)$, for all $i, 1 \leq i \leq \aleph$, making most "efficient" use of the optical resources available in the network. This is called the *routing problem* for the logical topology of a wavelength-routed network.

In recent years, the number of lightpaths on a single fiber has increased tremendously, accommodating hundreds of WDM channels [324, 287]. In most cases, WDM channels are no longer the scarce resources to be the primary targets of optimization. It is reasonable to assume that RWA may be solved in a separate step, and the issue of the RWA problem is ignored in this chapter. In other words, it is assumed that setting up any set of lightpaths, each lightpath connecting a pair of end nodes, is feasible. Later on, in Chapter 8, a technique for solving the combined problem — the RWA problem, the logical topology design problem, and the routing problem will be discussed.

Since the number of logical topologies, as well as the number of possible routing schemes for a given logical topology, is usually very large, the solution space for this problem is enormous and the computational cost of an optimum solution can be extremely high. In this chapter the following topics are covered:

- MILP-based logical topology design and routing,
- heuristic for the logical topology design problem,
- solving the routing problem in

---

[1] Most networks assume that the lightpaths satisfy the wavelength continuity constraint [324].

      – small- and medium-sized networks,
      – routing in large networks.

7.1 MILP-Based Solution of the Logical Topology Design and the Routing Problem

The problem of logical topology design is to find which pairs of end nodes are to be connected by a lightpath.[2] To represent the situation where a lightpath between any given pair of end nodes either exists or does not exist, variables are needed, one for each pair of end nodes, that are restricted to being either 0 or 1. If such a variable has a value 1 (0), it means that a lightpath exists (does not exist) between the corresponding pair of end nodes.

Since the objective is to design an optimum logical topology, an immediate question is exactly what network parameter should be optimized as the objective function of the MILP. One popular approach, in logical topology design, is to minimize the $\mathcal{L}$-congestion,[3] the maximum traffic load Λ_{max} on a logical link. Other objectives functions, such as minimizing the average number of hops or minimizing the average packet delay, have also been investigated [116]. The parameter $\mathcal{L}$-congestion is an important parameter to minimize for the following reasons:

- For a higher value of Λ_{max}, more data is transported by the lightpath carrying that traffic. To handle more data, more time and/or complex electronic hardware is needed [26, 287].
- The $\mathcal{L}$-congestion on a single lightpath can never exceed the capacity $c_{lightpath}$ of the lightpath (2.5 Gbps or 10 Gbps, depending on the technology). If, in a particular design, the $\mathcal{L}$-congestion is less than $c_{lightpath}$, all logical edges may be realized by one lightpath. A design that minimizes $\mathcal{L}$-congestion below $c_{lightpath}$ reduces the cost of the network, since each lightpath means additional cost due to the optical and the electronic hardware at the source and the destination of the lightpath.
- If $\Lambda_{max} \leq c_{lightpath}$, all logical edges may be realized by a single lightpath. The routing strategy remains viable as long as no entry in the traffic matrix increases by a factor greater than $c_{lightpath}/\Lambda_{max}$. Therefore, it is preferable to use a lower value of Λ_{max} to allow greater possibility of scaling up the traffic in the network without changing

---

[2] There is an assumption here that the number of lightpaths connecting a pair of end nodes will not exceed 1.

[3] The term congestion has been used by researchers to describe this parameter. Here the term $\mathcal{L}$-congestion is being used to make a distinction from the term congestion used in Chapters 4 and 5.

the routing strategy. A similar reasoning holds when $\Lambda_{max} > c_{lightpath}$ and some logical edges have to be realized by more than one lightpath.

The input to the formulation $FORM\ 3$ given below is the traffic matrix T giving the traffic between every pair of end nodes and the number of transmitters and receivers at each end node. In this discussion, the unit used to denote $t(i,j)$, the element in row i, column j of T, will be the capacity of a lightpath.

Notation used in $FORM\ 3$

$\mathcal{N}_E$: number of end nodes in the network.

T : a $\mathcal{N}_E \times \mathcal{N}_E$ matrix, called the traffic matrix, giving the traffic requirements for the network.

$t(i,j)$: entry in row i and column j of matrix T. It represents the traffic from source end node E_i to destination end node E_j.

Λ_{max} : the $\mathcal{L}$-congestion of the network.

b_{ij} : a binary variable such that
$$b_{ij} = \begin{cases} 1 \text{ if a lightpath exists from end node } E_i \text{ to end node } E_j, \\ 0 \text{ otherwise.} \end{cases}$$

x_{ij}^{sd} : a continuous variable to denote the portion of traffic $t(s,d) : t(s,d) > 0$, that is routed through the logical edge $E_i \Rightarrow E_j$.

$\Delta_{in}\ (\Delta_{out})$: a constant denoting the number of receivers (transmitters) at every end node. This defines the maximum number of lightpaths that may end at (start from) an end node.

The value of $\Delta_{in}\ (\Delta_{out})$ is determined by the number of input (output) lines available in the add-drop multiplexors at each end node. For simplicity, it is assumed that these numbers are the same for all end nodes.

The formulation for $FORM\ 3$

The following MILP minimizes the $\mathcal{L}$-congestion of the logical network by determining an optimal logical topology and a routing over the logical topology. After solving the following formulation, possibly using a mathematical programming package (such as the CPLEX [206]), for a particular network specification, each value of b_{ij}, such that $b_{ij} = 1$, corresponds to a logical edge $E_i \Rightarrow E_j$. In other words, the nonzero values of b_{ij} define the logical topology. For a given pair of end nodes, $(E_s, E_d) : t(s,d) \neq 0$, the values of x_{ij}^{sd} for different logical edges $E_i \Rightarrow E_j$ completely defines how the traffic $t(s,d)$ is to be routed using a number of paths from E_s to E_d.

Objective function:

$$\text{minimize } \Lambda_{max} \tag{7.1}$$

subject to

1. Ensure that Λ_{max} is the $\mathcal{L}$-congestion.

$$\sum_{s=1}^{\mathcal{N}_E} \sum_{d=1}^{\mathcal{N}_E} x_{ij}^{sd} \leq \Lambda_{max}, \quad \forall i, j, 1 \leq i, j \leq \mathcal{N}_E, i \neq j \tag{7.2}$$

2. Part of the traffic $t(s,d)$ may flow on the logical edge $E_i \Rightarrow E_j$ only if the logical edge exists. Also x_{ij}^{sd}, the part of the traffic $t(s,d)$ flowing on the logical edge $E_i \Rightarrow E_j$, cannot exceed the traffic $t(s,d)$.

$$x_{ij}^{sd} \leq b_{ij} \cdot t(s,d), \qquad \forall i,j,s,d, 1 \leq i,j,s,d \leq \mathcal{N}_E, i \neq j, s \neq d \tag{7.3}$$

3. Apply the flow conservation rules.

$$\sum_{j=1}^{\mathcal{N}_E} x_{ij}^{sd} - \sum_{j=1}^{\mathcal{N}_E} x_{ji}^{sd} = \begin{cases} t(s,d) & \text{if } s = i, \\ -t(s,d) & \text{if } d = i, \\ 0 & \text{otherwise.} \end{cases} \quad \begin{matrix} \forall i, s, d, 1 \leq s, d, i \leq \mathcal{N}_E, \\ s \neq d \end{matrix} \tag{7.4}$$

4. The number of lightpaths ending at (starting from) a given end node cannot exceed Δ_{in} (Δ_{out}).

$$\sum_{i=1}^{\mathcal{N}_E} b_{ji} \leq \Delta_{in}, \qquad \forall j, 1 \leq j \leq \mathcal{N}_E \tag{7.5}$$

$$\sum_{j=1}^{\mathcal{N}_E} b_{ij} \leq \Delta_{out}, \qquad \forall j, 1 \leq j \leq \mathcal{N}_E \tag{7.6}$$

Justification for *FORM* 3

1. The total traffic on the logical edge $E_i \Rightarrow E_j$, for all source-destination pairs (i,j) is $\sum_{s=1}^{\mathcal{N}_E} \sum_{d=1}^{\mathcal{N}_E} x_{ij}^{sd}$. The purpose of (7.2) is to ensure that Λ_{max} is greater than or equal to the traffic on any edge. Since the objective is to minimize Λ_{max}, the minimum value of Λ_{max} must be the $\mathcal{L}$-congestion, the value of the traffic on the edge carrying the maximum traffic.
2. Only if b_{ij} is 1 (i.e., there is a lightpath from end node E_i to end node E_j, so that there exists a logical edge $E_i \Rightarrow E_j$), part of the traffic $t(s,d)$ may be routed over the logical edge $E_i \Rightarrow E_j$. This is ensured by (7.3).

3. (7.4) corresponds to the flow balance equations (Appendix 3). It ensures that, for all $(E_s, E_d) : t(s,d) \neq 0$, the desired traffic $t(s,d)$ flows through the logical paths from E_s to E_d, each logical path passing through 0 or more intermediate end nodes. For a given pair of end nodes $(E_s, E_d), s \neq d$, the part of the traffic flowing on logical edge $E_i \Rightarrow E_j$ is x_{ij}^{sd}. Therefore, the total traffic due to $t(s,d)$ flowing out of end node E_i is $\displaystyle\sum_{j=1}^{\mathcal{N}_E} x_{ij}^{sd}$. Similarly, the total traffic due to $t(s,d)$ flowing into E_i is $\displaystyle\sum_{j=1}^{\mathcal{N}_E} x_{ji}^{sd}$. If E_i is the source E_s, there is no traffic flowing into E_s and the traffic flowing out of E_s is $t(s,d)$. Similarly, if E_i is the destination E_d, there is no traffic flowing out of E_d and the traffic flowing in is $t(s,d)$. For all other intermediate end nodes in a logical path, the inflow must be matched by the outflow so that the difference must be 0.

4. (7.5) and (7.6) ensures that the number of lightpaths terminating at (starting from) an end node do not exceed Δ_{in} (Δ_{out}), the number of receivers (transmitters) at each end node.

The complexity of *FORM 3*

The number of binary (0/1) variables $b_{ij}, 1 \leq i, j \leq \mathcal{N}_E, i \neq j$, is $\mathcal{N}_E(\mathcal{N}_E - 1)$. The number of continuous variables x_{ij}^{sd} is $\mathcal{N}_E^2(\mathcal{N}_E - 1)^2$ so that the total number of continuous variables, including Λ_{max}, is $\mathcal{N}_E^2(\mathcal{N}_E - 1)^2 + 1$. The number of constraints is $\mathcal{N}_E[\mathcal{N}_E^2(\mathcal{N}_E - 1) + \mathcal{N}_E + 1]$.

The computational cost of this type of MILP is dominated by the number of binary variables. Even for moderate values of $\mathcal{N}_E$, the time involved to solve such MILP is considerable. As a result, this type of MILP is not of much practical use except for small networks.

Exercise 7.1. Consider the traffic matrix given in (7.7) for the network shown in Figure 7.1. Define the MILP for this traffic matrix using the formulation described in this section. Assume that each end node in the network has two transmitters and receivers. Each of the transmitters and receivers may be tuned to any desired carrier wavelength.   □

Exercise 7.2. The formulation described here assumes that RWA is always feasible. Incorporate RWA into the MILP so that your formulation also ensures the feasibility of the RWA.   □

7.2 A Heuristic for Determining the Logical Topology

To make the problem tractable, it is necessary to dissociate the problem of logical topology design from the problem of routing over the logical topology. Researchers have proposed heuristics to determine a logical topology based on the traffic matrix T. A simple heuristic for logical topology, called the *heuristic logical topology design algorithm* (HLDA)[323], is given below. In this heuristic, it is assumed that each of the transmitters and receivers may be tuned to any desired carrier wavelengths. The network has $\mathcal{N}_E$ end nodes.

The inputs to the heuristic are the following:

- Δ_{in}^i denoting the number of receivers at end node $E_i, 1 \leq i \leq \mathcal{N}_E$.
- Δ_{out}^i denoting the number of transmitters at end node $E_i, 1 \leq i \leq \mathcal{N}_E$.
- The traffic matrix $T = [t(i,j)]$ giving the required volume of data so that $t(i,j)$ is the traffic from end node E_i to end node E_j in the traffic matrix T, for all $i, j, 1 \leq i, j \leq \mathcal{N}_E$.
- The physical topology of the network.
- The number of channels n_{ch} per fiber.

The steps of the heuristic are as follows:

Step 1: Find the entry $t(i_{max}, j_{max})$ in T having a maximum value among all the entries in T. If there is no nonzero entry in T, stop.

Step 2: Using any RWA technique (Chapters 4 and 5), check if it is possible to establish a lightpath from $E_{i_{max}}$ to $E_{j_{max}}$. If RWA is not possible, set $t(i_{max}, j_{max})$ to 0 and go to Step 1.

Step 3: If $\Delta_{in}^{j_{max}}$ is equal to the number of lightpaths ending at end node $E_{j_{max}}$, set $t(i_{max}, j_{max})$ to 0 and go to Step 1.

Step 4: If $\Delta_{out}^{i_{max}}$ is equal to the number of lightpaths starting from end node $E_{i_{max}}$, set $t(i_{max}, j_{max})$ to 0 and go to Step 1.

Step 5: Create a logical edge from end node $E_{i_{max}}$ to end node $E_{j_{max}}$. Set $t(i_{max}, j_{max})$ to $t(i_{max}, j_{max}) - c_{lightpath}$ or 0, whichever is greater. Here $c_{lightpath}$ is the capacity of a lightpath. Go to Step 1.

In this process, since the lightpaths are established sequentially, a simple heuristic for RWA is to consider the shortest route from end node i_{max} to end node j_{max}, and find whether any channel is free on every fiber in the route.

Example 7.1. A WDM network with four end nodes shown in Figure 7.1 (taken from Chapter 1) allows two channels c_1 and c_2 on each fiber and the capacity of a lightpath $c_{lightpath} = 1.0$. Let the user requirement be specified in traffic matrix T where

$$T = \begin{bmatrix} 0.00 & 0.30 & 0.10 & 0.50 \\ 0.20 & 0.00 & 0.30 & 0.20 \\ 0.35 & 0.10 & 0.00 & 0.30 \\ 0.00 & 0.20 & 0.10 & 0.00 \end{bmatrix} \tag{7.7}$$

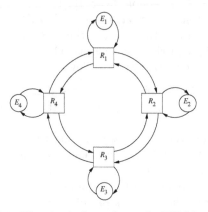

Fig. 7.1. The physical topology of a WDM network

Let each end node have two transmitters and two receivers.
The first iteration is as follows.

Step 1: The highest traffic is from end node E_1 to end node E_4 since $t(1,4) = 0.5$.

Step 2: It is possible to set up a lightpath from E_1 to E_4 using the route $E_1 \to R_1 \to R_4 \to E_4$ and channel c_1.

Step 3: There is a transmitter available in end node E_1.

Step 4: There is a receiver available in end node E_4.

Step 5: Create a logical edge $E_1 \Rightarrow E_4$.

At this stage, channel c_1 on edges $E_1 \to R_1$, $R_1 \to R_4$, $R_4 \to E_4$ is allotted to the lightpath just created. The new traffic matrix is T^{new} where

$$T^{new} = \begin{bmatrix} 0.00 & 0.30 & 0.10 & 0.00 \\ 0.20 & 0.00 & 0.30 & 0.20 \\ 0.35 & 0.10 & 0.00 & 0.30 \\ 0.00 & 0.20 & 0.10 & 0.00 \end{bmatrix} \tag{7.8}$$

In the next iteration, T^{new} will be used. The highest traffic now is $t(3,1) = 0.35$. The remaining steps of the heuristic are similar.   □

Exercise 7.3. Complete the design of the logical topology in Example 7.1.

Notice that all the entries in the traffic matrix could not be handled due to lack of transmitters and receivers. It is important to notice that when this design is finished, there is no guarantee that the traffic specified using the traffic matrix could be handled. The expectation is that, if the high values in the traffic matrix can be handled using a single lightpath, the routing algorithms, described later in Sections 7.4 and 7.5, will take care of the low values in the traffic matrix. □

Exercise 7.4. Write a program based on a greedy heuristic for the logical topology design problem using the following ideas.

During the RWA phase, when finding the route and the channel for a given ordered pair of end nodes E_s and E_d, only k routes from E_s to E_d will be considered. If possible, the k routes will be edge-disjoint. These k routes will be precomputed and saved in some routing table. Initially, there is no logical edge in the network. It is assumed that all the transmitters and receivers are tunable. Your program should be based on the following algorithm:

> **Step 1:** Pick the highest nonzero entry $t(s, d)$ in the traffic matrix. If no such entry exists, report success and stop.
>
> **Step 2:** Find, using some single commodity network flow algorithm [4], the maximum amount of traffic τ_{sd} that can be sent from E_s to E_d using the spare capacity on the logical edges already defined in the network.
>
> **Step 3:** If $\tau_{sd} \geq t(s, d)$, set $t(s, d) = 0$. Update the network flows so that traffic $t(s, d)$ is routed from E_s to E_d using the routes found in Step 2. Go to Step 1.
>
> **Step 4:** If $\tau_{sd} > 0$, set $t(s, d) = t(s, d) - \tau_{sd}$. Update the network flows so that traffic τ_{sd} is routed from E_s to E_d using the routes found in Step 2. Go to Step 1.
>
> **Step 5:** If there is a transmitter available at end node E_s and a receiver at end node E_d, check if RWA is possible using one of the k precomputed routes, already saved in the routing table. If so, set up a lightpath from E_s to E_d and set the traffic flowing on this logical edge to $t(s, d)$. Set $t(s, d) = 0$ and go to Step 1.
>
> **Step 6:** If the lightpath could not be set up in Step 5, report failure and stop.

Test your program using $k = 3$ and a number of physical topologies and traffic matrices. □

Exercise 7.5. In Step 5 of Exercise 7.4, only the case of setting up a single lightpath $E_s \Rightarrow E_d$ was considered. To handle the traffic $t(s, d)$, it is possible to use a number of logical paths, each of the form $E_s \Rightarrow E_{i_1} \Rightarrow E_{i_2} \Rightarrow \ldots \Rightarrow E_d$, where some of the logical edges correspond to new lightpaths.

Modify the heuristic to do so. Feel free to use reasonable simplifying assumptions. □

Exercise 7.6. In Chapter 6, a scalable topology, suitable for broadcast-and-select networks, was presented. Propose a heuristic for defining a logical topology for wavelength-routed networks, based on the topology discussed in Chapter 6, based on the traffic requirements. Some ideas are given below.

The graph G_S discussed in Chapter 6 defines a topology for $\mathcal{N}_S$ end nodes. There is considerable flexibility in mapping each of the end nodes of the wavelength-routed network to the vertices of G_S and this mapping may be done in $\mathcal{N}_S!$ ways. Your heuristic should take into account the traffic between end nodes. If there is high traffic between end node E_x and E_y, if possible, there should be a lightpath from E_x to E_y. A greedy heuristic is the simplest approach.

Propose a scheme for evaluating your heuristic and test it using a number of traffic matrices.   □

7.3 Routing in Wavelength-Routed Networks Viewed as an MCNF Problem

If the logical topology is already fixed, the routing strategy determines how the traffic $t(s,d), t(s,d) > 0$, for all $(s,d), 1 \leq s, d \leq \mathcal{N}_E$, may be handled in an optimal manner. This type of problem has been studied extensively by the operations research community and is known as the *Multi-Commodity Network Flow* (MCNF)[4] problem [4].

The routing problem in wavelength-routed networks may be viewed as an MCNF problem [273] by considering the traffic $t(s,d)$, $t(s,d) > 0$, from E_s to E_d, as a commodity, distinct from the commodity corresponding to the traffic for any other pair of end nodes. It is convenient to attach a distinct number k with each commodity, corresponding to each pair of end nodes, and designate the commodity K^k as that corresponding to the pair of end nodes (E_{s_k}, E_{d_k}) having $t(s_k, d_k) > 0$. Commodity K^k has a single source, E_{s_k}, and a single destination, E_{d_k}. The network, in this case, may be represented by the graph G_L, the logical topology of the network. If there are n_{sd}^{light} lightpaths from end node E_s to E_d and $c_{lightpath}$ is the capacity of a lightpath (2.5 Gbps or 10 Gbps depending on the technology used), the capacity of the logical edge from E_s to E_d is $n_{sd}^{light} \times c_{lightpath}$, representing the amount of data that may be handled by all the lightpaths that start from E_s and end at E_d. The notion of cost is not useful in the routing problem considered here.

Example 7.2. In the case of the traffic matrix T given in (7.9), there are ten nonzero entries, so that there are ten commodities. Commodity K^1 is for the

---
[4] A short description of some network flow problems, including multi-commodity network flow is given in Appendix 3.

pair (E_1, E_2), corresponding to row 1, column 2 of T, and has traffic 0.30. Commodity K^2 is for the pair (E_1, E_3), corresponding to row 1, column 3 of T, and has traffic 0.10. Commodity K^{10} is for the pair (E_4, E_3), corresponding to row 4, column 3 of T, and has traffic 0.10.

$$T = \begin{bmatrix} 0.00 & 0.30 & 0.10 & 0.00 \\ 0.20 & 0.00 & 0.30 & 0.20 \\ 0.35 & 0.10 & 0.00 & 0.30 \\ 0.80 & 0.00 & 0.10 & 0.00 \end{bmatrix} \tag{7.9}$$

□

7.4 Routing in Small- and Medium-Sized Networks

The routing in small- and medium-sized (less than 30 end nodes) networks is quite straightforward and may be solved by a formulation $FORM\ 1$ similar to that given in Section 7.1. Since the logical topology is defined, the values of $b_{ij}, 1 \leq i, j \leq \mathcal{N}_E$, are known, and that greatly simplifies the formulation $FORM\ 4$ given below. From a computational point of view, it is important to note that this formulation does not involve any integer variable and hence is a linear program (LP).

Notation used in $FORM\ 4$

q : number of commodities, i.e., the number of pairs of end nodes (E_{s_k}, E_{d_k}) having traffic $t(s_k, d_k) \neq 0$.

$src(k)$ $(dest(k))$: source s_k (destination d_k) of commodity $k, 1 \leq k \leq q$.

Υ^k : the traffic $t(s_k, d_k)$ for commodity k, $1 \leq k \leq q$.

$\mathcal{N}_E$: number of end nodes in the network.

$\widehat{x}_{ij}^k$: a continuous variable denoting the amount of traffic for commodity k flowing on the logical edge $E_i \Rightarrow E_j$.

Λ_{max} : a variable denoting the $\mathcal{L}$-congestion in the network.

L : the set of all pairs of indices of end nodes (i, j) such that $E_i \Rightarrow E_j$ is a logical edge (i.e., a lightpath exists from end node E_i to end node E_j).

The formulation for *FORM* 4

The problem of routing over a given logical topology to minimize $\mathcal{L}$-congestion is similar to the minimum-cost multi-commodity (MMCF) network flow problem [4] (Appendix 3) and can be formulated as follows:

Objective function:

$$\text{minimize} \quad \Lambda_{max} \tag{7.10}$$

Subject to

1. Ensure that Λ_{max} is the $\mathcal{L}$-congestion.

$$\sum_{k=1}^{q} \widehat{x}_{ij}^{k} \leq \Lambda_{max}, \forall (i,j) \in L \tag{7.11}$$

2. Apply the flow conservation rules.

$$\sum_{j:(i,j)\in L} \widehat{x}_{ij}^{k} - \sum_{j:(j,i)\in L} \widehat{x}_{ji}^{k} = \begin{cases} \Upsilon^k & \text{if src}(k) = i, \\ -\Upsilon^k & \text{if dest}(k) = i, \\ 0 & \text{otherwise.} \end{cases} \tag{7.12}$$

(7.12) has to be applied to node i, for all $i, 1 \leq i \leq N_E$, and has to be repeated for each commodity $k, 1 \leq k \leq q$.

Justification of *FORM* 4

(7.11) is very similar to (7.2). (7.12) serves the purpose of (7.4). The main difference between this formulation and that in Section 7.1 is that the set of logical edges L is known here, so that, instead of considering all pairs of end nodes $(i,j), 1 \leq i, j \leq N_E$, only the pairs $(i,j) \in L$ need to be considered.

The complexity of *FORM* 4

The number of continuous variables in the above formulation is $1 + mq$. The constraints of the LP is specified through (7.11) and (7.12). The computational cost of solving such problems is determined by the basis[5] size [377, 93]. In the case of the above LP, there are m constraints due to (7.11) and $q \times N_E$ constraints due to (7.12). The value of q is $O(N_E^2)$ since most, if not all, end nodes have some traffic to all other end nodes. Therefore, there are $O(N_E^3)$ constraints from (7.11) and (7.12).

This formulation is a linear program with no integer variable and, therefore, can be solved for small- and medium-sized networks, with the number of end nodes less than 30, using a commercial package such as the CPLEX.

---

[5] A short explanation of the revised simplex method, including an explanation of the term basis, is included in Appendix 1.

Exercise 7.7. Write a program that will read in a logical topology and a traffic matrix and generate an LP as described in this section. Test your program using the traffic matrix shown in (7.7) and the logical topology you designed in Exercise 7.3. Solve the resulting LP using a mathematical programming package (e.g., CPLEX [206]). □

7.5 Routing in Large Networks

Most multi-commodity network flow problems considered in the operations research community consider a relatively small number of commodities [273]. The analysis in Section 7.4 reveals an important aspect of the routing problem in WDM networks — the number of commodities is relatively large, $O(\mathcal{N}_E^2)$, and the number of constraints in the formulation given in Section 7.4 is $O(\mathcal{N}_E^3)$. In other words, the number of constraints rises rapidly as the number of end nodes $\mathcal{N}_E$ in the network increases. Table 7.1 shows how the basis size increases with the number of end nodes $\mathcal{N}_E$ using $\Delta_{in} = \Delta_{out} = 3$ for each end node so that the number of logical edges is $m = 3 \times \mathcal{N}_E$. It is assumed that each end node has some traffic to all other end nodes, so that the number of commodities $q = \mathcal{N}_E(\mathcal{N}_E - 1)$.

Table 7.1. Table showing the increase of the basis size with the network size

$\mathcal{N}_E$	m	q	$m + \mathcal{N}_E q$
10	30	90	930
20	60	380	7660
40	120	1560	62520
100	300	9900	990300

The problem of large basis size arises due to the fact that, for every commodity $k, 1 \le k \le q$, each end node in the network has to be considered as a potential intermediate end node in the logical path for the commodity in (7.12). Since $q = O(\mathcal{N}_E^2)$, the problem becomes intractable for larger value of $\mathcal{N}_E$.

The formulation $FORM\ 4$ discussed in Section 7.4 is based on the *node-arc* representation [4]. In the operations research community, there is an alternative representation for network flow problems, called the *arc-chain*[6] representation [152]. This representation may be used to handle the routing problem using a formulation which is more tractable for larger networks. The main advantage of the new formulation is that, for every commodity, there is only one constraint, so that the basis size is $O(\mathcal{N}_E^2)$.

---

[6] The terms *arc* and *chain* are used in the operations research literature rather than the terms edge and path used in the network literature.

In the formulation given below, there are three important aspects as follows:

  i) The notion of arc-chain representation, which helps in reducing the size of the basis from $O(\mathcal{N}_E^3)$ to $O(\mathcal{N}_E^2)$.
 ii) The concept of implicit column generation [385] to handle the problem of managing an enormous number of logical paths for each commodity that has to be considered in the arc-chain formulation.
iii) The application of the Generalized Upper Bounding (GUB) technique [93] to reduce the number of operations in each iteration of the revised simplex method.[7]

7.5.1 The Arc-Chain Representation

Definition 7.1. *A chain [35] from a source s to a destination d is a sequence of logical edges $[(s = i_0 \Rightarrow i_1), (i_1 \Rightarrow i_2), \ldots, (i_{p-1} \Rightarrow i_p = d)]$. The logical path described by the chain above is $[(s = i_0 \Rightarrow i_1 \Rightarrow i_2 \Rightarrow \ldots \Rightarrow i_{p-1} \Rightarrow i_p = d]$.*

In a network with m logical edges numbered $1, 2, \ldots, m$, a chain may be represented by a vector of m 1s and 0s so that, if the ith element in the chain is 1 (0), the ith logical edge does (does not) appear in the chain, for all $i, 1 \leq i \leq m$. For each commodity k, in general, there are many chains. In this discussion, $\overrightarrow{C_j^k}$ will represent the jth chain of the kth commodity — a vector of 1s and 0s.

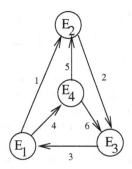

Fig. 7.2. A network with four nodes

Example 7.3. Figure 7.2 shows a simple logical topology with four end nodes and six logical edges. Each logical edge is assigned a number from 1 to 6 as

---

[7] The revised simplex method (including terms such as the basis, slack variables, and simplex multipliers) is reviewed briefly in Appendix 1.

shown. This network carries two commodities K^1 and K^2. The source and the destination for commodity K^1 (K^2) is E_1 (E_4) and E_2 (E_3). There are two logical paths in this network for commodity K^1. The first logical path consists of the logical edge $E_1 \Rightarrow E_2$ and the second path is $E_1 \Rightarrow E_4 \Rightarrow E_2$. The chain $\overrightarrow{C_1^1}$, representing the first logical path, is the vector $[1, 0, 0, 0, 0, 0]$ since the logical edge $E_1 \Rightarrow E_2$ has been assigned edge number 1. Similarly, the chain $\overrightarrow{C_2^1}$, corresponding to the second logical path $[E_1 \Rightarrow E_4 \Rightarrow E_2]$, is the vector $[0, 0, 0, 1, 1, 0]$. Commodity K^2 has two logical paths so that the chains $\overrightarrow{C_1^2}$ and $\overrightarrow{C_2^2}$ are the vectors $[0, 0, 0, 0, 0, 1]$ and $[0, 1, 0, 0, 1, 0]$.   □

A network having m edges and q commodities may be represented by an *arc-chain incidence matrix* AC. If there are $\hat{n}^k$ chains for the kth commodity K^k, for all $k, 1 \leq k \leq q$, AC is an $m \times \hat{n}$ matrix where $\hat{n} = \hat{n}^1 + \hat{n}^2 + \ldots + \hat{n}^q$ is the total number of chains for all commodities. Each chain of a commodity corresponds to a column in matrix AC so that the first $\hat{n}^1$ columns of AC correspond to chains for commodity K^1, the next $\hat{n}^2$ columns correspond to chains for commodity K^2, and so on.

Example 7.4. The arc-chain matrix AC, in the case of the network shown in Figure 7.2, with two commodities K^1 and K^2 is shown below.

$$AC = \begin{bmatrix} 1 & 0 & 0 & 0 \\ 0 & 0 & 0 & 1 \\ 0 & 0 & 0 & 0 \\ 0 & 1 & 0 & 0 \\ 0 & 1 & 0 & 1 \\ 0 & 0 & 1 & 0 \end{bmatrix} \tag{7.13}$$

In this example, the rows of arc-chain matrix AC corresponds to the edges $1, 2, \ldots, 6$ of the network. Columns 1 and 2 (3 and 4) correspond to the chains for commodity K^1 (K^2).   □

Exercise 7.8. Suppose there are two more commodities K^3 and K^4 in the logical topology shown in Figure 7.2. Commodity K^3 (K^4) is from E_3 (E_1) to E_2 (E_3). Show how the matrix AC will be modified.   □

7.5.2 An LP for the Routing Problem Using the Arc-Chain Representation

An LP for the routing problem using the arc-chain representation is best described using matrices and vectors rather than linear equations, used earlier in Section 7.4. This formulation is called $FORM$ 5 in the following description.

Notation used in *FORM 5*

m : the number of logical edges in the network.

q : the number of commodities, $K^1, K^2, \ldots, K^q$, in the network.

$\overrightarrow{C_j^k}$: the jth chain of the kth commodity — a vector of 1s and 0s.

$\hat{n}^k$: the number of chains for commodity K^k.

$\hat{n}$: the total number of chains for all commodities.

AC : arc-chain incidence matrix of size $m \times \hat{n}$.

AC^k : the $m \times \hat{n}^k$ sub-matrix of AC corresponding to all the chains of commodity K^k.

$\overrightarrow{AC_j^k}$: the column vector of AC corresponding to the jth chain of commodity K^k.

ac_{ij}^k : the ith element of the jth chain of commodity K^k so that

$$ac_{ij}^k = \begin{cases} 1 \text{ if edge } i \text{ is in chain } j \text{ of commodity } K^k, \\ 0 \text{ otherwise.} \end{cases}$$

Υ^k : the traffic $t(s_k, d_k)$ for commodity K^k, $1 \le k \le q$.

A : constraints matrix[8] of size $m \times (\hat{n} + m + 1)$.

$\overrightarrow{A^p}$: column p in the constraints matrix A.

$\overrightarrow{c}$: a vector of cost-coefficients where the pth element c_p represents the cost of the pth column of A.

x_j^k : a variable denoting the flow in the jth chain of commodity K^k, $0 \le x_j^k \le \Upsilon^k$.

$\overrightarrow{x^k}$: the column vector of variables, containing $\hat{n}^k$ elements $x_1^k, x_2^k, \ldots, x_{\hat{n}^k}^k$.

$\overrightarrow{e^k}$: a row vector $[1, 1, \ldots, 1]$ of $\hat{n}^k$ 1s.

Λ_{max} : a variable denoting the $\mathcal{L}$-congestion in the network.

$\overrightarrow{\Lambda}$: a column vector $[\Lambda_{max}, \Lambda_{max}, \ldots, \Lambda_{max}]$ with m occurrences of the variable Λ_{max}.

$\wp(s, d)$: the shortest path, in the logical topology, from s to d.

x_s^i : the ith slack variable.

---

[8] The constraints matrix will not be explicitly generated or stored in this formulation but will be referred to in the discussions.

$\overrightarrow{x_s}$: a column vector of m slack variables so that the elements of $\overrightarrow{x_s}$ are x_s^1, x_s^2, ..., x_s^m.

B : the basis for the revised simplex method, a matrix of size $(m+q) \times (m+q)$.

$\overrightarrow{c_B}$: a vector of $(m + q)$ cost coefficients, each coefficient corresponding to a variable in the basis.

$\overrightarrow{x_B}$: the vector corresponding to the variables in the basis.

$\overrightarrow{y}$: a vector of $(m + q)$ simplex multipliers.

The formulation for *FORM* 5

The linear programming formulation, using the arc-chain formulation, is as follows:

Objective function:
$$\text{minimize} \quad \Lambda_{max} \tag{7.14}$$

Subject to

$$AC^1\overrightarrow{x^1} + AC^2\overrightarrow{x^2} + \ldots + AC^q\overrightarrow{x^q} \leq \overrightarrow{\Lambda}$$
$$\overrightarrow{e^1}\overrightarrow{x^1} \qquad\qquad = \Upsilon^1$$
$$\overrightarrow{e^2}\overrightarrow{x^2} \qquad = \Upsilon^2 \tag{7.15}$$
$$\vdots$$
$$\overrightarrow{e^q}\overrightarrow{x^q} = \Upsilon^q$$

Justification of the LP Equation

The objective function is the same as in Section 7.4, minimizing the $\mathfrak{L}$-congestion.

The first line in (7.15) corresponds to the edges in the logical topology. For this line, the right hand side of the inequality represents the column vector $[\Lambda_{max}, \Lambda_{max}, \ldots, \Lambda_{max}]$ of size m. The left hand side is $\sum_{k=1}^{q} AC^k \cdot \overrightarrow{x^k}$. Here AC^k is a sub-matrix of AC having size $m \times \hat{n}^k$ and $\overrightarrow{x^k}$ is a column vector of size $\hat{n}^k$ so that the product $AC^k\overrightarrow{x^k}$ has size m. Considering the ith logical edge, the sum of the ith elements of all the products $AC^k\overrightarrow{x^k}$, for all commodities k, should be less than or equal to Λ_{max} so that $\sum_{k=1}^{q}\sum_{j=1}^{\hat{n}^k} ac_{ij}^k \cdot x_j^k \leq \Lambda_{max}$. In other words, if the logical edge i appears in chain j of commodity k, $ac_{ij}^k = 1$. Only then should the flow x_j^k through the jth chain for commodity k contribute to the total traffic on logical edge i. The sum of all these flows on logical edge i must be less than or equal to the $\mathcal{L}$-congestion Λ_{max}.

The lines 2 to $q+1$ of (7.15) correspond to the commodities, so that the constraint corresponding to the kth commodity is $\overrightarrow{e^k} \cdot \overrightarrow{x^k} = \varUpsilon^k$, for all $i, 1 \leq k \leq q$. Since $\overrightarrow{e^k}$ is a vector $[1, 1, \ldots, 1]$ of $\hat{n}^k$ 1s, the constraint becomes $\sum_{j=1}^{\hat{n}^k} x_j^k = \varUpsilon^k$. In other words, the sum of the flows for the kth commodity, using all the chains for commodity K^k, must be equal to the requirement $\varUpsilon^k$.

7.5.3 Solving the LP Specified Using the Arc-Chain Representation

The advantage of the formulation given in Section 7.5.2 is that the number of constraints, and hence the basis size, is $q + m$. The disadvantage is that the value of $\hat{n}$, the total number of chains for all commodities, is extremely large for all nontrivial networks, since the number of chains in such a network, for a given commodity, specified by its source and destination, is likely to be enormous. The number of commodities is usually $\mathcal{N}_E \times (\mathcal{N}_E - 1)$, since most end nodes normally communicate with each other. This is also a large number for nontrivial networks, so that storing the constraints in the form of a matrix A is not feasible. It is necessary to adapt some techniques from operations research to handle the problem.

The normal first step in solving an LP, using the revised simplex method,[9] is to remove the inequalities by adding slack variables. Using matrix notation, the column vector $\overrightarrow{x_s}$ of m slack variables x_s^1, x_s^2, $\ldots$, x_s^m may be used to remove the inequality constraints giving the following constraints:

$$
\begin{aligned}
AC^1\overrightarrow{x^1} + AC^2\overrightarrow{x^2} + \ldots + AC^q\overrightarrow{x^q} + \overrightarrow{x_s} &= \overrightarrow{\Lambda} \\
\overrightarrow{e^1}\overrightarrow{x^1} &= \varUpsilon^1 \\
\overrightarrow{e^2}\overrightarrow{x^2} &= \varUpsilon^2 \qquad (7.16) \\
&\cdots \\
\overrightarrow{e^q}\overrightarrow{x^q} &= \varUpsilon^q
\end{aligned}
$$

After taking the variable $\overrightarrow{\Lambda}$ to the left hand side, the constraints are as follows.

---

[9] A summary of the revised simplex method, with an explanation of the terms used in the revised simplex method, is given in Appendix 1.

$$-\overrightarrow{A} + AC^1\overrightarrow{x^1} + AC^2\overrightarrow{x^2} + \ldots + AC^q\overrightarrow{x^q} + \overrightarrow{x_s} = 0$$

$$\overrightarrow{e^1}\overrightarrow{x^1} \qquad\qquad\qquad = \Upsilon^1$$

$$\overrightarrow{e^2}\overrightarrow{x^2} \qquad\qquad = \Upsilon^2 \qquad (7.17)$$

$$\ldots$$

$$\overrightarrow{e^q}\overrightarrow{x^q} \qquad = \Upsilon^q$$

Example 7.5. For the network shown in Figure 7.2 and the two commodities K^1 and K^2 discussed in the previous example, let the traffic demand Υ^1 (Υ^2) for commodity K^1 (K^2) be 0.5 (0.3). The constraints described in (7.16), when specified in matrix form, are as follows:

$$
\begin{bmatrix}
1 & 0 & 0 & 0 & 1 & 0 & 0 & 0 & 0 & 0 \\
0 & 0 & 1 & 0 & 1 & 0 & 0 & 0 & 0 & 0 \\
0 & 0 & 0 & 0 & 0 & 0 & 1 & 0 & 0 & 0 \\
0 & 1 & 0 & 0 & 0 & 0 & 0 & 1 & 0 & 0 \\
0 & 1 & 0 & 1 & 0 & 0 & 0 & 0 & 1 & 0 \\
0 & 0 & 1 & 0 & 0 & 0 & 0 & 0 & 0 & 1 \\
1 & 1 & 0 & 0 & 0 & 0 & 0 & 0 & 0 & 0 \\
0 & 0 & 1 & 1 & 0 & 0 & 0 & 0 & 0 & 0
\end{bmatrix}
\times
\begin{bmatrix}
x_1^1 \\ x_2^1 \\ x_1^2 \\ x_2^2 \\ x_s^1 \\ x_s^2 \\ x_s^3 \\ x_s^4 \\ x_s^5 \\ x_s^6
\end{bmatrix}
=
\begin{bmatrix}
\Lambda_{max} \\ \Lambda_{max} \\ \Lambda_{max} \\ \Lambda_{max} \\ \Lambda_{max} \\ \Lambda_{max} \\ 0.5 \\ 0.3
\end{bmatrix}
\qquad (7.18)
$$

Here, the columns 1 and 2 (3 and 4) of the constraints matrix correspond to the two chains $\overrightarrow{C_1^1}$, $\overrightarrow{C_2^1}$ ($\overrightarrow{C_1^2}$, $\overrightarrow{C_2^2}$) for commodity K^1 (K^2). The variables x_1^1, x_2^1 (x_1^2, x_2^2) correspond to the flows for chains $\overrightarrow{C_1^1}$, $\overrightarrow{C_2^1}$ ($\overrightarrow{C_1^2}$, $\overrightarrow{C_2^2}$) for commodity K^1 (K^2). The variables x_s^1, x_s^2, ..., x_s^6 are the slack variables. The first m (here $m = 6$) rows of the constraints matrix A correspond to the first line in (7.16) and the last q (here $q = 2$) rows correspond to the constraints given in the remaining lines in (7.16). The identity matrix in rows 1–6 and columns 5–10 of the constraints matrix A correspond to the slack variables.

After moving Λ_{max} to the left hand side, as done in (7.17), the resulting equation, in matrix form, is as follows:

$$
\begin{bmatrix}
-1 & 1 & 0 & 0 & 0 & 1 & 0 & 0 & 0 & 0 & 0 \\
-1 & 0 & 0 & 0 & 1 & 0 & 1 & 0 & 0 & 0 & 0 \\
-1 & 0 & 0 & 0 & 0 & 0 & 0 & 1 & 0 & 0 & 0 \\
-1 & 0 & 1 & 0 & 0 & 0 & 0 & 0 & 1 & 0 & 0 \\
-1 & 0 & 1 & 0 & 1 & 0 & 0 & 0 & 0 & 1 & 0 \\
-1 & 0 & 0 & 1 & 0 & 0 & 0 & 0 & 0 & 0 & 1 \\
0 & 1 & 1 & 0 & 0 & 0 & 0 & 0 & 0 & 0 & 0 \\
0 & 0 & 0 & 1 & 1 & 0 & 0 & 0 & 0 & 0 & 0
\end{bmatrix}
\times
\begin{bmatrix}
\Lambda_{max} \\ x_1^1 \\ x_2^1 \\ x_1^2 \\ x_2^2 \\ x_s^1 \\ x_s^2 \\ x_s^3 \\ x_s^4 \\ x_s^5 \\ x_s^6
\end{bmatrix}
=
\begin{bmatrix}
0.0 \\ 0.0 \\ 0.0 \\ 0.0 \\ 0.0 \\ 0.0 \\ 0.5 \\ 0.3
\end{bmatrix}
\qquad (7.19)
$$

□

Exercise 7.9. Considering that the network now supports the two additional commodities defined in Exercise 7.8, show how (7.19) will be modified.   □

From (7.16), it should be noted that, in the constraints matrix, the values in the first m positions in a column corresponding to a flow variable x_j^k comes from chain $\overrightarrow{C_j^k}$ in AC^k. The remaining q elements in the column are all 0s except for element number $m+k$ which is 1. The column for Λ_{max} has -1 in the first m positions and 0 in the remaining q positions. If a column corresponds to the slack variable x_s^i, then the column has 0s in all positions except i, where there is a 1. Since the total number of constraints is $m + q$, the basis is an $(m + q) \times (m + q)$ matrix.

Finding an initial feasible solution for *FORM 5*

For this problem, the normal method of using artificial variables and the "big M" method [377] to find an initial feasible solution (IFS) in an LP with equality constraints is time consuming. To create the basis B of size $(m + q) \times (m + q)$, and $\overrightarrow{x_B}$ denoting the vector corresponding to the basis variables, the following procedure may be used for the IFS.

The idea is to have only one chain in the basis B for the IFS, for each commodity, and adjust the slack variables accordingly. The formal description is tedious and is illustrated through an example.

Example 7.6. In the case of the network shown in Figure 7.2 with two commodities K^1 and K^2 having traffic demands 0.5 and 0.3, the basis is an 8×8 matrix. The steps for creating the basis corresponding to an initial feasible solution are as follows:

> **Step 1)** Choose any one chain for each of the commodities K^1 and K^2. Send the entire traffic Υ^1 (Υ^2) for K^1 (K^2) on the selected chain for K^1 (K^2). Figure 7.3 shows the situation where the chain [0, 0, 0, 1, 1, 0] ([0, 1, 0, 0, 1, 0]) corresponding to logical path $E_1 \Rightarrow E_4 \Rightarrow E_2$ ($E_4 \Rightarrow E_2 \Rightarrow E_3$) for commodity K^1 (K^2) has been selected. The flows for the commodities K^1 and K^2 on the edges in the chains $\overrightarrow{C_2^1}$ and $\overrightarrow{C_2^2}$ are shown in Figure 7.3. For example, edge 5 has a flow of 0.5 for commodity K^1 and 0.3 for the commodity K^2. In Figure 7.3, beside edge 5, this is shown by the pair (0.5, 0.3).
>
> **Step 2)** Calculate the sum of the flows on each logical edge and find the logical edge carrying the maximum flow. The total traffic on this logical edge is Λ_{max}. In the case shown in Figure 7.3, logical edge number 5, from E_4 to E_2, carries the maximum flow of 0.8. The value of $\Lambda_{max} = 0.8$ in this example.
>
> **Step 3)** Create the columns in the basis for Λ_{max} and for the chains selected for K^1 and K^2. In this example, this step creates the first

three columns of the basis, one each for Λ_{max}, K^1, and K^2. The values
of the corresponding basis variables are 0.8, 0.5, and 0.3 respectively.
Step 4) Since edge 5, from E_4 to E_2, is carrying the flow Λ_{max}, no
slack variable is needed for this edge. For each remaining edge $i, i \neq 5$,
a slack variable x_s^i is needed. The values of the slack variable x_s^i, $i \neq 5$,
need to be adjusted to satisfy the constraint in the first line of (7.17).

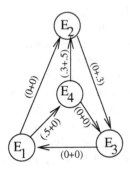

Fig. 7.3. A network with four nodes having some flows

The resulting equation using matrix notation is given below.

$$
\begin{bmatrix}
-1 & 0 & 0 & 1 & 0 & 0 & 0 & 0 \\
-1 & 0 & 1 & 0 & 1 & 0 & 0 & 0 \\
-1 & 0 & 0 & 0 & 0 & 1 & 0 & 0 \\
-1 & 1 & 0 & 0 & 0 & 0 & 1 & 0 \\
-1 & 1 & 1 & 0 & 0 & 0 & 0 & 0 \\
-1 & 0 & 0 & 0 & 0 & 0 & 0 & 1 \\
0 & 1 & 0 & 0 & 0 & 0 & 0 & 0 \\
0 & 0 & 1 & 0 & 0 & 0 & 0 & 0
\end{bmatrix}
\times
\begin{bmatrix}
0.8 \\
0.5 \\
0.3 \\
0.8 \\
0.5 \\
0.8 \\
0.3 \\
0.8
\end{bmatrix}
=
\begin{bmatrix}
0.0 \\
0.0 \\
0.0 \\
0.0 \\
0.0 \\
0.0 \\
0.5 \\
0.3
\end{bmatrix}
\qquad (7.20)
$$

□

Exercise 7.10. Considering that the network now supports the two additional commodities defined in Exercise 7.8, show how (7.20) will be
modified.   □

Finding the entering column

As outlined in Appendix 1, the standard approach in the revised simplex
method is to find, if possible, an entering column by

- computing the simplex multipliers $\overrightarrow{y} = \overrightarrow{C_B} B^{-1}$,

- finding, if possible, a column $\overrightarrow{A^p}$ in the constraints matrix A, such that $\overrightarrow{y} \cdot \overrightarrow{A^p} > c_p$, where c_p is the cost associated with the column p.

To handle this problem, it is possible to adapt Tomlin's approach for solving minimum cost multi-commodity flow problems [385]. Instead of explicitly storing the constraints as done in an LP solver, Tomlin's approach implicitly keeps track of the constraints and generates a chain, only when it is established that the chain should be part of the column entering the basis. The central idea is that, in each iteration, the algorithm checks only one potential chain/commodity. For each commodity, the algorithm checks only the "shortest" chain from the source to the destination of that commodity where each logical edge is assigned a distance having an appropriate value. Such a shortest chain may be created, on the fly, using, for instance, Djikstra's algorithm [4, 108]. The method then checks if the chain satisfies the condition $\overrightarrow{y} \cdot \overrightarrow{A^p} > c_p$ to be part of an entering column. If so, the algorithm creates the entering column using the shortest chain. The details of the algorithm, tailored for the routing problem, are given below.

In this description, there will be references to the pth column $\overrightarrow{A^p}$ of the constraints matrix A. This constraints matrix, as mentioned earlier, will never be explicitly stored.

For convenience in describing the algorithm, let $\pi_1, \pi_2, \ldots, \pi_m$ be the first m simplex multipliers (the first m elements of vector $\overrightarrow{y}$), corresponding to the m arcs of the network, and let $\alpha_1, \alpha_2, \ldots, \alpha_q$ be the remaining q simplex multipliers (the last q elements of vector $\overrightarrow{y}$), corresponding to the q commodities.

Theorem 7.1.

a) *If $\pi_i > 0$, for any $i, 1 \le i \le m$, slack variable x_s^i is a candidate to enter the basis.*

b) *If the sum of the first m simplex multipliers is less than -1, Λ_{max} is a candidate to enter the basis.*

c) *If, for chain j of commodity k, $\displaystyle\sum_{i=1}^{m}(-\pi_i)a_{ij}^k < \alpha_k$, then the variable x_j^k corresponding to the chain $\overrightarrow{C_j^k}$ is a candidate to enter the basis.*

Proof. a) Let p be the column corresponding to the ith slack variable x_s^i. The cost coefficient for any slack variable is 0, so $c_p = 0$, and the vector $\overrightarrow{A^p}$ consists of 0s except in position i. Therefore, for x_s^i, $\overrightarrow{y} \cdot \overrightarrow{A^p} - c_p = \pi_i$. Since π_i is positive, x_s^i is a candidate to enter the basis.

b) The cost coefficient for variable Λ_{max} is 1 and the corresponding column $\overrightarrow{A^p}$ in the constraints matrix consists of -1s in the first m positions and 0s in the remaining q positions. This means

$$\vec{y} \cdot \overrightarrow{A^p} - c_p = [\pi_1 \ldots \pi_m \alpha_1 \ldots \alpha_q] \cdot \begin{bmatrix} -1 \\ -1 \\ \vdots \\ -1 \\ 0 \\ 0 \\ \vdots \\ 0 \end{bmatrix} - 1 = -\sum_{i=1}^{m} \pi_i - 1$$

Since $\sum_{i=1}^{m} \pi_i < -1$, Λ_{max} is a candidate to enter the basis.

c) The column $\overrightarrow{A^p}$ corresponding to the chain $\overrightarrow{C_j^k}$ has value a_{ij}^k in row i, for all $i, 1 \le i \le m$. The values in $\overrightarrow{A^p}$ in the remaining q positions are all 0s except for a 1 in position $m + k$. It is therefore possible to represent $\overrightarrow{A^p}$ by $[a_{1j}^k a_{2j}^k \ldots a_{mj}^k 0 \ldots 010 \ldots 0]$. Since $c_p = 0$,

$$\vec{y} \cdot \overrightarrow{A^p} - c_p = \vec{y} \cdot \overrightarrow{A^p} = [\pi_1 \ldots \pi_m \alpha_1 \ldots \alpha_{k-1} \alpha_k \alpha_{k+1} \ldots \alpha_q] \cdot \begin{bmatrix} a_{1j}^k \\ a_{2j}^k \\ \vdots \\ a_{mj}^k \\ 0 \\ \vdots \\ 0 \\ 1 \\ 0 \\ \vdots \\ 0 \end{bmatrix}$$

$$= \sum_{i=1}^{m} \pi_i a_{ij}^k + \alpha_k .$$

If $\vec{y} \cdot \overrightarrow{A^p} - c_p$ is positive, $\overrightarrow{C_j^k}$ may be used to create an entering column. Therefore the condition for creating an entering column is $\sum_{i=1}^{m} \pi_i a_{ij}^k + \alpha_k > 0$. This means that $\sum_{i=1}^{m} (-\pi_i) a_{ij}^k < \alpha_k$. This completes the proof.   □

Theorem 7.1 may be used for determining a column to enter the basis. If any π_i is positive, the corresponding slack variable can be entered into the basis (Theorem 7.1, part a). The column corresponding to slack variable x_s^i

can be constructed immediately since it contains a 0 in all positions except in position i where there is a 1.

To use Theorem 7.1 part c, it is important to note that the case of $\pi_i > 0$ has been already considered in Theorem 7.1 part a. Therefore, it is safe to assign $(-\pi_i)$ as the length of edge i and guarantee that the length of any edge i is 0 or positive, since $\pi_i \leq 0$ for all $i, 1 \leq i \leq m$. Then $\sum_{i=1}^{m}(-\pi_i)a_{ij}^{k}$

is the length of chain $\overrightarrow{C_j^k}$. If this length is less than α_k, this chain satisfies the condition given in Theorem 7.1 and may be used to create the entering column. In the arc-chain formulation, in any nontrivial network, there are, in general, an enormous number of chains from any given source to any given destination. To avoid listing these chains explicitly, a mechanism is needed to quickly find a suitable chain.

Finding the shortest path for each commodity can be done very efficiently using well-known techniques such as Dijkstra's algorithm [4, 108] if all the edges have nonnegative lengths. Let E_{s_k} (E_{d_k}) be the source (destination) of commodity K^k and let $\wp_{s_k d_k}$ be the shortest path (for commodity K^k) from E_{s_k} to E_{d_k}. This shortest path $\wp_{s_k d_k}$ corresponds to some chain, say the jth chain $\overrightarrow{C_j^k}$. Then, the length of the shortest path is $= \sum_{i=1}^{m}(-\pi_i)a_{ij}^{k}$. If this length is less than α_k, $\overrightarrow{C_j^k}$ is a valid chain to enter the basis. In this case, the path $\wp_{s_k d_k}$ found by the shortest-path algorithm immediately gives the chain $\overrightarrow{C_j^k}$ and, therefore, the first m elements in the entering column. The remaining q elements in the entering column are all 0, except in position k, where there is a 1.

Furthermore, if the length of path $\wp_{s_k d_k} \geq \alpha_k$, this means that all other chains ($\overrightarrow{C_r^k}$) for commodity K^k have a length $\geq \alpha_k$ so that $\sum_{i=1}^{m}(-\pi_i)a_{ir}^{k} \geq \alpha_k$.

Therefore, no chain $\overrightarrow{C_r^k}$ for commodity K^k can be a valid candidate for an entering column. It is therefore sufficient to consider only the shortest path $\wp(s_k, d_k)$ for each commodity.

The following steps use Theorem 7.1 to determine a column to enter the basis:

Step 1) Repeat Step 2 for all $i, 1 \leq i \leq m$.

Step 2) If $\pi_i > 0$, create an entering column corresponding to slack variable x_s^i and stop.

Step 3) Assign length $-\pi_i$ to edge i for all $i, 1 \leq i \leq m$, in the logical topology.

Step 4) Repeat Steps 5–6 for all $k, 1 \leq k \leq q$.

Step 5) Find the shortest path $\wp(s_k, d_k)$ from E_{s_k} to E_{d_k}.

Step 6) If the length of the path $\wp(s_k, d_k) < \alpha_k$, create the entering column from the path $\wp(s_k, d_k)$, and k, the commodity number, and stop.

Step 7) No entering column exists. Stop.

This algorithm does not include any step to enter the column corresponding to Λ_{max}. The initial feasible solution automatically inserts Λ_{max} in the basis. Once Λ_{max} is in the basis, it is never a candidate for leaving the basis. The process for finding the leaving column involves the standard ratio test of a revised simplex algorithm[10] and is omitted.

Use of generalized upper bounding and eta-factorization

The most time-consuming components of an iteration in the revised simplex method, as outlined in Appendix 1, are the following steps:

Step 1) Solve the equation $\overrightarrow{y} \cdot B = \overrightarrow{c_B}$.

Step 4) Solve the equation $B \cdot \overrightarrow{d} = \overrightarrow{A^j}$.

Both these steps involve inverting the basis B, a matrix of size $(m + q) \times (m + q)$. Generalized upper bounding (GUB) [93, 101] is a technique to compute and update y and d when the basis has a special structure. A basis B, for some m and q, exhibits a GUB structure if

- q is relatively large compared to m and
- each column of the last q rows has at most one nonzero entry, say e, e being equal to 1.

It is well known [93] that, if the basis exhibits a GUB structure, it is possible to compute $\overrightarrow{y}$ and $\overrightarrow{d}$ using operations on matrices of smaller size, with the crucial operation of inverting a matrix needed only on a matrix of size m [93]. In cases where q is much larger than m, the GUB technique can dramatically improve the time to compute $\overrightarrow{y}$ and $\overrightarrow{d}$. It is also well-known that, if the basis exhibits a GUB structure, the use of eta-factorization [93] may also be used profitably to speed up the process of computing y and d.

The columns in the constraint matrix A corresponding to the arc-chain formulation (7.17) belong to one of three categories as follows:

Category 1) The column corresponding to Λ_{max}. In this column, the last q elements are 0.

Category 2) A column corresponding to some chain $\overrightarrow{C_j^k}$. In such a column, there is a 1 only in row $m+k$; the remaining elements in rows $m+i, i \neq k, 1 \leq i \leq q$, have a value 0.

Category 3) A column corresponding to some slack variable x_s^i. In such a column, all elements in rows $m, \ldots, m + q$ are 0s.

---

[10] briefly reviewed in Appendix 1.

Therefore the basis B in the arc-chain representation satisfies the GUB structure. Thus the routing problem can be easily handled even for relatively large networks.

Exercise 7.11. Write a program to implement the algorithm given in this section. The input to the algorithm is a logical topology and a traffic matrix. Your program should generate a routing for minimum $\mathcal{L}$-congestion. Run the program for WDM networks with 10–50 nodes. Compare the execution time of this program with that of the program you wrote in Exercise 7.7 using the same logical topology and traffic matrix.   □

Exercise 7.12. Analyze your program to find the part which is most time consuming. Can you improve the performance of the program in that part? An outline of one way to improve is given below.

Do not run Dijkstra's algorithm in each iteration of the LP. When running the Dijkstra's algorithm, the weight of the ith logical edge in the network was $-\pi_i$. In each iteration, the straightforward approach used in the algorithm given above is to assign the weight obtained in the last iteration and run Dijkstra's algorithm to find the shortest path.

It is possible to simply keep the shortest paths obtained using a previous iteration of Dijkstra's algorithm, reassign the weights in each iteration, and check whether the condition for the entering column is still satisfied. Of course, the paths used are no longer the shortest paths. However, they are valid paths and, so long as the condition for an entering column is satisfied, these paths can be used.

A stage will be reached when no further improvement is possible using the shortest paths found in a previous iteration, for all the commodities. Only at this stage, should Dijkstra's algorithm be run again.   □

7.6 Bibliographic Notes

The complete logical topology design and traffic routing problem has been formulated in [236] as an MILP. This joint problem is then decomposed into two subproblems:

    i) the connectivity problem and
    ii) the flow assignment problem.

A heuristic is used for the first problem and the second problem is cast and solved as a multi-commodity flow problem. The overall algorithm then iterates from the current solution by applying branch-exchange operations to the connectivity diagram. [323] is a milestone paper that studies the design of logical topologies, over a wavelength-routed optical network physical topology. The authors first formulate the topology design and routing problem as an MILP, and then propose splitting the combined problem into two subproblems. A number of heuristics for topology design are also presented. The topology design problem has been formulated in [289] as an optimization problem that either minimizes the average packet delay or maximizes the factor by which the traffic matrix can be scaled up. The formulation also considers constraints on the number of available wavelengths per fiber. A heuristic based on simulated annealing and flow deviation is also presented for larger networks.

In [116] the static virtual topology design problem has been formulated as an MILP. It then describes and compares a number of important theoretical results, algorithms, and heuristics in the area. A tabu-search-based meta-heuristic for optimal logical topology design, to minimize the $\mathcal{L}$-congestion of the network, has been presented in [177]. The paper assumes that the physical topology and a stochastic description of the traffic pattern is given, and does not consider constraints on the number of wavelengths per fiber.

The concept of "absorption probability", which can be used to express the grade of service of an optical network, has been introduced in [293]. Exact calculation of absorption probability is computationally intensive. A method for approximation of absorption probabilities, together with the lower and upper bounds, for arbitrary topologies also appears in [293].

The minimum number of wavelengths required to support all possible logical topologies on a bidirectional ring physical topology was considered in [292]. This was extended to general two-connected and three-connected physical topologies as well. The design of logical topologies over bidirectional ring and linear physical topologies has been considered in [12]. The objective of the design in [12] is to minimize the processing delays for the worst-case traffic flow, which is equivalent to minimizing the diameter of the logical topology for uniform traffic. A number of logical topology design algorithms are proposed, along with theoretical lower bounds for the traffic delay.

An MILP for optimal logical topology design and an evaluation of a number of existing heuristics for logical topology design, in terms of both performance and complexity, has been given in [241]. The complete logical topology design problem including traffic routing, lightpath selection, and RWA has been considered in [232]. The problem is formulated as an MILP, where the objective is to minimize $\mathcal{L}$-congestion. For large networks, a solution based on LP-relaxation of the integer variables is presented. Topology design, under a static traffic model, for multi-fiber WDM networks has been studied in [386]. Two ILPs have been presented, for wavelength-continuous and wavelength-convertible networks respectively, to solve the problem.

Load balancing algorithms for networks where the average traffic between nodes is changing with time is an important research topic that has been investigated by many researchers. An overview of reconfiguration issues in virtual topology design is given in [116]. Algorithms have been proposed in [378] to iteratively reconfigure the logical topology to minimize the $\mathcal{L}$-congestion in response to changes in the traffic pattern. Since each change to the logical topology is small, the disturbance to the network is small. It was shown that the algorithms perform near-optimally.

Reconfiguration issues in single-hop lightwave networks has been considered in [22]. The reconfiguration problem is formulated in [22] as a Markovian decision process and a number of retuning strategies are also presented. The effects of reconfiguration, in terms of different parameters such as packet delay and packet loss, have been studied in [22] through simulations.

The logical topology design problem has been formulated as a linear program in [311]. The approach is used as a reconfiguration algorithm for updating the topology as needed. A two-stage heuristic has been proposed in [366], which provides a tradeoff between the objective function value and the number of changes to the network. An analytical model for logical topology reconfiguration in optical networks has been proposed in [420]. A case study on reconfiguration policies, using the analytical model, was also conducted in [420]. Reconfiguring the logical topology causes disruption in traffic, so the time to trigger reconfiguration is an important issue. The conflicting objectives of efficient resource utilization and minimizing traffic disruption has been considered in [357] using an information theoretic point of view.

A number of important issues in IP/WDM networks and reconfigurable IP/WDM networks has been presented in [402]. A new traffic engineering framework based on MPLS-based IP routing and MPLS-based WDM reconfiguration architecture and an IP/WDM network test bed is also presented in [402].

For the routing problem in logical topologies, discussed in Section 7.5, a fast approximation algorithm has been proposed in [10].

8

Faults in Optical Networks

With the rapid deployment of optical networks on a global scale, involving millions of kilometers of optical fibers, the technological improvements in the number of channels on a fiber and the resulting increases in the volume of data carried by optical networks, the vulnerability of such networks to failures of different types has become vitally important. In recent years, the design of robust WDM networks has become a major topic [408, 409, 106, 266, 107, 155, 264, 168, 167, 208, 329, 316, 317]. In this chapter, common types of faults in WDM networks and ways to handle the faults are discussed.

In a WDM network, failure may occur in every component of the network. This includes link failures, node failures, channel failures, and software failures. Link failure is the most common type of fault, where the fiber constituting a link between two nodes in the network does not permit data transmission. This happens, for example, when, due to human errors, a backhoe cuts through a fiber during earthmoving operations. The optical/electronic component within a node can fail. The failure of a transmitter or a receiver results in the failure of a channel. An entire node can fail, for example, due to fire or flooding. Such a fault means that all traffic flowing through the failed node is disrupted.

In view of the fact that each lightpath is carrying data at the rate of a few Gbps (2.5 Gbps or 10 Gbps typically, depending on the technology used) and each fiber can carry 100 or more lightpaths, it is clear that disruption of this traffic, even for a brief period of time, is a serious event. Many schemes have been proposed to take care of faults in WDM networks. The simplest scheme is to deal with the problem at a higher layer. The IP protocol, for instance, already has the ability to deal with failure. However, recovery at the IP layer is time consuming (several seconds), too long considering the data transfer rate on a fiber (many lightpaths, each carrying data at several Gbps). The alternative that has attracted the most attention is that of dealing quickly with faults at the optical layer so that the potential of disruption to traffic is minimized.

To handle faults, it is important to make a distinction between single faults and multiple faults. The time to repair a fault ranges from a few hours to a few days. For every 10 km of fiber, a cut is expected once in 12 years [318]. The probability of two faults occurring simultaneously is normally taken to be extremely small. Here, simultaneous failure is not necessarily the failure of two components at exactly the same time; it means that a second fault occurs before the first fault is repaired. In view of the negligibly small probability of multiple failures, most work on faults assume that there can be only a single fault in the network. Recently, some attention has been paid to the fact that several fibers may be sharing the same duct [434]. If there is a cut through the duct, many fibers may be cut at the same time. Multiple faults may also occur due to a natural calamity such as an earthquake affecting an entire region. The following discussions will focus mainly on single-link faults — the most common type of faults.

The basic idea of handling link faults at the optical level is simple. As discussed in Chapter 7, for networks having no provision for handling faults, once the logical topology has been determined with m logical edges, each logical edge has to be realized using one of the RWA algorithms discussed in Chapters 4 and 5. To create the ith edge in the logical topology, from s_i to d_i, the selected RWA algorithm defines a lightpath L_i, from s_i to d_i, using a route ρ_i through the physical topology. This process has to be done for all $i, 1 \leq i \leq m$. Figure 8.1 shows the physical topology of a WDM network where the squares represent nodes (which have add/drop multiplexers as well as OXCs so that they can function both as end nodes and as routers). A bidirectional arrow ($N_i \leftrightarrow N_j$) between node N_i and node N_j represents links in the physical network that allow N_i to communicate directly with N_j and viceversa. In the network shown in Figure 8.1 a lightpath L_i is defined from node N_1 to node N_5 and passes through nodes N_2, N_3, and N_4 using the route $\rho_i = N_1 \rightarrow N_2 \rightarrow N_3 \rightarrow N_4 \rightarrow N_5$.

If the possibility of link faults in an optical network is allowed, any one of the fibers in the physical topology may become faulty. In the graph model described here, a failure in a link (or fiber) means that an edge in the graph corresponding to the physical network does not exist any longer. If the route ρ_i of lightpath L_i includes an edge that has become faulty, the lightpath L_i no longer remains operational. One straightforward way to handle the fault is to replace the lightpath L_i by some other lightpath L_i^{new} with its own route and channel assignment. For instance, any of the fibers $N_1 \rightarrow N_2$, $N_2 \rightarrow N_3$, $N_3 \rightarrow N_4$, or $N_4 \rightarrow N_5$, used by lightpath L_i in Figure 8.1, may develop a link fault. When such a fault occurs, a lightpath L_i^{new} must be set up to carry the traffic earlier communicated by L_i. To illustrate this, let there be a link failure involving the fiber $N_2 \rightarrow N_3$ (Figure 8.2). The route $N_1 \rightarrow N_2 \rightarrow N_3 \rightarrow N_4 \rightarrow N_5$ must be replaced by a route from N_1 to N_5 that does not include the faulty edge $N_2 \rightarrow N_3$. In the case of this specific fault, fault management may be done in a variety of ways as follows:

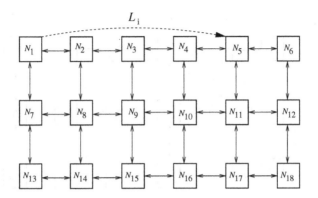

Fig. 8.1. A lightpath in a fault-free WDM network

I) **Replace the faulty route by another route.** One possible route for L_i^{new} is $N_1 \to N_7 \to N_8 \to N_9 \to N_{10} \to N_{11} \to N_5$ (Figure 8.3).

II) **Replace a faulty link by a subpath around it.** For instance, the original route may be replaced by the new route $N_1 \to N_2 \to N_8 \to N_9 \to N_3 \to N_4 \to N_5$ (Figure 8.4) for L_i^{new}.

III) **Replace a subpath that contains the faulty link.** For instance, the original route may be replaced by the new route $N_1 \to N_2 \to N_8 \to N_9 \to N_{10} \to N_4 \to N_5$ for L_i^{new}.

To make a distinction between the lightpath L_i, to be used when there is no fault, and the lightpath L_i^{new} that replaces it when there is a fault in the route ρ_i of lightpath L_i, from now on, the following notation will be used. The lightpath used in the absence of faults will be called the *primary lightpath* and the ith primary lightpath will be denoted by L_i^P. The route of L_i^P through the physical topology will be called the *primary path* (also called the *service path* or the *working path*) and will be denoted by ρ_i^P. If there is a faulty link in ρ_i^P, the lightpath replacing L_i^P will be called the *backup lightpath* for L_i^P and will be denoted by L_i^B. The route of L_i^B through the physical topology is called the *backup path* (also called the *alternate path*, the *restoration path*, or the *protection path*) and will be denoted by ρ_i^B. In order to be able to use the lightpath L_i^B in place of L_i^P, when a fault occurs, it is essential to ensure that the faulty edge in route ρ_i^P cannot appear in route ρ_i^B. Depending on the fault management technique used, it is quite possible that

- different backup paths may be used to take care of faults in different components of the primary path ρ_i^P or

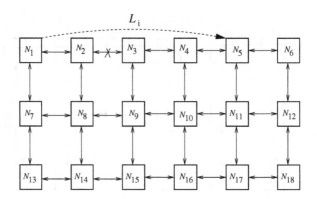

Fig. 8.2. A faulty WDM network

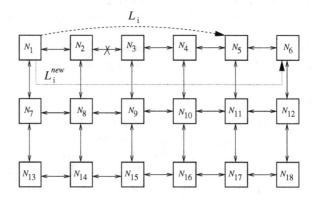

Fig. 8.3. Replacing a faulty route in a WDM network

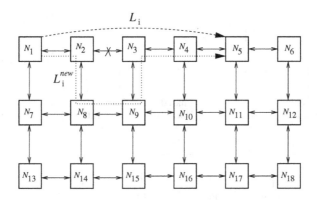

Fig. 8.4. Replacing a faulty link in a WDM network

- the same backup path may be used to take care of a fault anywhere in the primary path.

8.1 Categorization of Faults

A broad categorization of the main schemes for the management of edge failure, following [318], is given below (Figure 8.5). In a *fault protection scheme* scheme, for every primary lightpath L_i^P, a backup lightpath L_i^B is determined at the same time as L_i^P. Either both L_i^P and L_i^B are simultaneously operational in fault-free networks carrying the same traffic or the backup lightpath is set up quickly (within ten to one hundred milliseconds after a fault occurs) when L_i^P fails in order to carry the traffic originally carried by L_i^P.

In a *restoration* scheme [252, 281], only after a fault occurs, for every lightpath L_i^P that become inoperative, the search for a route and channel assignment for a backup lightpath L_i^B is carried out. Restoration schemes take longer time but are more efficient in utilizing capacity due to the multiplexing of the spare capacity requirements and provide resilience against different types of failure [318].

In most cases, protection (or restoration) is based on protecting the entire primary path or on protecting (or restoring) a failed edge of the primary path. The idea of *path protection* (or *restoration*) is that, in the case of a failed edge in the primary path, a backup path, edge-disjoint from the primary path, will be used. Since the edge in the physical topology is a fiber, this means that

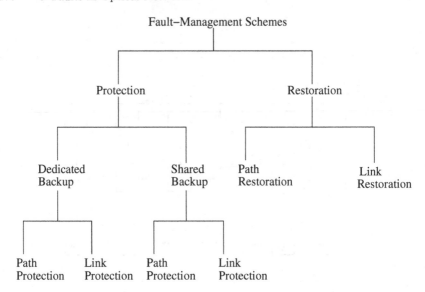

Fig. 8.5. A categorization of fault management schemes

the primary and backup paths are fiber-disjoint. Figure 8.6 shows a network satisfying the wavelength continuity constraint with three primary (backup) lightpaths L_1^P, L_2^P, and L_3^P (L_1^B, L_2^B, and L_3^B) using primary (backup) paths $\rho_1^P = N_8 \rightarrow N_7 \rightarrow N_6$, $\rho_2^P = N_4 \rightarrow N_6$, and $\rho_3^P = N_4 \rightarrow N_2 \rightarrow N_1$ ($\rho_1^B = N_8 \rightarrow N_4 \rightarrow N_6$, $\rho_2^B = N_4 \rightarrow N_7 \rightarrow N_6$, and $\rho_3^B = N_4 \rightarrow N_6 \rightarrow N_1$). Since ρ_1^B was deliberately chosen to be edge-disjoint with respect to ρ_1^P, the route ρ_1^B may be used to avoid a fault in *any edge* in the original route ρ_1^P. A similar situation holds for routes ρ_2^P and ρ_2^B and for routes ρ_3^P and ρ_3^B.

The idea of *link protection* (or *link restoration*) is illustrated in Figure 8.7. In the case of a failed edge, say from $N_4 \rightarrow N_6$, in the primary path $\rho_i^P = N_8 \rightarrow N_4 \rightarrow N_6 \rightarrow N_1$, all the remaining edges of the primary path are fault-free. The backup path ρ_i^B is the same as ρ_i^P except that the faulty edge $N_4 \rightarrow N_6$ is replaced by an alternate route $N_4 \rightarrow N_7 \rightarrow N_6$ from N_4 to N_6 that does not involve the edge $N_4 \rightarrow N_6$ so that the backup path is $\rho_i^B = N_8 \rightarrow N_4 \rightarrow N_7 \rightarrow N_6 \rightarrow N_1$. It is to be noted that, except for the subpath $N_4 \rightarrow N_7 \rightarrow N_6$ (which replaces the failed edge $N_4 \rightarrow N_6$), the backup path uses exactly the same resources used in the primary path. In other words, the link protection scheme has the potential of saving resources particularly in situations where the primary path is relatively long.

Path protection schemes may be classified into two categories — *dedicated backup* or *shared backup* . In the dedicated backup scheme, any lightpath L_i^P, using the primary path ρ_i^P, has an associated lightpath L_i^B using the backup path ρ_i^B such that the network resources (i.e., transmitters, receivers, and channels on fibers) are reserved for L_i^B from the very outset. This scheme was used in Figure 8.6. The resources for the backup lightpath L_1^B consists of

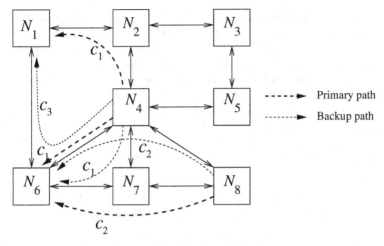

Fig. 8.6. Path protection scheme

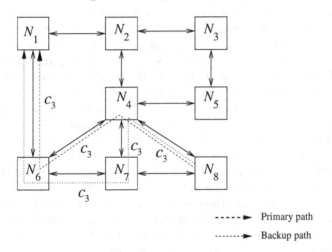

Fig. 8.7. Link protection scheme

the transmitter at N_8 tuned to the wavelength corresponding to channel c_2, the receiver at N_6 tuned to the same wavelength, and the channel c_2 on the edges $N_8 \to N_4$ and $N_4 \to N_6$. These resources are exclusively dedicated to this backup lightpath and cannot be used by any other lightpath — primary or backup. It may be readily verified that this is the case for all the lightpaths illustrated in Figure 8.6. In particular, it is to be noted that the backup paths $\rho_1^B = N_8 \to N_4 \to N_6$ and $\rho_3^B = N_4 \to N_6 \to N_1$ share the edge $N_4 \to N_6$ and, to ensure that L_1^B and L_2^B are given exclusive resources, are allotted different resources (channels c_2 and c_3 on each edge in their respective backup paths).

Dedicated protection schemes may be further classified as a 1+1 *protection scheme* or a 1:1 *protection scheme*. In a 1+1 protection scheme, the data is sent *simultaneously* using both the primary and the backup lightpaths. If there is a fault in the primary path ρ_i^P, the destination simply continues to get the data from the lightpath L_i^B. The recovery of the network from faults in such a scheme is very fast since the destinations affected by a fault do not need to communicate with the corresponding sources or have to wait for any corrective action.

In a 1:1 protection scheme, when the network is fault-free, the data is sent using only the primary lightpath L_i^P, for all $i, 1 \leq i \leq m$, where m is the number of logical edges in the network. If there is a fault in the primary path ρ_i^P, the destination node informs the corresponding source node to

- stop communicating using the primary path ρ_i^P,
- set up the backup lightpath using the backup path ρ_i^B,
- continue data transmission using the lightpath L_i^B.

The recovery of the network from faults in such a scheme is slower than the 1+1 scheme since the destinations affected by a fault need to communicate with the corresponding sources, wait for the source to set up the backup lightpath, and then communicate using the backup lightpath.

There is one advantage of the 1:1 scheme over the 1+1 scheme. Since the lightpath L_i^B is not set up until there is a fault and the probability of a fault is low, the resources for the backup lightpaths are not needed most of the time. These resources may be used for low-priority traffic until there is a fault and there is a need to set up a number of backup lightpaths . When a fault occurs, affecting a number of existing primary lightpaths, backup lightpaths will be set up by preempting resources (channels on fibers) used by low-priority traffic. The low-priority traffic will be communicated most of the time since the probability of a fault is small. One scheme involving low-priority traffic will be informally discussed later on in Section 8.3.3, including Exercise 8.3. So far as the design is concerned, there is no difference between a 1+1 scheme and a 1:1 scheme. The difference is in the operation of the network. As mentioned above, in the 1+1 scheme, for each logical edge, both the primary and the backup lightpath carry the same traffic concurrently.

In dedicated protection approximately 50% of network resources are allotted to backup lightpaths . Since faults occur rarely, these resources are "wasted", in the sense they are not used, most of the time. *Shared-path protection* (also called *backup multiplexing* or *1:N protection*) reduces this wastage to some extent. The idea is that, if two primary lightpaths L_i^P and L_j^P use edge-disjoint primary paths ρ_i^P and ρ_j^P, then, under the single fault assumption, both ρ_i^P and ρ_j^P can never contain a faulty edge simultaneously. Therefore it is never necessary to use the respective backup lightpaths L_i^B and L_j^B at the same time. In shared-path protection, if the primary paths ρ_i^P and ρ_j^P are edge-disjoint, the backup lightpaths L_i^B and L_j^B are allowed to share resources. In such circumstances, ρ_i^B and ρ_j^B may share one or more fibers and L_i^B and L_j^B may be allotted the same channel number.

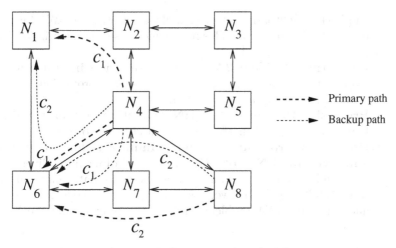

Fig. 8.8. A network illustrating shared-path protection

Shared-path protection is illustrated in Figure 8.8 with the same primary and backup routes as used in Figure 8.6. The difference is that the backup lightpaths L_1^B and L_3^B are given the same channel c_2 even though they share the edge $N_4 \rightarrow N_6$ because the paths ρ_1^P and ρ_3^P are edge-disjoint.

8.2 Important Problems in Protection and Restoration

As discussed in Chapters 4 and 5, the optimization techniques used in determining an optimum routing and wavelength assignment (RWA) in a WDM network are quite complex. The problem becomes more complex when implementing a protection scheme, since a backup path has to be determined whenever a primary path between a source-destination pair is determined. The problem becomes even more complicated when the logical topology and the routing scheme for a given traffic matrix have to be solved at the same time.

In such a design, the characteristics of the WDM network are important. This includes whether or not wavelength converters are present, how many transmitters and receivers are available at each node, and how many channels are supported on each fiber.

In protection, the following problems are important:

- **Protection Problem 1A — 1:1 (or 1+1) protection in static allocation.** Given a network and a set of logical edges, how to carry out static RWA to ensure 1:1 protection.
- **Protection Problem 1B — 1:1 (or 1+1) protection in dynamic allocation.** Given a network and an existing set of primary

and backup lightpaths, how to handle, efficiently, the RWA for a new request for a connection. This new request also needs a primary lightpath and a backup lightpath.

- **Protection Problem 2A — 1:N protection in static allocation.** Same as problem 1A, except that shared protection is needed to make more efficient use of resources.
- **Protection Problem 2B — 1:N protection in dynamic allocation.** Same as problem 1B, except that shared protection is needed.
- **Protection Problem 3 — determining the logical topology and RWA for 1:N protection in static allocation**. Given a network and a traffic matrix, specifying the static data communication to be handled by the network, how to determine the logical topology and solve the RWA problem for both the primary and the backup paths.

Protection problem 1A (1B) is important when dedicated path protection (using either 1:1 or 1+1 protection) is being used in a network having static (dynamic) lightpath allocation. Protection problem 2A (2B) is important when shared-path protection is being used in a network having static (dynamic) lightpath allocation. In these problems, the logical topology has been predetermined. Protection problem 3 is the most complicated problem, involving the design of the logical topology, and, at the same time, finding the optimum primary and backup paths for optimum performance.

As presented in [110], the three basic problems in restoration are as follows:

- **Restoration Problem 1A — Maximum restoration.** Given a network, a set of demands for data communication, a set of logical edges, the primary paths to realize each logical edge, and spare capacities for lightpaths on each fiber in the network, how to find backup paths for as many demands as possible to handle a given fault involving any one node or edge in the network.
- **Restoration Problem 1B — Minimum capacity restoration.** Given a network, a set of logical edges, and their primary paths to realize each logical edge, how to find backup paths for all possible faults so that the total capacity requirements for the network is minimized.
- **Restoration Problem 2 — Joint optimization.** Given a network and a set of logical edges, for each logical edge, how to find both the primary path and the backup path so that the total capacity requirements for the network is minimized.

Restoration Problem 1A is a very practical problem of a network where it may not always be possible to find backup paths and channels to restore all affected lightpaths whose primary paths use the faulty edge or node. In this formulation, the idea is to search the network to find, if possible, for each affected lightpath L_i^P, a backup path ρ_i^B that bypasses the faulty edge. Every edge on the route ρ_i^B must have a spare channel that may be used to set up a backup lightpath L_i^P. If there is no wavelength converter in the network, the search is successful only if the same spare channel is available on every edge in the route ρ_i^B. Since the search may not always succeed for each affected lightpath, the idea is to maximize the number of lightpaths that may be restored.

Restoration Problem 1B is useful in making sure that 100% restorability is possible without changing the primary paths. In a network where static lightpath allocation is used, it is useful to ensure that complete restorability is guaranteed. If it turns out that a particular network cannot handle all faults, a solution to this problem is useful in identifying where additional transmitters and receivers may be installed to ensure complete restorability.

Restoration Problem 2 is very useful in situations where

- existing demands on a legacy[1] network are being transferred to a WDM network,
- periodic cleanup is being done to optimize the optical resources for realizing the logical topology of the network,
- long-term planning is needed to determine future capacity requirements of the system.

Each of these problems may be handled using mathematical optimization, or using a heuristic giving quick solutions. The heuristic may be implemented using a central server or a distributed algorithm executed on the nodes of the network.

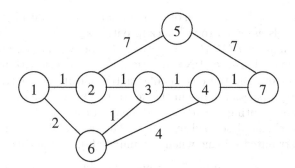

Fig. 8.9. A physical topology of a network illustrating the trap situation

The proper choice of primary path in a path protection scheme is very important. It is tempting to use heuristics to determine a primary path and only then determine a backup path, edge-disjoint from the primary path. Using the shortest path as the primary path is an obvious option in such a heuristic. Selecting the primary path, without taking into account the backup

---

[1] A legacy network is an existing network, possibly based on old and outmoded technology.

path, however may lead to a situation where it is impossible to find a backup path, edge-disjoint from the primary path. This is called a *trap* situation [353]. The following example is taken from [353]. Figure 8.9 shows the physical topology of a network where each edge represents a bidirectional link. The label of each edge in Figure 8.9 is the length of the edge. If the shortest-path routing is used for the primary path, the shortest path from 1 to 7 is $1 \rightarrow 2 \rightarrow 3 \rightarrow 4 \rightarrow 7$. If this is chosen, there is no remaining path, edge-disjoint from the primary path, to be used as the backup path.

8.3 Schemes for Handling Faults

In the following discussions, formulations to handle a number of protection and restoration schemes are covered[2] as follows:

- Formulation *FORM* 6 in Section 8.3.1 finds optimal routings for the primary and the backup paths using the 1:1 or 1+1 path protection scheme. This formulation is for a network where every node is equipped with a wavelength converter at each node, for a specified set of logical edges for the network.
- Formulation *FORM* 7 in Section 8.3.2 sets up, using dynamic lightpath allocation, a primary lightpath and a backup lightpath. This is for a network using 1:N protection in a network with no wavelength converter.
- An informal description of a possible approach for utilizing the unused channels assigned to the backup paths is given in Section 8.3.3.
- Formulation *FORM* 8 in Section 8.3.4 determines an optimal logical topology. This uses 1:N protection, for a network with no wavelength converter , and there is a traffic matrix specifying the traffic demands between pairs of nodes in the network. This formulation also finds an optimal routing scheme over the logical topology.
- The scheme described in Section 8.3.5 is useful for restoration using a distributed scheme when dynamic lightpath allocation is used.

As mentioned in Chapter 5, formulations where each node in the network is equipped with a wavelength converter are relatively easier, since the channel numbers assigned to a primary (or backup) lightpath in a wavelength-convertible network are allowed to differ from one fiber to the next. This means that only the route of each primary and backup path through the physical topology is important. For any edge $N_i \rightarrow N_j$ in the physical topology, as long as the total number of primary and backup lightpaths using the edge $N_i \rightarrow N_j$ does not exceed n_{ch} (the maximum number of channels that a fiber

---

[2] For ease of understanding, the simpler formulations are given before the more difficult cases.

supports), each of the primary and backup lightpaths on the edge $N_i \rightarrow N_j$ in the physical topology may be assigned a unique channel number.

In the following discussions, the WDM network has $\mathcal{N}_E$ end nodes and $\mathcal{N}_R$ router nodes. Each node (either an end node or a router node) is assigned a unique number $n, 1 \leq n \leq \mathcal{N}_E + \mathcal{N}_R$.

8.3.1 1:1 Path Protection in Wavelength-Convertible Networks Using Static Allocation

In formulation $FORM$ 6, the logical topology, specified by graph G_L, is already determined, possibly using one of the schemes given in Chapter 7. This formulation may also be used for 1+1 protection. Since path protection is being used, the primary path must use a route, through the physical topology, that is edge-disjoint with respect to that used by the backup path. The problem is to find the primary and backup paths, corresponding to each of the logical edges in G_L, ensuring that no fiber carries more than the maximum number n_{ch} of lightpaths, so that the total number of channels used by all the primary and backup lightpaths is minimum.

Notation used in $FORM$ 6

x_{ij}^k : an integer variable having a value of 0 or 1 where

$$x_{ij}^k = \begin{cases} 1 \text{ if the primary path of the } k^{\text{th}} \text{ logical edge uses the edge} \\ \quad N_i \rightarrow N_j \text{ in the physical topology} \\ 0 \text{ otherwise.} \end{cases}$$

y_{ij}^k : an integer variable having a value of 0 or 1 where

$$y_{ij}^k = \begin{cases} 1 \text{ if the backup path of the } k^{\text{th}} \text{ logical edge uses the edge} \\ \quad N_i \rightarrow N_j \text{ in the physical topology} \\ 0 \text{ otherwise.} \end{cases}$$

E : the set of all pairs of indices (i, j) of nodes such that $N_i \rightarrow N_j$ is an edge in the physical topology (i.e., there is a fiber from node N_i to N_j in the network).

$\mathcal{N}$: the number of nodes (router nodes and end nodes) in the network.

m : number of edges in the logical topology. There is an assumption that no edge in the logical topology requires more than one lightpath.

source(k) : the source of the kth logical edge.

destination(k) : the destination of the kth logical edge.

The formulation for *FORM* 6

Formulation *FORM* 6 minimizes the total amount of optical resources used. Here the number of channels on each fiber is the resource being minimized.

Objective function

$$\text{minimize} \sum_{k=1}^{m} \sum_{(i,j)\in E} (x_{ij}^k + y_{ij}^k) \tag{8.1}$$

subject to

1. Define the primary path and the backup path ($\forall k, 1 \le k \le m, \forall i, 1 \le i \le N$).

$$\sum_{j:(i,j)\in E} x_{ij}^k - \sum_{j:(j,i)\in E} x_{ji}^k = \begin{cases} 1 & \text{if source}(k) = i, \\ -1 & \text{if destination}(k) = i, \\ 0 & \text{otherwise.} \end{cases} \tag{8.2}$$

$$\sum_{j:(i,j)\in E} y_{ij}^k - \sum_{j:(j,i)\in E} y_{ji}^k = \begin{cases} 1 & \text{if source}(k) = i, \\ -1 & \text{if destination}(k) = i, \\ 0 & \text{otherwise.} \end{cases} \tag{8.3}$$

2. Ensure that the primary path and the backup path are edge-disjoint.

$$x_{ij}^k + y_{ij}^k \le 1, \qquad \forall i,j : (i,j) \in E, \forall k, 1 \le k \le m. \tag{8.4}$$

3. Ensure that the total number of channels used by primary paths and the backup paths on any edge is less than the maximum number of channels n_{ch} permitted on a fiber.

$$\sum_{k=1}^{m} x_{ij}^k + \sum_{k=1}^{m} y_{ij}^k \le n_{ch}, \qquad \forall i,j : (i,j) \in E. \tag{8.5}$$

Justification of *FORM* 6

The objective of *FORM* 6 is to minimize the optical resources used in establishing lightpaths for all requests for communication from source s_i to destination $d_i, 1 \le i \le m$. Each edge in the primary path as well as the backup path requires one channel — a resource of the network. Therefore in (8.1), the sum of the lengths of the primary and the backup paths have to be minimized. (8.2) ((8.3)) is the standard flow conservation equation[3] [4] to define the primary (backup) path from a source to a destination. Since the primary path must be edge-disjoint with respect to the backup path, (8.4) enforces that x_{ij}^k and y_{ij}^k cannot both be 1. If $x_{ij}^k = 1$, the primary lightpath

---

[3] Briefly discussed in Appendix 3.

for the kth logical edge is using the edge $N_i \rightarrow N_j$ in the physical topology. $\sum_{k=1}^{m} x_{ij}^k \left(\sum_{k=1}^{m} y_{ij}^k \right)$ gives the number of primary (backup) paths that use the edge $N_i \rightarrow N_j$. Therefore, (8.5) limits the number of lightpaths on edge $i \rightarrow j$ to n_{ch}.

The complexity of *FORM* 6

For all $i, j, (i, j) \in E$, and for all $k, 1 \leq k \leq m$, binary (0/1) variables x_{ij}^k, y_{ij}^k are needed. Therefore, the formulation has $2|E|m$ binary variables. The number of constraints is $2Nm + |E|(m+1)$.

Exercise 8.1. A WDM network with four end nodes is shown in Figure 8.10 (taken from Chapter 1). Specify formulation *FORM* 6 to define the logical edges shown in Figure 8.11. This logical topology also appears in Chapter 1. If you generate the lp file [206] corresponding to *FORM* 6 and execute it using CPLEX, you will obtain the routes for the primary and the backup paths. □

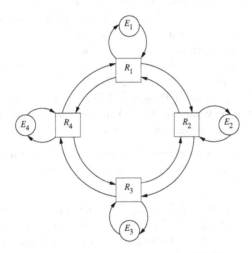

Fig. 8.10. The physical topology of a WDM network

8.3.2 Dynamic Wavelength Allocation with Wavelength Continuity Constraint

Formulation *FORM* 6 described in Section 8.3.1 had the advantage that wavelength converters were available in every node, so that the channel used by

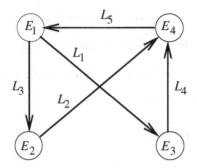

Fig. 8.11. Required logical topology

any lightpath was allowed to change from fiber to fiber. This is not a realistic scenario in existing networks. Formulation $FORM$ 7, described in this section, assumes that no node in the network is equipped with a wavelength converter and uses dynamic lightpath allocation using the shared-path protection scheme.

In $FORM$ 7, there are M source-destination pairs (s_i, d_i), such that source s_i is communicating with destination d_i, for all $i, 1 \leq i \leq M$. Network resources have already been allocated to maintain a primary lightpath and a backup lightpath from s_i to d_i,[4] for all $i, 1 \leq i \leq M$. The problem is to process, efficiently, a new request for a communication from source S to destination D to find, if possible, a primary path and a backup path for the new request, minimizing the additional optical resources needed for the connection.

Formulation $FORM$ 7, like formulation $FORM$ 6 in Section 8.3.1, will involve a Mixed Integer Linear Program (MILP) with binary variables (integer variables restricted to values of 0 or 1) and continuous variables. There are two important features of this formulation:

- the requirement of wavelength continuity constraint, and
- the replacement of integer variables by continuous variables whenever possible.

The time to solve an MILP using P binary variables is $\Theta(2^P)$ [294]. The replacement of binary variables by continuous variables, to the extent possible, will have a major impact on the efficiency of this formulation.

The objective of $FORM$ 7 is to minimize, if a new request for communication can be handled, the amount of additional channels needed for the primary and the backup lightpaths for the request, without disturbing existing lightpaths, such that:

i) there exists an unused channel number and a route for the primary lightpath,

---

[4] The route of each of these primary lightpaths, of course, has to be edge-disjoint with respect to that of the corresponding backup lightpath.

ii) there exists an available channel number and a route for a backup lightpath which is edge-disjoint with respect to the primary path for this connection.

Notation used in *FORM 7*

$\mathcal{N}$: number of nodes (routers and end nodes) in the network.

M : number of requests that have been handled already, for which primary and backup lightpaths are currently established.

E : the set of all pairs (i, j) of nodes N_i, N_j such that $N_i \rightarrow N_j$ is an edge in the physical topology representing a fiber in the network.

n_{ch} : number of channels that each edge in this network can accommodate.

P^1 : list of routes for the primary lightpaths already established, $[\rho_1^P, \rho_2^P, \ldots, \rho_M^P]$, where ρ_k^P is the kth primary path, for all $k, 1 \leq k \leq M$.

P^2 : list of routes for the backup lightpaths already established, $[\rho_1^B, \rho_2^B, \ldots, \rho_M^B]$, where ρ_k^B is the kth backup path, for all $k, 1 \leq k \leq M$.

ρ_{M+1}^P : route for the primary lightpath to be set up from source S to destination D.

ρ_{M+1}^B : route for the backup lightpath to be set up from source S to destination D.

$\tilde{A}$: the primary edge-path incidence matrix[5] $(\tilde{a}_{ij}^k)$ such that
$$\tilde{a}_{ij}^k = \begin{cases} 1 \text{ if the edge } N_i \rightarrow N_j \text{ is in the } k^{\text{th}} \text{ primary path,} \\ 0 \text{ otherwise.} \end{cases}$$

$\tilde{B}$: the backup edge-path incidence matrix $(\tilde{b}_{ij}^k)$ such that
$$\tilde{b}_{ij}^k = \begin{cases} 1 \text{ if the edge } N_i \rightarrow N_j \text{ is in the } k^{\text{th}} \text{ backup path,} \\ 0 \text{ otherwise.} \end{cases}$$

c_P^k (c_B^k) : the channel assigned to the kth primary (backup) lightpath, for all $k, 1 \leq k \leq M$.

c_P^{M+1} (c_B^{M+1}) : the channel to be assigned to the new primary (backup) lightpath from S to D.

---

[5] An edge-path incidence matrix is one where there is one row for each edge and one column for each path. Such a matrix is useful in recording which edge of a physical topology appears in which route of a primary (backup) lightpath.

$\boldsymbol{\delta_k}$: a non-negative continuous[6] variable for each existing backup path, $\rho_k^B, 1 \leq k \leq M$, whose values are restricted by the constraints such that

$$\delta_k = \begin{cases} 1 \text{ if the } k^{\text{th}} \text{ backup path shares some edge } i \to j \text{ with route } \rho_{M+1}^B \\ \quad \text{of the new backup lightpath and has the same channel number} \\ \quad c_P^{M+1} \text{ used by the new backup lightpath,} \\ 0 \text{ otherwise.} \end{cases}$$

$\boldsymbol{x_{ij}}$: a binary variable for each edge $i \to j$ such that

$$x_{ij} = \begin{cases} 1 \text{ if the new primary path } \rho_{M+1}^P \text{ uses the edge } N_i \to N_j, \\ 0 \text{ otherwise.} \end{cases}$$

$\boldsymbol{y_{ij}}$: a binary variable for each edge $i \to j$, such that

$$y_{ij} = \begin{cases} 1 \text{ if the new backup path } \rho_{M+1}^B \text{ uses the edge } N_i \to N_j \\ 0 \text{ otherwise.} \end{cases}$$

$\boldsymbol{w_k \, (z_k)}$: a binary variable, for all k, $1 \leq k \leq n_{ch}$, such that

$$w_k \, (z_k) = \begin{cases} 1 \text{ if the new primary (backup) lightpath is assigned} \\ \quad \text{channel number } k, \\ 0 \text{ otherwise.} \end{cases}$$

Once the formulation $FORM$ 7 finds values for binary variables x_{ij} (y_{ij}), the primary (backup) path ρ_{M+1}^P (ρ_{M+1}^B) from the source S to the destination D can be obtained easily.[7]

At the same time, the value of k such that $w_k \, (z_k) = 1$ gives the unused (available) channel number c_P^{M+1} (c_B^{M+1}) to be used on the primary (backup) path.

The formulation for $FORM$ 7

Formulation $FORM$ 7 minimizes the sum of the number of channels used in establishing the primary and the backup path.

Objective function:

$$\text{minimize} \sum_{(i,j)\in E} x_{ij} + \sum_{(i,j)\in E} y_{ij} \tag{8.6}$$

subject to:

1. Define the primary and the backup path.

---

[6] It is important to note that δ_k is a continuous variable but with a value restricted to 0 or 1. The use of such variables rather than binary variables can have a dramatic effect on the time required to solve integer linear programs.

[7] If $x_{ij} = 1$ $(y_{ij} = 1)$, the edge $N_i \to N_j$ is in the primary (backup) path.

$$\sum_{j:(i,j)\in E} x_{ij} - \sum_{j:(j,i)\in E} x_{ji} = \begin{cases} 1 & \text{if } i = S, \\ -1 & \text{if } i = D, \\ 0 & \text{otherwise.} \end{cases} \qquad \forall i, 1 \leq i \leq N, \quad (8.7)$$

$$\sum_{j:(i,j)\in E} y_{ij} - \sum_{j:(j,i)\in E} y_{ji} = \begin{cases} 1 & \text{if } i = S, \\ -1 & \text{if } i = D, \\ 0 & \text{otherwise.} \end{cases} \qquad \forall i, 1 \leq i \leq N, \quad (8.8)$$

2. The primary path and the backup path must be edge-disjoint.

$$x_{ij} + y_{ij} \leq 1, \qquad \forall (i,j) \in E \qquad (8.9)$$

3. The primary (backup) path must have exactly one channel number associated with it.

$$\sum_{k=1}^{n_{ch}} w_k = 1 \qquad (8.10)$$

$$\sum_{k=1}^{n_{ch}} z_k = 1 \qquad (8.11)$$

4. The primary path must not share an edge as well as the channel number with any other path, primary or backup.

$$x_{ij} + w_{c_P^k} \leq 1, \qquad \forall (i,j) \in E : \tilde{a}_{ij}^k = 1, k = 1, 2, \ldots, M \qquad (8.12)$$

$$x_{ij} + w_{c_B^k} \leq 1, \qquad \forall (i,j) \in E : \tilde{b}_{ij}^k = 1, k = 1, 2, \ldots, M \qquad (8.13)$$

5. The backup path cannot share an edge and a channel number with another existing primary path.

$$y_{ij} + z_{c_P^k} \leq 1, \qquad \forall (i,j) \in E : \tilde{a}_{ij}^k = 1, k = 1, 2, \ldots, M \qquad (8.14)$$

6. The backup path may share a channel number and an edge with the kth backup path only if the primary path is edge-disjoint with respect to the kth primary path.

$$z_{c_B^k} + y_{ij} - \delta_k \leq 1, \qquad \forall (i,j) \in E : \tilde{b}_{ij}^k = 1, k = 1, 2, \ldots, M \quad (8.15)$$

$$\delta_k - z_{c_B^k} \leq 0, \qquad k = 1, 2, \ldots, M \qquad (8.16)$$

$$\delta_k - \sum_{\forall (i,j) \in E : b_{ij}^k = 1} y_{ij} \leq 0, \qquad k = 1, 2, \ldots, M \qquad (8.17)$$

$$x_{ij} + \delta_k \leq 1, \qquad \forall (i,j) \in E : \tilde{a}_{ij}^k = 1, k = 1, 2, \ldots, M \qquad (8.18)$$

Justification for _FORM_ 7

The objective of _FORM_ 7, as stated in (8.6), is to minimize the sum of the lengths of the primary and the backup paths. The justification is the same as that for (8.1). The flow conservation equations are stated in (8.7) and (8.8) and are similar to (8.2) and (8.3). The purpose of (8.9) (similar to (8.4)) is to ensure that the primary path must be edge-disjoint with respect to the backup path. The purpose of (8.10) (resp. (8.11)) is to ensure that exactly one channel number is reserved for the primary (backup) path, since w_k (z_k) must be 1 for exactly one k. In other words, (8.10) (resp. (8.11)) enforces the wavelength continuity constraint for the primary (backup) lightpath.

If $\tilde{a}_{ij}^k, 1 \leq k \leq M$, is 1, the kth primary path (which is one of the existing paths) passes through the edge $N_i \to N_j$ in the physical topology. This k^{th} primary path has already been allotted channel number c_P^k. If $w_{c_P^k}$ is 1, it means that the new primary path is also assigned the same channel number as the existing k^{th} primary path. The purpose of (8.12) is to prohibit the situation where the new primary path is assigned the same channel number as the kth primary path and shares the same edge $N_i \to N_j$ in the physical topology with the kth primary path. In exactly the same way, (8.13) (resp. (8.14)) prohibits the situation where the new primary (backup) path is assigned the same channel number as the kth backup (primary) path and shares the same edge (i, j), with the kth backup (primary) path.

If the kth backup path uses the same channel as used by the new backup path to be built, $z_{c_B^k} = 1$. Then (8.16) indicates that $\delta_k \leq 1$. Now if the new backup path shares an edge (i, j) with the kth backup path, $y_{ij} = 1$. The purpose of (8.15) is to state that $\delta_k \geq 1$, so that the only value of δ_k that satisfies both (8.15) and (8.16) is $\delta_k = 1$.

If the kth backup path does not share any edge with the new backup path, (8.17) states that $\delta_k \leq 0$. Since all variables must be greater than or equal to 0, δ_k must be 0. If the kth backup path shares edge $N_i \to N_j$ in the physical topology with the new backup path, $y_{ij} = 1$. If the kth backup path does not share the same channel number as the new backup path, $z_{c_B^k} = 0$. In this situation, (8.15) states that $\delta_k \geq 0$. Since (8.16) states that $\delta_k \leq 0$, the only solution that satisfies both (8.15) and (8.16), in this situation, is $\delta_k = 0$.

In summary, (8.15), (8.16), and (8.17) ensure that, for all k, the continuous variable δ_k has a value 1 if and only if

    I. the kth backup path shares some edge $N_i \to N_j$ in the physical topology with the new backup path and

    II. the kth backup lightpath shares the same channel number as the new backup lightpath.

Otherwise δ_k has a value 0.

The complexity of _FORM_ 7

For all $(i,j) \in E$ and for all $k, 1 \leq k \leq n_{ch}$, the variables x_{ij}, y_{ij}, w_k and z_k are integer variables having binary (0/1) values. For each $k, 1 \leq k \leq M$, δ_k are M nonnegative continuous variables. Therefore, the formulation has $2 \times (|E| + n_{ch})$ binary variables and M continuous variables. Since there are $\mathcal{N}$ nodes in the physical topology, no route through the physical topology will have more than $(\mathcal{N} - 1)$ edges. Therefore, the number of constraints in this formulation is bounded by $2\mathcal{N} + |E| + 2 + 2(\mathcal{N} - 1)M + 3(\mathcal{N} - 1)M + 2M = 2\mathcal{N} + |E| + 2 + (5\mathcal{N} - 3)M$.

Exercise 8.2. Propose another formulation of the same problem using binary variables in place of continuous variables $\delta_k, 1 \leq k \leq M$. Write a program that will generate an MILP based on the formulation given here and an MILP based on your formulation. The program should read in all the existing primary and backup paths, the physical topology of the network, and a source S and destination D. Considering a number of networks of various sizes and, for each network, various sets of primary and backup paths, compare the time needed in your formulation to the time needed for the formulation given here.   □

8.3.3 Utilizing the Channels Used by Backup Paths When There Is No Fault in the Network

The backup paths in protection schemes represent resources that are "wasted", most of the time. In discussing the 1:1 protection scheme, it was pointed out that these backup paths may be utilized for low-priority data communication which may be interrupted, if needed. This possibility is illustrated below through an informal discussion of a possible scenario, followed by Exercises 8.3, 8.4, and 8.5.

The networks discussed so far support only requests having the same priority level. In real life, different communication requests may have different priority levels. When downloading a video, for instance, the priority level could be low while the priority level would be high in business-to-business urgent communication involving, for instance, distributed databases. In a _differentiated service environment_, the network supports traffic having multiple priority levels with different levels of service. To illustrate the idea, a simplified model is considered here.

Let a network support requests for connections having one of two possible priority levels — type 1 connections with high-priority or type 2 connections with low-priority. Lightpaths are allocated dynamically in response to requests for connections. Each type 1 connection in this model has a primary path and a backup path using the 1:1 protection scheme. Each type 2 connection has only a primary path with no backup path so that, if there is a fault in a route used by a type 2 connection, the communication will be interrupted and possibly

will be attempted later. The primary path of a type 2 connection is allowed
to overlap with the backup path of a type 1 connection. As discussed earlier,
in a protection scheme, the channels reserved for backup lightpaths are not
used until a fault occurs. Since faults are rare events, the channels reserved
for backup lightpaths represent unutilized resources. The scheme here tries to
utilize the resources devoted to the backup paths of high-priority traffic to
carry low-priority traffic.

In response to a request for a type 1 connection, the request will be successful if both the primary and backup paths can be established. In response to a
request for a type 2 connection, the request will be successful if the primary
path can be established.

Once the lightpath(s) has (have) been established, the data communication
begins using the designated primary path. Since this is a dynamic lightpath
allocation scheme, when a data communication is over, the resources allocated to the connection will be reclaimed and reused for other requests for
connections.

In the event of a link failure, one or more primary paths of type 1 connec-
tion(s) may become faulty. This scheme allows the lightpaths of relevant type
2 connections to be preempted to establish the backup path of each disrupted
high-priority connection if they overlap. In other words, if the backup path
of a type 1 connection overlaps the primary path of a type 2 connection, the
low-priority type 2 connection will be interrupted when there is a link failure
involving the primary path of a high-priority connection, in order to establish
the backup path of the high-priority connection.

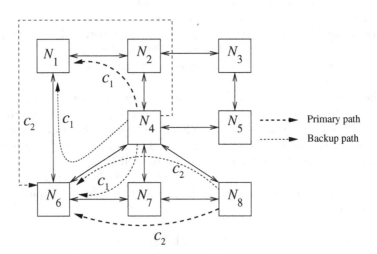

Fig. 8.12. A network with differentiated services

For example, let each fiber in the network shown in Figure 8.12 be able
to carry only two channels c_1 and c_2 and let the 1:1 protection scheme be

used for type 1 communication. Let all communications in progress at time t, as shown in Figure 8.12, be type 1 connections. If there is a request for a type 2 connection from N_4 to N_6, such a request can be satisfied by using the primary path $N_4 \to N_6$ and channel c_2. This type 2 connection using the path $N_4 \to N_6$ may continue as long as the network is fault-free. In the event of a failure of the link $N_8 \to N_7$, the high-priority lightpath from N_8 to N_6 will be affected and the backup path $N_8 \to N_4 \to N_6$ must be used for communication from N_8 to N_6 using the channel c_2. In this case, the low-priority lightpath from N_4 to N_6 must be interrupted so that the channel c_2 used by that lightpath may be reclaimed to establish the backup path for the request for communication from N_8 to N_6.

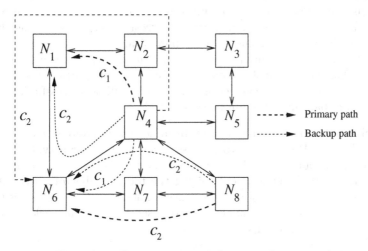

Fig. 8.13. A network with differentiated services

It is also possible to use shared-path protection to reduce further the "wastage" due to backup paths [342]. For instance, in the network shown in Figure 8.13, the backup paths $N_4 \to N_6 \to N_1$ and $N_8 \to N_4 \to N_6$ may both use the same channel c_2 as done in the scheme shown in Figure 8.8. Once again, if there is a request for a type 2 connection from N_4 to N_6, such a request can be satisfied by using the primary path $N_4 \to N_6$ and the channel c_2. The difference from the situation where the 1:1 protection scheme is used is that, here the channel c_1 is still free for use in servicing some other request.

Exercise 8.3. Consider a network where each node (router node or end node) in the network has a full wavelength converter and the 1:1 protection scheme is being used.

In this problem, there are M_1 (source-destination) end node pairs such that source s_i^1 is communicating with destination d_i^1, for all $i, 1 \leq i \leq M_1$, using type 1 communication. This means that network resources have been already allocated to maintain a primary lightpath and a backup lightpath from s_i^1 to d_i^1, for all $i, 1 \leq i \leq M_1$. At the same time, there are M_2 (source-destination) end node pairs such that source s_i^2 is communicating with destination d_i^2, for all $i, 1 \leq i \leq M_2$, using type 2 communication. Network resources have been already allocated to maintain a primary lightpath from s_i^2 to d_i^2, for all $i, 1 \leq i \leq M_2$.

Develop MILP equations to process a new request for a type 1 communication from source S to destination D to find, if possible, a primary path and a backup path for the new type 1 request, minimizing the total number of previously unused channels needed to establish the new connection. Here, the idea is that each channel used by the primary path must be a previously unused channel. A channel used by the backup path from S to D may be currently used in some type 2 communication. If the backup path from S to D uses such a channel, this channel does not "cost" anything since the channel is already being used in some type 2 communication.

When allocating a route to a lightpath, it must be ensured that no fiber in the route has more than the maximum number n_{ch} of lightpaths. In response to a request for a type 1 connection, the request will be successful if both the primary and backup paths can be established.   □

Exercise 8.4. Consider the same problem defined in Exercise 8.3.

Develop MILP equations to process a new request for a type 2 communication from source S to destination D to find, if possible, a primary path for the new type 2 request, minimizing the total number of previously unused channels needed to establish the new connection. Here the idea is to use, as much as possible, the channels already used to maintain backup paths of existing type 1 connections.   □

Exercise 8.5. Consider the problems defined in Exercises 8.3 and 8.4. In this exercise, you have to use shared protection scheme.

Develop MILP formulations to establish a new type 1 connection and a new type 2 connection.   □

8.3.4 1:N Protection Using Static Allocation with Wavelength Continuity Constraint

In formulations $FORM\ 6$ and $FORM\ 7$, described in Sections 8.3.1 and 8.3.2, the source(s) and the destination(s) of the lightpath(s) was (were) specified and the problem was that of route and channel assignment. This made the formulations in Sections 8.3.1 and 8.3.2 somewhat simpler. In this section, the logical topology design is discussed .

Given the user requirements for data communication, specified in the form of a traffic matrix $T = [t(s, d)]$ and the characteristics of the network, the objective of formulation $FORM\ 8$ [9] is to minimize the $\mathfrak{L}$-congestion, as in

Section 7.1. The logical topology design problem without the RWA aspect and the establishment of a protection path has been discussed in Section 7.1. In formulation $FORM$ 8 the primary and the backup lightpath for each logical edge must also be determined.

Similar to the approach in Chapter 7, formulation $FORM$ 8 must optimize network performance by

1) determining the set of logical edges E_L in graph G_L,
2) ensuring that, for each logical edge in E_L, both the primary and backup lightpath have an assigned
   - route through the physical topology,
   - channel number on each fiber in its route,
3) determining the strategy for routing the traffic T over the logical topology, giving minimum $\mathcal{L}$-congestion.

The difference between the problem discussed in Chapter 7 and this problem is the fact that each logical edge here represents two lightpaths — the primary lightpath and the backup lightpath, whose routes must be edge-disjoint. Therefore, the RWA here is more complex than that needed in Chapter 7.

Since the problem is very complex from a computational point of view, it is important to make some simplifying assumptions. One important aspect is finding the routes for the primary and backup paths for RWA. If every possible route has to be considered in the formulation, the problem becomes computationally intractable. Here, an assumption is made that, when looking for a primary or a backup path from a source end node E_s to a destination end node E_d, it is sufficient to consider $\mathcal{R}$ edge-disjoint routes from end node E_s to E_d, for some predetermined value of $\mathcal{R}$. From these $\mathcal{R}$ routes, one will be selected for the primary path and another one, distinct from the route selected for the primary path, will be selected for the backup path. This means that for each pair of end nodes (E_s, E_d), edge-disjoint routes $\rho^1_{sd}, \rho^2_{sd}, \ldots, \rho^{\mathcal{R}}_{sd}$ will be precomputed before solving the formulation given below.

Djikstra's shortest-path algorithm [4] may be used to precompute the edge-disjoint routes $\rho^1_{sd}, \rho^2_{sd}, \ldots, \rho^{\mathcal{R}}_{sd}$. It has been shown that, when searching for a route for a lightpath, it is sufficient to search three edge-disjoint routes from the source to the destination of the lightpath [314] so that $\mathcal{R} = 3$ may be used as a pragmatic way to find a solution within reasonable time. One important question is whether it will be possible to get $\mathcal{R}$ edge-disjoint routes between all possible ordered pairs of end nodes. Since path protection is being used, the physical topology must be 2-edge-connected. There is no guarantee that more than two edge-disjoint paths will always exist. To simplify the discussions, it is assumed below that $\mathcal{R}$ edge-disjoint routes between all possible ordered pairs of end nodes exist.[8]

The idea of the formulation $FORM$ 8 is as follows:

---

[8] Since formulation $FORM$ 8 is quite complex as it is, and will work only for relatively small networks, the primary purpose of such a formulation is to be used as a benchmark for heuristics useful in practical networks. When used as

- start with a list $\mathcal{L}$ of all possible logical edges in the network. In a network with $\mathcal{N}_E$ end nodes, the list $\mathcal{L}$ has $\mathcal{N}_E(\mathcal{N}_E-1)$ elements, one for each pair of end nodes.
- determine G_L by selecting logical edges for the network, making sure that, for each logical edge in G_L, it is possible to establish a primary and a backup lightpath.
- find an optimum routing over G_L for the specified traffic matrix.

Notation used in *FORM* 8

$\mathcal{N}_E$: number of end nodes in the network.

E : the set of all pairs (i, j) of indices of nodes such that $N_i \rightarrow N_j$ is an edge in the physical topology.

n_{ch} : the number of channels permitted on each edge of the network.

$\mathcal{R}$: the number of edge-disjoint routes, numbered $1, 2, \ldots, \mathcal{R}-1, \mathcal{R}$ over the physical topology, between every pair of end nodes.

$\mathcal{L}$: the list $[\ell_1, \ell_2, \ldots, \ell_{\mathcal{N}_E(\mathcal{N}_E-1)}]$ of potential, directed logical edges. Since there is one potential logical edge from each source s to each destination d, the number of elements in $\mathcal{L}$ is $\mathcal{N}_E(\mathcal{N}_E - 1)$.

ℓ_p : the pth element of $\mathcal{L}, 1 \leq p \leq \mathcal{N}_E(\mathcal{N}_E - 1)$.

source(p) (destination(p)) : end node E_X (E_Y) when ℓ_p is a directed logical edge $E_X \Rightarrow E_Y$ from end node E_X to end node E_Y.

d_{pr}^e : a precomputed binary coefficient for each pair of end nodes (i, j), $1 \leq i, j \leq \mathcal{N}_E, i \neq j$, identifying a lightpath p, each edge $e \in E$ and each route number $r, r \leq \mathcal{R}$, defined as follows:

$$d_{pr}^e = \begin{cases} 1 \text{ if the } r^{\text{th}} \text{ route from source}(p) \text{ to destination}(p) \text{ uses edge } e, \\ 0 \text{ otherwise.} \end{cases}$$

Λ_{max} : a continuous variable denoting the $\mathcal{L}$-congestion of the network.

b_p : a binary variable relevant when ℓ_p is selected, $1 \leq p \leq \mathcal{N}_E(\mathcal{N}_E - 1)$, defined as follows:

$$b_p = \begin{cases} 1 \text{ if } \ell_p \text{ is selected to be included in the logical topology,} \\ 0 \text{ otherwise.} \end{cases}$$

$\widehat{x}_{pr}$ $(\widehat{y}_{pr})$: a binary variable for the primary (backup) path, used if ℓ_p is selected, $\forall p, 1 \leq p \leq \mathcal{N}_E(\mathcal{N}_E - 1), \forall r, 1 \leq r \leq \mathcal{R}$, defined as follows:

---

a benchmark, synthetic networks will be used and the assumption that $\mathcal{R}$ edge-disjoint routes between all possible ordered pairs of end nodes exist is reasonable.

$$\widehat{x}_{pr} \ (\widehat{y}_{pr}) = \begin{cases} 1 \text{ if } \ell_p \text{ is selected, and the } r^{\text{th}} \text{ route from source}(p) \\ \quad \text{to destination}(p) \text{ is chosen for the primary (backup)} \\ \quad \text{lightpath,} \\ 0 \text{ otherwise.} \end{cases}$$

$w_{kp} \ (z_{kp})$: a binary variable relevant when ℓ_p is selected, $\forall p, 1 \leq p \leq \mathcal{N}_E(\mathcal{N}_E - 1), \forall k, 1 \leq k \leq n_{ch}$, defined as follows:

$$w_{kp} \ (z_{kp}) = \begin{cases} 1 \text{ if } \ell_p \text{ is selected, and the primary (backup) lightpath for} \\ \quad \ell_p \text{ is allocated channel } k, \\ 0 \text{ otherwise.} \end{cases}$$

$\delta_e^{kp}, \gamma_e^{kp}, \alpha_{ke}, \theta_{e_1 e_2}^{kp}$ and η_e^{kp} : continuous variables which are restricted[9] to have a value of either 0 or 1 as follows:

$$\delta_e^{kp} \ (\gamma_e^{kp}) = \begin{cases} 1 \text{ if } \ell_p \text{ is selected, and the primary (backup) lightpath for} \\ \quad \ell_p \text{ is allocated channel } k \text{ and assigned a route containing} \\ \quad \text{edge } e, \\ 0 \text{ otherwise.} \end{cases}$$

$$\alpha_{ke} = \begin{cases} 1 \text{ if there is at least one backup lightpath that uses edge } e \text{ and} \\ \quad \text{the backup lightpath is allocated channel } k, \\ 0 \text{ otherwise.} \end{cases}$$

$$\theta_{e_1 e_2}^{kp} = \begin{cases} 1 \text{ if } \ell_p \text{ is selected, and the primary lightpath for } \ell_p \text{ is assigned} \\ \quad \text{a route containing edge } e_2, \text{ the corresponding backup light-} \\ \quad \text{path is assigned a route containing edge } e_1, \text{ and the backup} \\ \quad \text{lightpath is allocated channel } k, \\ 0 \text{ otherwise.} \end{cases}$$

$$\eta_e^{kp} = \begin{cases} 1 \text{ if } \ell_p \text{ is selected, the route for the primary lightpath for } \ell_p \text{ uses} \\ \quad \text{edge } e, \text{ and the corresponding backup lightpath is allocated} \\ \quad \text{channel } k, \\ 0 \text{ otherwise.} \end{cases}$$

f_p^{sd} : a continuous variable denoting the traffic on logical edge ℓ_p from source end node s to destination end node d.

$t_u^k \ (r_u^k)$: the number of transmitters (receivers) at end node u tuned to the wavelength corresponding to channel k.

ϵ : a very small constant.

The formulation for *FORM 8*

Formulation *FORM 8* determines a logical topology and finds a routing over the logical topology to give minimum $\mathfrak{L}$-congestion and the primary and

---

[9] These variables are just like the continuous variable δ_k in Section 8.3.2 and are useful in reducing the number of integer variables significantly.

backup lightpaths for every edge in the logical topology. For the ease of understanding, the constraints for the following formulation have been described in related groups.

Objective function:

$$\text{minimize } \Lambda_{max} \tag{8.19}$$

subject to:

a) Path creation and channel allocation constraints

1. If logical edge ℓ_p is selected, select a primary path and a backup path for ℓ_p.

$$\sum_{r=1}^{\mathcal{R}} \widehat{x}_{pr} - b_p = 0, \qquad \forall p, 1 \leq p \leq \mathcal{N}_E(\mathcal{N}_E - 1) \tag{8.20}$$

$$\sum_{r=1}^{\mathcal{R}} \widehat{y}_{pr} - b_p = 0, \qquad \forall p, 1 \leq p \leq \mathcal{N}_E(\mathcal{N}_E - 1) \tag{8.21}$$

2. Enforce the wavelength continuity constraint for the primary and backup lightpaths.

$$\sum_{k=1}^{n_{ch}} w_{kp} - b_p = 0, \qquad \forall p, 1 \leq p \leq \mathcal{N}_E(\mathcal{N}_E - 1) \tag{8.22}$$

$$\sum_{k=1}^{n_{ch}} z_{kp} - b_p = 0, \qquad \forall p, 1 \leq p \leq \mathcal{N}_E(\mathcal{N}_E - 1) \tag{8.23}$$

3. If logical edge ℓ_p is selected, ensure that the primary and backup paths for ℓ_p are edge-disjoint.

$$\widehat{x}_{pr} + \widehat{y}_{pr} - b_p \leq 0, \quad \forall p, 1 \leq p \leq \mathcal{N}_E(\mathcal{N}_E - 1), \forall r, 1 \leq r \leq \mathcal{R} \tag{8.24}$$

b) Routing of the primary and the backup lightpaths

4. Define the continuous variables δ_e^{kp} and γ_e^{kp}.

$$\sum_{r=1}^{\mathcal{R}} d_{pr}^e \cdot \widehat{x}_{pr} + w_{kp} - \delta_e^{kp} \leq b_p, \quad \begin{array}{l} \forall p, 1 \leq p \leq \mathcal{N}_E(\mathcal{N}_E - 1), \\ \forall e \in E, \forall k, 1 \leq k \leq n_{ch} \end{array} \tag{8.25}$$

$$\delta_e^{kp} - \sum_{r=1}^{\mathcal{R}} d_{pr}^e \cdot \widehat{x}_{pr} \leq 0, \quad \begin{array}{l} \forall p, 1 \leq p \leq \mathcal{N}_E(\mathcal{N}_E - 1), \\ \forall e \in E, \forall k, 1 \leq k \leq n_{ch} \end{array} \tag{8.26}$$

$$\delta_e^{kp} - w_{kp} \leq 0, \quad \forall p, 1 \leq p \leq \mathcal{N}_E(\mathcal{N}_E - 1), \quad (8.27)$$
$$\forall e \in E,$$
$$\forall k, 1 \leq k \leq n_{ch}$$

$$\sum_{r=1}^{\mathcal{R}} d_{pr}^e \cdot \widehat{y}_{pr} + z_{kp} - \gamma_e^{kp} \leq b_p, \quad \forall p, 1 \leq p \leq \mathcal{N}_E(\mathcal{N}_E - 1), \quad (8.28)$$
$$\forall e \in E,$$
$$\forall k, 1 \leq k \leq n_{ch}$$

$$\gamma_e^{kp} - \sum_{r=1}^{\mathcal{R}} d_{pr}^e \cdot \widehat{y}_{pr} \leq 0, \quad \forall p, 1 \leq p \leq \mathcal{N}_E(\mathcal{N}_E - 1), \quad (8.29)$$
$$\forall e \in E,$$
$$\forall k, 1 \leq k \leq n_{ch}$$

$$\gamma_e^{kp} - z_{kp} \leq 0, \quad \forall p, 1 \leq p \leq \mathcal{N}_E(\mathcal{N}_E - 1), \quad (8.30)$$
$$\forall e \in E,$$
$$\forall k, 1 \leq k \leq n_{ch}$$

5. Define continuous variable α_{ke}.

$$\gamma_e^{kp} - \alpha_{ke} \leq 0, \quad \forall p, 1 \leq p \leq \mathcal{N}_E(\mathcal{N}_E - 1), \quad (8.31)$$
$$\forall e \in E,$$
$$\forall k, 1 \leq k \leq n_{ch}$$

$$\alpha_{ke} \leq 1, \quad \forall e \in E, \forall k, 1 \leq k \leq n_{ch} \quad (8.32)$$

$$\alpha_{ke} - \sum_{p=1}^{\mathcal{N}_E(\mathcal{N}_E-1)} \gamma_e^{kp} \leq 0, \quad \forall e \in E, \forall k, 1 \leq k \leq n_{ch} \quad (8.33)$$

6. Ensure that each primary lightpath is edge-channel disjoint[10] with respect to all other primary as well as backup lightpaths.

$$\alpha_{ke} + \sum_{p=1}^{\mathcal{N}_E(\mathcal{N}_E-1)} \delta_e^{kp} \leq 1, \quad \forall e \in E, \forall k, 1 \leq k \leq n_{ch} \quad (8.34)$$

7. Define the continuous variable $\theta_{e_1 e_2}^{kp}$.

$$\gamma_{e_1}^{kp} + \sum_{r=1}^{\mathcal{R}} d_{pr}^{e_2} \cdot \widehat{x}_{pr} - \theta_{e_1 e_2}^{kp} \leq b_p, \quad \forall p, 1 \leq p \leq \mathcal{N}_E(\mathcal{N}_E - 1), \quad (8.35)$$
$$\forall e_1, e_2 \in E, e_1 \neq e_2,$$
$$\forall k, 1 \leq k \leq n_{ch}$$

---

[10] In other words, a primary lightpath cannot share an edge with another lightpath (primary or backup) and use the same channel number.

$$\theta_{e_1 e_2}^{kp} - \gamma_{e_1}^{kp} \leq 0, \quad \forall p, 1 \leq p \leq \mathcal{N}_E(\mathcal{N}_E - 1), \tag{8.36}$$
$$\forall e_1, e_2 \in E, e_1 \neq e_2,$$
$$\forall k, 1 \leq k \leq n_{ch}$$

$$\theta_{e_1 e_2}^{kp} - \sum_{r=1}^{\mathcal{R}} d_{pr}^{e_2} \cdot \widehat{x}_{pr} \leq 0, \quad \forall p, 1 \leq p \leq \mathcal{N}_E(\mathcal{N}_E - 1), \tag{8.37}$$
$$\forall e_1, e_2 \in E, e_1 \neq e_2,$$
$$\forall k, 1 \leq k \leq n_{ch}$$

8. Ensure that, if two backup lightpaths share a common edge and a channel, the corresponding primary lightpaths are edge-disjoint.

$$\sum_{p=1}^{\mathcal{N}_E(\mathcal{N}_E - 1)} \theta_{e_1 e_2}^{kp} \leq 1, \qquad \forall e_1, e_2 \in E, e_1 \neq e_2, \forall k, 1 \leq k \leq n_{ch} \tag{8.38}$$

c) **Constraints corresponding to the transmitters and the receivers**

9. Ensure that, in a fault-free network, the number of primary lightpaths starting from (terminating at) an end node is limited by the number of transmitters (receivers) at that end node.

$$\sum_{p:source(p)=u} w_{kp} \leq t_u^k, \qquad \forall k, 1 \leq k \leq n_{ch}, \forall u, 1 \leq k \leq \mathcal{N}_E \tag{8.39}$$

$$\sum_{p:destination(p)=u} w_{kp} \leq r_u^k, \quad \forall k, 1 \leq k \leq n_{ch}, \forall u, 1 \leq k \leq \mathcal{N}_E \tag{8.40}$$

10. Define the continuous variable η_e^{kp}.

$$\sum_{r=1}^{\mathcal{R}} d_{pr}^e \cdot \widehat{x}_{pr} + z_{kp} - \eta_e^{kp} \leq 1, \quad \forall p, 1 \leq p \leq \mathcal{N}_E(\mathcal{N}_E - 1), \tag{8.41}$$
$$\forall e \in E,$$
$$\forall k, 1 \leq k \leq n_{ch}$$

$$\eta_e^{kp} - \sum_{r=1}^{\mathcal{R}} d_{pr}^e \cdot \widehat{x}_{pr} \leq 0, \quad \forall p, 1 \leq p \leq \mathcal{N}_E(\mathcal{N}_E - 1), \tag{8.42}$$
$$\forall e \in E,$$
$$\forall k, 1 \leq k \leq n_{ch}$$

$$\eta_e^{kp} - z_{kp} \leq 0, \quad \forall p, 1 \leq p \leq \mathcal{N}_E(\mathcal{N}_E - 1), \tag{8.43}$$
$$\forall e \in E,$$
$$\forall k, 1 \leq k \leq n_{ch}$$

11. Ensure that, in a faulty network, the number of primary lightpaths starting from (terminating at) an end node is limited by the number of transmitters (receivers) at that end node.

$$\sum_{p:source(p)=u} w_{kp} - \delta_e^{kp} + \eta_e^{kp} \le t_u^k, \quad \forall e \in E, \tag{8.44}$$
$$\forall k, 1 \le k \le n_{ch},$$
$$\forall u, 1 \le u \le \mathcal{N}_E$$

$$\sum_{p:destination(p)=u} w_{kp} - \delta_e^{kp} + \eta_e^{kp} \le r_u^k, \quad \forall e \in E, \tag{8.45}$$
$$\forall k, 1 \le k \le n_{ch},$$
$$\forall u, 1 \le u \le \mathcal{N}_E$$

d) Traffic flow constraints

12. Ensure that the total traffic, corresponding to source s and destination d, flowing on the logical edge ℓ_p, when ℓ_p is selected, does not exceed the traffic $t(s, d)$, the traffic from s to d. When ℓ_p is not selected, the traffic on the logical edge ℓ_p must be zero.

$$f_p^{sd} - b_p t_{sd} \le 0, \quad \forall p, 1 \le p \le \mathcal{N}_E(\mathcal{N}_E - 1), \tag{8.46}$$
$$\forall s, d, 1 \le s, d \le \mathcal{N}_E, s \ne d$$

13. Eliminate logical edges that carry very little traffic.

$$\epsilon b_p - \sum_{(s,d):s \ne d} f_p^{sd} \le 0, \quad \forall p, 1 \le p \le \mathcal{N}_E(\mathcal{N}_E - 1), \tag{8.47}$$
$$\forall s, d, 1 \le s, d \le \mathcal{N}_E$$

14. Route the traffic, from a given source s to a given destination d, over the logical edges.

$$\sum_{p:destination(p)=i} f_p^{sd} - \sum_{p:source(p)=i} f_p^{sd} = \begin{cases} t(s,d) & \text{if } s = i, \\ -t(s,d) & \text{if } d = i, \\ 0 & \text{otherwise.} \end{cases} \tag{8.48}$$

(8.48) is to be repeated for all $i, 1 \le i \le \mathcal{N}_E, \forall s, d, s \ne d, 1 \le s, d \le \mathcal{N}_E$.

15. Ensure that the total amount of traffic on any lightpath does not exceed Λ_{max}, the congestion.

$$f_p^{sd} - \Lambda_{max} \le 0, \quad \forall p, 1 \le p \le \mathcal{N}_E(\mathcal{N}_E - 1) \tag{8.49}$$

16. Ensure that the total amount of traffic on any lightpath does not exceed 1, its maximum capacity. (In this formulation, all traffic values are normalized with respect to the maximum capacity of a lightpath — 10 Gbps or 2.5 Gbps today depending on the technology).

$$f_p^{sd} \le 1, \quad \forall p, 1 \le p \le \mathcal{N}_E(\mathcal{N}_E - 1) \tag{8.50}$$

Justification for *FORM* 8

1. If ℓ_p is selected, $b_p = 1$. Only then, from the possible $\mathcal{R}$ routes from source(p) to destination(p), exactly one route must be selected for the primary path for ℓ_p and exactly one route for the backup path. If $b_p = 0$, no route through the physical topology should be selected ((8.20) and (8.21)).

2. (8.22) and (8.23) serve the same purpose as (8.10) and (8.11). The allocation of a channel is done only if ℓ_p is selected.

3. If ℓ_p is selected, (8.24) ensures that the primary path and the backup paths for ℓ_p are edge-disjoint. If the rth route is used by the primary lightpath for ℓ_p, then the rth route cannot also be used by the backup lightpath for ℓ_p. Since the different precomputed routes, for a given source-destination pair, are all edge-disjoint, the constraint ensures that any primary lightpath is edge-disjoint with respect to the corresponding backup lightpath.

4. The purpose of (8.25) – (8.27) is to define the continuous variable δ_e^{kp} so that $\delta_e^{kp} = 1$ if

   a) ℓ_p is selected and
   b) the primary lightpath for ℓ_p is allocated channel k and
   c) uses edge e.

   If the condition is not satisfied, $\delta_e^{kp} = 0$.
   If ℓ_p is selected, $b_p = 1$. If the primary lightpath for ℓ_p is allocated channel k, $w_{kp} = 1$. If edge e is used by the primary lightpath, (8.20) ensures that $\widehat{x}_{pr} = 1$ for some r. The precomputed constant $d_{pr}^e = 1$ for this p and r. In this situation, (8.25), (8.26), and (8.26) become $\delta_e^{kp} \geq 1$, $\delta_e^{kp} \leq 1$ and $\delta_e^{kp} \leq 1$ respectively. The only solution compatible with these constraints is $\delta_e^{kp} = 1$.
   If ℓ_p is not selected, $b_p = 0$. In this case, (8.20) and (8.22) state that $\widehat{x}_{pr} = 0$ and $w_{kp} = 0$ for all r and k. The result is that (8.25), (8.26), and (8.27) become $\delta_e^{kp} \geq 0$, $\delta_e^{kp} \leq 0$, and $\delta_e^{kp} \leq 0$ respectively. The solution compatible with these constraints is $\delta_e^{kp} = 0$.
   If the primary lightpath for ℓ_p is not allocated channel k, $w_{kp} = 0$. If the primary lightpath does not use edge e, $\sum_{r=1}^{\mathcal{R}} d_{pr}^e \cdot \widehat{x}_{pr} = 0$. If $b_p = 1$, and $w_{kp} = 0$, (8.25) and (8.26) become $\sum_{r=1}^{\mathcal{R}} d_{pr}^e \cdot \widehat{x}_{pr} \leq 1 + \delta_e^{kp}$ and $\delta_e^{kp} \leq 0$.
   Since all variables are nonnegative, the only possible solution is $\delta_e^{kp} = 0$.
   The remaining cases are similar.
   The purpose of (8.28) – (8.30) is to define the continuous variable γ_e^{kp} so that $\gamma_e^{kp} = 1$ if

   a) ℓ_p is selected and
   b) the backup lightpath for ℓ_p is allocated channel k and
   c) uses edge e.

   (8.28) – (8.30) are just like (8.25) – (8.27).

5. (8.31) – (8.33) define a continuous variable α_{ke}, for each channel k and edge e, such that $\alpha_{ke} = 1$ if there is at least one backup lightpath that uses edge e and is assigned channel k. Otherwise $\alpha_{ke} = 0$.

   If logical edge ℓ_p is selected, and the backup lightpath for ℓ_p uses edge e and is allocated channel k, $\gamma_e^{kp} = 1$. In this case, (8.31) and (8.32) then state $\alpha_{ke} \geq 1$ and $\alpha_{ke} \leq 1$ respectively. The only compatible solution is $\alpha_{ke} = 1$.

   If there is no backup lightpath that uses edge e and is assigned channel k, $\gamma_e^{kp} = 0$ for all p. In this situation, (8.31) and (8.33) then state $\alpha_{ke} \geq 0$ and $\alpha_{ke} \leq 0$ respectively, leading to $\alpha_{ke} = 0$.

6. If $\alpha_{ke} = 1$, there must exist a backup lightpath that uses edge e and is allocated channel k. Therefore, no primary path can use edge e and be allocated channel k. If $\delta_e^{kp} = 1$, some primary lightpath for logical edge ℓ_p is allocated channel k and assigned a route containing edge e. Since at most one primary lightpath may use edge e and be allocated channel k, $\displaystyle\sum_{p=1}^{\mathcal{N}_E(\mathcal{N}_E-1)} \delta_e^{kp}$ is at most 1. The following two purposes are served by (8.34):

   a) each primary lightpath must be edge-channel disjoint with respect to all other primary lightpaths,

   b) each primary lightpath must be edge-channel disjoint with all backup lightpaths.

7. If logical edge ℓ_p is selected, $b_p = 1$. If the primary lightpath for ℓ_p is assigned a route containing edge e_2, for exactly one route r from source(p) to destination(p), $\widehat{x}_{pr} = 1$ and $d_{pr}^{e_2} = 1$. If the backup lightpath for ℓ_p is allocated channel k and passes through edge e_1, $\gamma_{e_1}^{kp} = 1$. Therefore, if

   a) the logical edge ℓ_p is selected,

   b) the primary lightpath is assigned a route containing edge e_2,

   c) the backup lightpath is allocated channel k,

   (8.35), (8.36), and (8.37) state that $\theta_{e_1 e_2}^{kp} \geq 1$, $\theta_{e_1 e_2}^{kp} \leq 1$, and $\theta_{e_1 e_2}^{kp} \leq 1$, leading to $\theta_{e_1 e_2}^{kp} = 1$. It may be readily verified that if any one of the conditions is false, $\theta_{e_1 e_2}^{kp} = 0$.

8. For a given channel k, edge e_1, and edge e_2 in the physical topology, if $\theta_{e_1 e_2}^{kp} = 1$, for some logical edge ℓ_p, the primary lightpath for ℓ_p passes through edge e_2. If the backup lightpath for another logical edge $\ell_q, p \neq q$, shares the edge e_1 and the channel k with the backup lightpath for logical edge ℓ_p, then $\theta_{e_1 e_i}^{kq} = 1$ for any edge e_i in the primary path for logical edge ℓ_q. For such sharing of edge e_1 and channel k by two backup lightpaths, the primary paths for ℓ_p and ℓ_q must be distinct. Therefore edge e_i can never be the edge e_2. The purpose of (8.38) is to stipulate that $\displaystyle\sum_{p=1}^{\mathcal{N}_E(\mathcal{N}_E-1)} \theta_{e_1 e_2}^{kp}$ cannot exceed 1. In other words, if two backup lightpaths share a common

edge and a channel number, the corresponding primary lightpaths must be edge-disjoint.

9. When there is no fiber failure, the number of primary lightpaths originating (terminating) at node u and using channel k should not exceed the number of transmitters (receivers) at node u tuned to channel k. The number of primary lightpaths starting from (terminating at) a node u is $\displaystyle\sum_{p:source(p)=u} w_{kp}$ $\Big(\displaystyle\sum_{p:destination(p)=u} w_{kp}\Big)$. This number must be limited by the number of transmitters (receivers) t_u^k (r_u^k) at node u. This constraint is enforced by (8.39) and (8.40).

10. If the logical edge ℓ_p is selected, and the corresponding primary lightpath uses edge e, $d_{pr}^e \cdot \widehat{x}_{pr}$ must be 1 for exactly one r so that $\displaystyle\sum_{r=1}^{\mathcal{R}} d_{pr}^e \cdot \widehat{x}_{pr} = 1$. If the backup lightpath corresponding to ℓ_p uses channel k, $z_{kp} = 1$. In this situation, (8.41) and (8.42) state that $\eta_e^{kp} \geq 1$ and $\eta_e^{kp} \leq 1$, leading to $\eta_e^{kp} = 1$. If $\displaystyle\sum_{r=1}^{\mathcal{R}} d_{pr}^e \cdot \widehat{x}_{pr} = 0$ $(z_{kp} = 0)$, (8.42) and (8.43) force η_e^{kp} to the value 0. In summary, for the case of a single fiber failure on edge e, the continuous variable η_e^{kp} is defined, using (8.41), (8.42), and (8.43), so that $\eta_e^{kp} = 1$ iff ℓ_p is selected, the primary lightpath for ℓ_p uses edge e, and the corresponding backup lightpath uses channel k, and $\eta_e^{kp} = 0$ otherwise.

11. The number of primary lightpaths at node u, using transmitters tuned to the wavelength corresponding to channel k, before any fiber failures is $\displaystyle\sum_{p:source(p)=u} w_{kp}$. If edge e fails, $\displaystyle\sum_{p:source(p)=u} \delta_e^{kp}$ provides the number of those primary lightpaths originating from node u and using channel k which are now destroyed. Similarly, $\displaystyle\sum_{p:source(p)=u} \eta_e^{kp}$ provides the number of new backup lightpaths, replacing the destroyed primary lightpaths originating from node u, which use channel k. Therefore, the total number of lightpaths (including the working primary paths and the new backup lightpaths replacing the destroyed primary paths) originating from node u and using channel k is $\displaystyle\sum_{p:source(p)=u} w_{kp} - \delta_e^{kp} + \eta_e^{kp}$. This number must not exceed t_u^k, the number of transmitters at node u tuned to channel k. This explains (8.44).

    The explanation for (8.45) is almost identical, considering receivers, with lightpaths terminating at node u.

12. The variable f_p^{sd} denotes the traffic on logical edge ℓ_p from source end node s to destination end node d. If ℓ_p is selected (i.e., $b_p = 1$), then f_p^{sd} cannot exceed $t(s,d)$, the traffic from s to d. If ℓ_p is not selected the amount of traffic on ℓ_p must 0, so that f_p^{sd} must be 0. This is enforced with (8.46).

13. To eliminate logical edges that carry very little traffic, (8.47) states that, when $b_p = 1$, the total traffic f_p^{sd}, from source s to destination d over logical edge ℓ_p, is greater than or equal to a small constant ϵ. When $b_p = 0$, (8.46) ensures that $f_p^{sd} = 0$.
14. Flow balance equations [4] are used in (8.48) to route the traffic, from a given source s to a given destination d, over the logical edges. At any intermediate node i in a logical path from s to d, the outflow $\sum\limits_{p:source(p)=i} f_p^{sd}$

must be equal to the inflow $\sum\limits_{p:destination(p)=i} f_p^{sd}$. At source s, there is no inflow, only the outflow $t(s,d)$. At destination d, there is no outflow, only the inflow $t(s,d)$.
15. (8.49) and (8.50) are obvious.

Exercise 8.6. In this formulation the objective was to design a network to minimize the congestion. Another important problem is that of designing the network so that a minimum number of lightpaths are needed. Show how the formulations may be modified so that the objective function is to minimize the total number of lightpaths in the network.   □

Exercise 8.7. Formulation *FORM* 8 will not work for networks of practical size. Suggest a greedy heuristic to solve the problem discussed in this section.   □

8.3.5 Restoration in Networks That Support Dynamic Lightpath Allocation

In restoration schemes, as explained earlier, there is no guarantee that the scheme will be able to find an alternative route if there is a fault in the network. A scheme is given below for dynamically allocating the route and the channel number in response to a request for a new lightpath in networks with a faulty edge or node. When determining the route and the channel for the lightpath, the faulty edge/node has to be avoided.

The scheme may also be adapted for use in situations where

I) static lightpath allocation is used so that all the lightpaths are already allocated and are being used for data communication when some node or edge fails,
II) dynamic lightpath allocation where some lightpaths are already allocated and are being used for data communication when some node or edge fails

The scheme described here is a distributed algorithm, quite similar to the scheme for dynamic RWA in fault-free networks, as described in Section 5.4.2.

An informal description of the scheme is given here since the formal description, given in [344], is a little tedious.

In networks where this scheme is used, each node, which may be either a router node or an end node, stores a route to all the end nodes in the network. This is a little different from the algorithm described in Section 5.4.2 where only the end nodes store the routes to all other end nodes. For communication from source end node E_s to destination end node E_d, if there is no fault in the network, the route ρ_{sd} from E_s to E_d, saved in the source E_s, is used and the scheme will be identical to that described in Section 5.4.2. However, any node or any edge in the route ρ_{sd} may become faulty. If that happens, the route ρ_{sd} is no longer usable and it is necessary to find an alternative route for the lightpath from E_s to E_d. The following example explains the idea.

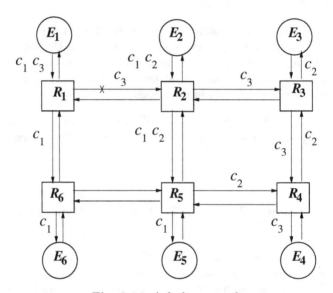

Fig. 8.14. A faulty network

Let there be a request for a lightpath from E_1 to E_5 in the network shown in Figure 8.14. The network in Figure 8.14 is identical to that shown in Figure 5.6 except for the faulty edge $R_1 \rightarrow R_2$. As in Section 5.4.2, each fiber can support five channels $c_1, \ldots, c_5$. The steps followed in determining a fault-free route from E_1 to E_5 are as follows:

Step 1: At node E_1, the stored route ρ_{15} for communication from E_1 to E_5 is $E_1 \rightarrow R_1 \rightarrow R_2 \rightarrow R_5 \rightarrow E_5$. Node E_1 is not aware that the edge $R_1 \rightarrow R_2$ is faulty. The initial set of channels that may be used for communication is the set $C_0 = \{c_1, \ldots, c_5\}$. The initial set of channels C_1 that may be used for communication from E_1 to R_1, as explained in Section 5.4.2, is $C_1 = \{c_2, c_4, c_5\}$. As in Section 5.4.2, a T-lock is placed on all the channels c_2, c_4, and c_5 for the edge $E_1 \rightarrow R_1$.

Step 2: At node R_1, the next edge to be used is the edge $R_1 \rightarrow R_2$. However, node R_1 is aware of the fact that the edge $R_1 \rightarrow R_2$ has a fault. Therefore, node R_1 looks for another node to create a fault-free route for the lightpath. In this case, E_1 is an end node and the router node R_2 is not reachable due to the fault. Therefore, the only remaining option for router R_1 is the route through router R_6 using the edge $R_1 \rightarrow R_6$. The set of channels is $C_2 = \{c_2, c_4, c_5\} \setminus \{c_1\} = \{c_2, c_4, c_5\}$. This set of channels, the destination E_5, and the partial route $E_1 \rightarrow R_1 \rightarrow R_6$ are sent to node R_6 and T-locks are placed on the channels in C_2, for the edge $R_1 \rightarrow R_6$.

Step 3: At node R_6, since the destination is E_5 and the route is a partial route, the first task is to compute a route to E_5. Let the route ρ_{65}, saved in node R_6, from R_6 to E_5 be $R_6 \rightarrow R_5 \rightarrow E_5$. This route will be appended to the route found so far from E_1, giving the route $\rho_{15}^{new} = E_1 \rightarrow R_1 \rightarrow R_6 \rightarrow R_5 \rightarrow E_5$. The set of channels that may be used for a lightpath from E_1 to R_5 is $C_3 = C_2 \setminus \emptyset = \{c_2, c_4, c_5\}$. The set C_3 and the route $E_1 \rightarrow R_1 \rightarrow R_6 \rightarrow R_5 \rightarrow E_5$ are sent to node R_5. T-locks are placed on the channels in C_3 on the edge $R_6 \rightarrow R_5$.

Step 4: At node R_5, the set $C_4 = \{c_2, c_4, c_5\} \setminus \{c_1\} = \{c_2, c_4, c_5\}$ of channel numbers that may be used to set up a lightpath to E_5 is computed. The set C_4 and the route ρ_{15}^{new} is sent to node E_5. T-locks are placed on the channels in C_4 on the edge $R_5 \rightarrow E_5$.

Step 5: Since node E_5 is the destination node, and the set C_4 is not empty, any of the channels in C_4 may be used to set up a lightpath from E_1 to E_5 using the route ρ_{15}^{new}. Node E_5 arbitrarily picks any channel, say c_4, from the set C_4. The setup phase is now over and the response phase begins.

The response phase works exactly as described in Section 5.4.2. There are several complications that did not arise in this example. If there are multiple faults in the network or if the route ρ_{15}^{new} includes the same faulty edge $R_1 \rightarrow R_2$, the scheme may not give a valid route through the physical topology. One way to deal with this is to allow the source end node E_s to wait for a certain time for the response signal to arrive. If the response signal does not arrive within a "reasonable" time, the source may draw the conclusion that a viable route through he physical topology does not exist and the connection will simply be blocked.

The above scheme avoids faulty edges/nodes. The same scheme may be readily used to restore after a failure. When a failure occurs, the lightpath(s) passing through the faulty node/edge will fail. For each failed lightpath, this recovery procedure will start as soon as a source node E_s becomes aware that a lightpath from itself to some other end node E_d is no longer operational.

The steps followed to replace each affected lightpath by a new lightpath are identical to those outlined above.

Exercise 8.8. Suggest another algorithm for the problem discussed in Section 8.3.5 based on the following ideas:

i) Before the network starts operating, for each node N_i, two edge-disjoint routes ρ_1^{ij} and ρ_2^{ij} from itself to each end node E_j are determined. Here, ρ_1^{ij} is the shortest path from N_i to E_j.

ii) Node N_i, for all nodes N_i in the network, will store two nodes, for each end node E_j,

   a) the next node N_1^{ij}, using route ρ_1^{ij}, from itself to E_j and

   b) the next node N_2^{ij}, using route ρ_2^{ij}, from itself to E_j.

iii) When there is a request for a lightpath from E_s to E_d, the route will be determined dynamically. An intermediate node N_i receives a route from E_s to itself and a set of channels that may be used for communication from E_s to itself on the supplied route. Node N_i selects, as the next node in the path from E_s to E_d,

   • N_1^{id} if the edge $N_i \rightarrow N_1^{id}$ is fault-free and there is an available channel that may be used to communicate from E_s to N_1^{id},

   • N_2^{id} otherwise.

iv) Node N_i sends the path from E_s to N_1^{id} or to N_2^{id} (as may be the case) and the set of channels that may be used for communication.

v) The process is over either when the end node E_d is reached or when it is decided that no feasible path exists.

$\square$

Exercise 8.9. Compare, through simulation, the performance of the algorithm illustrated in this section and the algorithm you have proposed. In this simulation you should consider various sizes of networks and various topologies and use 32 as the number of channels on each fiber in the network. Each request for a lightpath should specify a source end node, a destination end node, and a duration. Use some convenient unit for time. Determine, through simulation, the average blocking probability and the average setup time for each size of network.

$\square$

8.4 Bibliographic Notes

8.4.1 Review Papers on Fault

A review of many research topics in WDM networks, including protection and restoration in WDM networks, appears in [287]. A survey on restoration schemes, for dynamic as well as static traffic requests, appears in [281]. An overview of protection/restoration schemes is given in [424]. A protection scheme for multi-fiber physical links and hierarchical OXC structures is also given in [424]. The paper also presents a vision of protection/restoration schemes that are closely coordinated at both the IP and WDM layers. A review of protection and restoration methods in WDM networks, including multiple faults, appears in [435]. A collection of research results on the nature of faults in optical networks and fault handling techniques is available in [361].

8.4.2 Link and Path Protection Schemes

The factors affecting the complexity of implementing different protection schemes in the optical layer have been discussed in [167] and [168] — two landmark papers. The factors considered in [168] include requirements for fast signaling, dealing with multiple types of failures, such as node and link failures, and multiple concurrent failures, and supporting low-priority traffic. Different protection schemes have been compared in terms of their cost and efficiency. In [168], the advantages and disadvantages of protection at the optical layer have been discussed.

An ILP and a heuristic, based on simulated annealing, have been proposed in [64] for resource allocation to set up the primary and the backup paths in survivable WDM networks. The influence of the cost function, the potential benefits of wavelength conversion, and the tunability of the laser sources on network design have also been studied in [64]. A heuristic for RWA to handle a set of static connection requests is given in [5]. A distributed algorithm for shared-path protection in dynamic WDM mesh networks has been proposed in [346]. The algorithm determines, in a distributed manner, the extent to which information may be shared and uses a "retry" scheme to improve the blocking probability. The algorithm can be modified to design restorable networks capable of handling a given set of failures.

Approaches for dynamic lightpath provisioning in optical networks and analysis of the performances of dynamic provisioning algorithms are given in [313]. The paper considers unprotected, 1+1 protected, as well as mesh-restored lightpaths and concludes that mesh-restored lightpaths lead to significant capacity savings.

A centralized and a distributed algorithm for routing and wavelength allocation of lightpaths have been compared in [315]. In [341] a novel way of choosing the primary and the backup paths has been presented to ensure that the lengths of both the primary and the backup paths are small. This problem turns out to be NP-complete for both the edge-disjoint and the node-disjoint versions of the problem. An approximation algorithm for the problem with a guaranteed performance bound and a mathematical programming formulation for the exact solution of the problem has been given in [341].

A framework for jointly assigning wavelengths to multiple classes of service, including 100% protected, unprotected, and low-priority connections has been proposed in [335]. The problem has been formulated in [335] as a vertex coloring problem on a graph. A pool-based channel reservation scheme for shared-path protection has been proposed in [103]. It was shown that the approach is optimal when the set of primary and backup paths are given. The proposed scheme also offers advantages since it requires less overhead for restoration capacity, less data storage at nodes, and easier access to control and monitoring parameters. It is assumed in [103] that a primary and a link-disjoint backup path is set up for each connection request.

Algorithms for network capacity design that take into account the existence of protection paths and the use of shared protection have been proposed in [189]. The problem has been formulated in [189] as a generalized maximum concurrent flow problem and then two polynomial approximation algorithms have been developed to solve the problem.

An efficient framework for dynamic shared-path protection, under distributed control with partial information, has been studied in [306]. The complete information is partitioned in [306] so that each node only maintains and uses partial information. An ILP for shared-path protection and logical topology design has been proposed in [332], solving the same problem discussed in Section 8.3.4. Since the ILP formulation is not suitable for practical networks, a heuristic to find efficient solutions in typical networks has also been proposed in [332]. A comparison of the protection-switching time for WDM shared-path protection with the restoration time for IP restoration also appears in [332].

A stochastic algorithm to determine the extent to which protection channels may be shared (for shared-path protection) has been presented in [61]. It was shown in [61] that the stochastic approach is faster, more scalable, and requires less information to determine sharability, compared to existing deterministic models. An ILP formulation as well as a fast heuristic has been presented. Simulation results show that the heuristic can achieve almost the same performance as the ILP-based scheme. An ILP for shared-path protection was proposed in [415] which looks not only at resource requirements (wavelengths, links) for selecting backup paths but also considers restoration time guarantees. Formulations for both a centralized scheme and a distributed scheme with partial information have been proposed in [415].

Dynamic allocation of restorable connections in response to requests for connections cannot assume knowledge of future requests for connection. In path protection schemes, the overhead, in the form of additional network resources needed for the backup path, is considerable [228]. Shared-path protection, as discussed in Section 8.3.2 above, assumes that the routings used in each existing primary and backup lightpaths are known to the routing algorithm at the time of allocating a new pair of primary and backup lightpaths in response to a new request. In distributed routing such complete global information is not easily available. It was shown in [228] that an aggregate information allows an algorithm given in [228] to perform quite efficiently.

Apart from the trap problem, mentioned in Section 8.2, the choice of a shortest path as the primary path may result in the length of a link-disjoint backup path becoming unacceptably large. It was shown in [296] that the problem of finding an eligible pair of primary and backup paths for a new lightpath request requiring shared-path protection and using dynamic lightpath allocation is NP-complete. A heuristic to compute a feasible solution and an algorithm to optimize resource consumption for a given solution has been proposed in [296].

A distributed constraint-based path selection method for dynamic RWA in WDM networks has been discussed in [215]. Multiple, possibly conflicting, constraints have been considered for selecting a feasible path. Simulations on different topologies such as the ring, the mesh, and interconnection networks have been used to validate the approach.

In a network where the wavelength continuity constraint must be satisfied and connections for communication, using path protection scheme, are set up on request have been studied in [8]. When there is a request for connection, the problem is to set up, if possible, a primary lightpath and a backup lightpath, taking into account the lightpaths set up in response to earlier requests that are still in progress. The route of the backup lightpath must be edge- (node-)disjoint with respect to the route of the primary lightpath to ensure the ability to handle single-link (node) failure. Both the edge-disjoint and node-disjoint problems are NP-complete [8]. For the edge-disjoint version of the problem, an approximation algorithm and an exact algorithm have been presented in [8].

8.4.3 Schemes for Restoration

Three problems for restoration have been investigated in [110]. Problem 1 is to find the restoration paths of as many lightpaths as possible, given the primary paths of these lightpaths. Problem 2 is to find the restoration paths of all lightpaths, given the primary paths of these lightpaths, to minimize the network capacity. Problem 3 is to find both the restoration and the primary paths so that network capacity is minimized.

The performance issues of several restoration algorithms, using either centralized or distributed control, have been discussed in [281]. A distributed algorithm for dynamic RWA in WDM networks has been presented in [100]. The approach takes into consideration multiple, possibly conflicting, metrics such as physical layer impairments, reliability, and traffic conditions for distributed discovery of routes for lightpaths. An efficient algorithm for distributed path selection to achieve end-to-end path-based connection restoration in a GMPLS framework has been proposed in [248]. The paper assumes that the restoration path is preselected, but the channels are not assigned until a failure occurs. Therefore, resources can be shared by connections whose primary paths are not susceptible to simultaneous failure.

ILPs for path/link protection and distributed protocols for path/ link restoration have been proposed in [318]. The protection-switching time and the restoration time for each of these schemes have been studied in [318], and it was shown that the path restoration has a better restoration efficiency than link restoration, and that link restoration has faster restoration compared with path restoration. In [91] it was shown that fast restoration times depend on the ability to pre-cross-connect protection paths. The concept of a *pre-cross-connect trail* (PXT), for both path-based and link-based protection schemes, has been introduced in [91]. The PXT scheme achieves fast restoration speeds and bandwidth efficiency, comparable to shared mesh protection.

A number of approximation formulae have been proposed in [235] to quickly estimate the size of an optical network with limited inputs. These formulae may be used to determine the amount of traffic that can be carried over a given network, specified by the number of nodes, the average fiber connectivity, the traffic, and the technique for handling faults. Also, for a given traffic to be supported, the approach can determine the characteristics of the topology required as well as the quality of restoration possible.

8.4.4 Attacks on All-Optical Networks

In this chapter the focus was on faults caused by failure of components, mainly link failures. In [407], the special characteristics of malicious attempts to disrupt all-optical networks have been discussed. Such attacks on all-optical networks are different, both from the problems created by component failures in all-optical networks and from attacks on traditional networks. A model for such attacks has been proposed in [407]. It was shown that placing a relatively small number of attack monitors on a selected set of nodes in a network is sufficient to achieve the required level of performance. A diagnosis method to localize the attack has also been proposed in [407].

8.4.5 Reducing the Overhead for Fault Tolerance

A new technique for dynamic lightpath allocation, called *primary-backup multiplexing*, has been proposed in [282] to reduce the overhead required for path protection. In this approach, a primary lightpath and one or more backup lightpaths share the same channel. By doing so, the number of lightpaths that can be established is increased, thus reducing the overhead for protection and hence increasing the network resource utilization. However, when a link fails, there is no guarantee that a backup lightpath can be set up. This technique is useful in the case of dynamic lightpath allocation, in response to requests for short-term communication, where the lightpaths are relatively short-lived. A connection between a source and a destination, called a *D-connection* in [282] loses its recoverability when a channel on its backup lightpath is used by some other primary lightpath. It regains its recoverability when the other primary lightpath terminates. When a primary lightpath fails due to a fault, the source node sends control messages along the route of the backup path to configure the routers for the backup path. If any channel in the backup route is used by some other primary lightpath, the establishment process fails. Since faults are infrequent and every connection does not necessarily require fault tolerance, the backup lightpaths with 100% guarantee are needed only for relatively few communications.

The concept of generalized loop-back, a new scheme for performing loop-back in optical mesh networks, has been introduced in [271, 272]. The idea is to have a primary digraph with each directed edge corresponding to a unidirectional fiber and an associated secondary digraph with edges that correspond to the unidirectional fibers in the reverse direction of the edges in the primary digraph. When a failure of the unidirectional fiber $x \rightarrow y$ occurs, the data to be carried by the failed edge $x \rightarrow y$ in the primary digraph floods the secondary digraph, possibly along multiple paths. When the data reaches node y along the edges of the secondary digraph, node y directs the data to the original outgoing edge in the primary digraph from y. Algorithms to recover from both link and node failures have been discussed in [271, 272]. The proposed approach demonstrates advantages in terms of speed, transparency, and flexibility. An algorithm for loop recovery, based on the same idea of generalized loopback, has been proposed in [263]. Additional spare capacity has been used in [263] to carry unprotected traffic. The effect of two-link failures on several different networks has also been studied in this paper. Three loop-back methods for handling double-link failures in WDM mesh networks have been discussed in [90]. Simulation results indicate that it is possible to achieve nearly 100% recovery from double-link failures with only a small increase in backup capacity.

The concept of *shared segment protection* has been introduced in [416]. Each primary path is divided into a set of segments, each segment is protected, and the problem is to determine the optimal set of segments for the primary paths. This idea can be visualized as a generalization of link protection (where a faulty link is bypassed) and path protection (where the entire path is avoided). An integer linear program (ILP) model to determine an optimal set of segments to protect a given primary path and a fast heuristic algorithm to obtain a near-optimal set of segments has been described in [416]. Simulation results indicate that this approach is much faster and has similar bandwidth efficiency to shared-path protection schemes.

A *partial path protection* (PPP) scheme selecting backup paths based on local information about network failures has been discussed in [395]. PPP allows operational segments of the primary path to be used in the protection path. Blocking probability has been used in [395] as the primary performance metric, rather than capacity-efficiency. The approach in [395] can handle multiple faults as well.

In [184], a number of distributed real-time applications (e.g., medical imaging, air traffic control, and video conferencing) have been identified where it is important to guarantee the message delivery time, so that the delay in recovery from failures must be controllable and low. In [184] the concept of *segmented backups* has been introduced . A segmented backup consists of multiple backup paths, each corresponding to a contiguous portion of the primary path. In [184] it was shown that the segmented backup scheme offers improved network resource utilization, a higher average call acceptance rate, better quality-of-service guarantees on propagation delays, and failure-recovery times and increased flexibility to control the level of fault tolerance of each connection separately.

The concept of *subpath protection* has been proposed in [295] to generalize shared-path protection. In subpath protection the optical network is conceptually partitioned into several smaller sub-networks, and each lightpath spanning more than one sub-network is conceptually split into components, each within a single sub-network. Shared-path protection is applied to each component of the lightpath. An ILP and a heuristic have been proposed in [295] to achieve subpath protection. It was shown in [295] that subpath protection achieves improved survivability, much higher scalability, and significantly reduced fault-recovery time.

A service-guaranteed shared protection scheme for dynamic lightpath allocation, called *Short Leap Shared Protection* (SLSP), has been introduced in [196]. The idea of SLSP is to divide each primary path into several overlapped protection "domains", each consisting of a primary and backup subpath pair. It was shown in [196, 197] that the scheme has better capacity utilization compared to shared-path protection. A study, in [216], compares through simulation three restoration techniques — path, subpath, and link restoration — in terms of desirable features, such as high availability for the entire network and low restoration time for each connection disrupted by the faults. It was reported in [216] that both subpath and link restoration can provide faster restoration time compared to path restoration. Subpath and link restoration have a lower restoration success rate compared to path restoration [216].

A way to handle single faults by designing a logical topology with sufficient redundancy for data communication was explored in [11]. The idea is that logical topologies, in general, have multiple logical paths for each source-destination pair. When there is a link failure, all logical paths, using any lightpath that uses the failed link, are no longer operational. However, in a properly designed logical topology with sufficient redundancy, it is possible to reroute the data, which was originally routed over logical paths that are now inoperative, using other logical paths. It was shown in [23] that this scheme requires considerably less network resources in terms of the number of channels compared to dedicated path protection (DPP) and shared-path protection (SPP).

8.4.6 Handling Multiple Faults

In this chapter the focus was on single-link failure, considered to be the most common cause for faults. Other types of failures are also becoming quite important. Double-link failure can be important if there is considerable time lapse between the first fault and the time to repair the fault since another fault may occur before the first fault can be repaired. In a shared-path protection scheme, the first fault may affect the primary (backup) paths of a number of source-destination pairs. If the second fault affects any of the backup (primary) paths of these source-destination pairs, the lightpath connecting the source-destination pair will be disrupted. The idea in [436] is to identify such "vulnerable" source-destination pairs and set up (or reprovision) a new backup path in anticipation of a future failure. In [262] two-link failures have been studied and a quantitative measure of the ability of a network to recover from two-link failures has been proposed. Restoration paths that embed rings within mesh networks fare poorly for failure localization. Determining the restoration paths ahead of time has been shown in [262] to be worse than determining a restoration path at the time of failure. *Preemptive reprovisioning* has been introduced in [312] to create a backup path in advance of a second failure. It was shown that preemptive reprovisioning reduces the restoration time in the case of a second failure.

Since fibers are typically laid in bundles, inside a *duct* (also called *conduits*), cutting through all or many fibers in a duct, either during earth-moving operations or due to a calamity, is quite possible. The argument that the primary path and the backup path should be duct-disjoint is made in [434]. An ILP and a heuristic algorithm have been presented in [434] for handling duct-layer constraints.

Shared Risk Link Group (SRLG) describes the relationship between links with a shared vulnerability [372]. An SRLG is a group of links that are subject to a common risk, such as a conduit cut. Therefore, finding a pair of diverse paths at the optical layer involves computing a pair of SRLG-diverse paths. Each link at the OXC layer may be related to several SRLGs. Although the concept of SRLG was originally proposed to deal with conduit cuts, it can be extended to include general risks. For example, all the fiber links located in an area may be assigned the same SRLG, considering the risk of earthquakes. A highly readable description of the fault scenarios that a designer must take into account also appears in [372]. Three kinds of SRLG-diverse path protection schemes — dedicated-path protection, shared-path protection, and "no protection" — have been considered in [353]. When service providers consider a number of requests for connections, a reasonable objective is to maximize the revenue, possibly by rejecting some requests. This is the *revenue maximization problem* in [365]. When all the connection requests can be handled, the objective becomes minimizing the network resources — the *capacity minimization* problem in [365]. For the revenue maximization and capacity minimization problems, an ILP and two heuristics have been proposed in [353]. A heuristic for avoiding traps in networks with shared link risk groups (SLRG) has been proposed in [417]. The performance of the heuristic is comparable to ILP-based approaches, but requires much less time than the ILPs. A shared SLRG protection scheme based on trap avoidance has also been proposed.

In [414] it was shown that, in order to avoid the trap problem, mentioned in Section 8.2, it is much more desirable to determine the primary and backup path together. A fast and efficient approximation algorithm, termed *successive survivable routing* (SSR), has been proposed in [259] to reduce the total spare capacity to "near optimality" for arbitrary single as well as multiple link failures or all single node failures. Numerical results comparing several SCA algorithms show that SSR has the best trade-off between solution optimality and computation speed. Fault management is viewed in [259] as a *spare capacity allocation* (SCA) problem — a multi-constraint optimization problem. In [259] an efficient approximation algorithm has been developed to minimize the cost of total spare capacity The approach can be applied to multiple faults (e.g., SRLG, single node failures). The trap problem, mentioned in Section 8.2, has been discussed in [259].

8.4.7 Survivable Routing

In a logical topology design, the routing for lightpaths must be designed in such a way that a single-link failure does not disconnect the logical topology. Such a routing is termed *survivable* in [280]. The survivable routing problem is shown to be NP-complete. Necessary and sufficient conditions for survivable routing have been developed in [280]. An ILP formulation for survivable routing as well as an LP-relaxation technique to handle practical networks have been proposed in [280]. The testing of the necessary and sufficient condition for even a ring logical topology to withstand the failure of a single physical link (in other words, the survivable routing problem for a ring) was shown in [338, 339] to be an NP-complete problem. An algorithm for testing the necessary and sufficient condition for survivable routing was also presented in [339]. The problem of determining the minimum number of links that has to be added to a physical topology so that the survivable routing of a logical ring is possible was shown to be NP-complete in [351]. An ILP formulation for the optimal solution of the problem and an algorithm for finding an approximate solution was given in [351].

8.4.8 Miscellaneous Topics

In [224], the concept of *Streams* as a new path protection scheme has been introduced. Two backup paths B_1 and B_2 can form a stream as long as there is no divergence between these two paths. For example, backup path $B_1 = v_1 \rightarrow v_2 \rightarrow v_3$ and $B_2 = v_2 \rightarrow v_3 \rightarrow v_4$ can form a stream $S = v_1 \rightarrow v_2 \rightarrow v_3 \rightarrow v_4$. However, backup path $B_3 = v_1 \rightarrow v_2 \rightarrow v_5$ and B_1 given above cannot form a stream as these two paths diverge at node v_2. The stream scheme can be conceptualized as a member of a virtually shared DPP scheme. In this scheme, all the routers are preconfigured (just like DPP schemes) and, in the case of a failure, the traffic previously communicated using the primary lightpaths is sent through the backup path that forms a part of a stream. This scheme results in fast recovery as the routers are not required to spend any time for decision making at the time of a failure since no signalling or configuration of routers is needed after a failure [224]. The resource utilization in this scheme is much better than that of DPP schemes as it allows sharing of resources on the backup paths between multiple connections. Thus the scheme retains advantages of both DPP and SPP. In addition to introducing the concept of stream, algorithms for the formation of streams are also given in [224]. The concept of *gStream*, a more generalized and improved form of stream, was introduced in [340]. Finding the set of gStreams that maximizes capacity utilization is NP-complete. Efficient heuristics were presented in [340] for maximizing the capacity utilization without sacrificing the benefit of fast recovery of streams [224].

Both IP and WDM layers can handle faults and a number of schemes (including those for WDM protection discussed in this chapter) have been proposed. A review of IP restoration techniques and WDM protection is available in [156]. A heuristic, using simulated annealing, has been proposed in [156] to use IP restoration and WDM protection schemes concurrently in the same network. The proposed heuristic can trade solution optimality for computational time, giving useful design choices.

The use of *lightpath diversity* (multiple disjoint lightpaths at the optical layer) to achieve reliable communication between any source-destination pair has been explored in [404]. The network has been optimized either by minimizing the error probability for a given transmitter power or by minimizing the transmitted power for a target error probability.

An efficient algorithm for locating multiple hard failures (unexpected events that suddenly interrupt the established channels) and soft failures (events that progressively degrade the quality of transmission) has been proposed in [269].

9

Traffic Grooming

One straightforward way to design a WDM network is to use a single-hop model for data communication where each individual request for communication from a source E_s to a destination E_d is mapped to a lightpath from E_s to E_d. Each request for communication is mapped to a different lightpath. Each lightpath in a WDM network, as pointed out in Chapter 2, carries today 10 Gbps or 2.5 Gbps depending on the technology used. Individual requests for connection are typically for a much lower data communication rate, of the order of Mbps. Defining a separate lightpath for each individual request for communication is clearly very wasteful and expensive in terms of the number of lightpaths and hence the number of optical components used in the network.

Traffic Grooming in WDM networks can be defined as a family of techniques for combining a number of low-speed traffic streams from users so that the high capacity of each lightpath may be used as efficiently as possible. Traffic grooming minimizes the network cost in terms of the number of transmitters, receivers, and optical switches [86, 117, 170, 202, 438, 444, 447]. Chapters 1, 3, and 7 have already discussed the idea of considering, at the same time, the communication requests between all the ordered pairs of end nodes when optimizing the logical topology and the routing strategy. In this chapter, additional aspects of this topic will be discussed in detail.

In discussing traffic grooming, the convention is to specify the data communication rate using the Optical Carrier level notation (OC-n)[1]. If the capacity of a lightpath is 10 Gbps (2.5 Gbps), it is specified as OC-192 (OC-48). Similarly the unit for user data rate is typically OC-3, OC-6, OC-12, or OC-48.

In general, the traffic grooming problem is to find the optimum strategy for data communication to handle a set R of requests for data communication. Each request $r \in R$ is characterized by its

1) data communication rate, s_r, using the Optical Carrier level OC-n notation,

---

[1] The base rate (OC-1) is 51.84 Mbps, and OC-n means $n \times 51.84$ Mbps (Chapter 1).

2) source end node,

3) destination end node.

There are two basic approaches [445] in traffic grooming:

Approach i) for a given set of traffic requests, minimize the total network cost, with the condition that all traffic requests are satisfied.

Approach ii) for given resource limitations and traffic demands, maximize the network throughput, measured by the total amount of traffic that is successfully carried by the network.

These two approaches may be considered in a variety of situations, similar to those given in Chapters 7 and 8, as follows:

1. static traffic, where the requests for communication are known in advance.

2. dynamic traffic, where the requests come one at a time and have to be processed, with no advance knowledge of the pattern of requests to arrive in the future.

3. single-hop communication, where all logical paths have a length of exactly 1.

4. multi-hop communication, where the lengths of the logical paths are not restricted.

5. blocking networks, where the objective is to successfully process as many requests for communication as possible, but it is possible that some of the requests for communication may not succeed.

6. non-blocking networks, where all requests must be satisfied and the objective is to minimize the cost of the network.

7. handle faults in the network, using, for example, schemes for fault avoidance discussed in Chapter 8.

An example of static traffic grooming is given below. The physical topology shown in Figure 9.1 was discussed in Section 1.1.2. For this example, the data communication capacity of a lightpath is taken as OC-48 (i.e., 2.5 Gbps).

Let the requests for data communication be as shown in Table 9.1.

A possible logical topology to support all these requests for communication is shown in Figure 9.2. For instance, the lightpath L_1 may be selected to service the following requests:

- OC-12 from E_1 to E_3,
- OC-3 from E_1 to E_3,
- OC-6 from E_1 to E_3,
- OC-6 from E_1 to E_4,
- OC-3 from E_1 to E_4,
- OC-3 from E_2 to E_3,

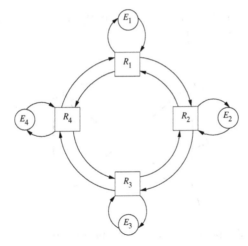

Fig. 9.1. The physical topology of a typical WDM network

	E_1	E_2	E_3	E_4
E_1		OC-12 OC-24	OC-12 OC-3 OC-6	OC-6 OC-3 OC-3
E_2	OC-6		OC-3	OC-6 OC-3 OC-3
E_3	OC-12	OC-3 OC-6		OC-6
E_4	OC-3	OC-3		

Table 9.1. Traffic requests

In this example, the routing strategy was to use a single logical path from its source to its destination[2] to handle each request. In this static traffic grooming problem, for instance, the OC-6 request from E_1 to E_4 used the logical path $E_1 \Rightarrow E_3 \Rightarrow E_4$. It is important to note that there may be other choices for a valid logical path to handle this OC-6 request from E_1 to E_4.

Each logical edge, after the allocation of requests to logical edges, will have a *residual capacity*, representing the difference between the capacity of a lightpath and the sum of the data communication rates of each of the requests which uses this logical edge in its logical path from its source to its destination. For instance, the logical edge L_1 has a capacity of 48 (in OC-n notation) and

---

[2] The requirement of selecting only one logical path for each request is important. The repercussions of either requiring a single logical path or allow the possibility of multiple logical paths to handle each request will be discussed later on in this section.

is carrying a total of 33 units (in OC-n notation). Logical edge L_1, therefore, has a residual capacity of 15.

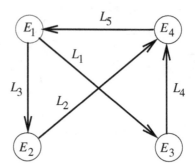

Fig. 9.2. A logical topology defined on the physical topology shown in Figure 9.1

Exercise 9.1. Find, using the logical topology shown in Figure 9.2, a routing to support all the requests in Table 9.1. In this routing, you should use only one logical path for each request for data communication.   □

In Chapter 7, the logical topology design problem was presented as a network optimization problem to determine

a) the set of lightpaths to be created,
b) for each lightpath in E_L,
   - its route through the physical topology,
   - its channel number on each fiber in its route,
c) the strategy for routing the traffic T over the logical topology.

In terms of the OC-n notation used for discussing traffic grooming, given a set R of requests for data communication, it is easy to compute a traffic matrix T, where the element $t(s, d)$, in row s and column d of matrix T, is the sum of the data communication rates of all requests from E_s to E_d in R. This gives $t(s, d)$ in OC-n notation, which can be easily converted to the Mbps data communication rate since OC-1 is 51.84 Mbps.

In the above formulation of the logical topology design problem, as discussed in Chapter 7, the traffic $t(s, d)$ from end node E_s to end node E_d may be divided into as many components as necessary to optimize the network performance. This means that an individual request for data communication from E_s to E_d, which constitutes a part of the total traffic $t(s, d)$ from end node E_s to end node E_d, may be split into a number of components and the different components of that request may be communicated using different logical paths from E_s to E_d.

Some researchers term this approach as the *bifurcated model of traffic grooming* [118]. In the other approach, called the *non-bifurcated model of traffic grooming*, the data stream corresponding to each request is communicated using a *single* logical path from the source of the data stream to its destination. In Exercise 9.1, the non-bifurcated model has been used. In this model, the data to be communicated from an end node E_i has the following sources:

1. data from users who are using the end node E_i as the entry point for data communication to other end nodes in the network. Each user data stream is typically several Mbps, that has to be transmitted to some specified end node.

2. data carried as payload by lightpaths whose destination is E_i. The payload of such a lightpath consists of many individual data streams, each data stream with a specified end node as its destination. Some of these data streams have E_i itself as their destination. These data streams will terminate at E_i. Each of the remaining data streams is using E_i as an intermediate end node in a multi-hop logical path from the source to the destination of that data stream.

In a non-bifurcated traffic grooming scenario, at end node E_i, each data stream that has to be communicated from end node E_i becomes part of the payload of a *single* lightpath whose source is E_i.

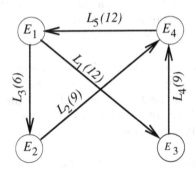

Fig. 9.3. A logical topology defined on the physical topology shown in Figure 9.1

It is important to note that the requirement that the data stream corresponding to each request be communicated, using a single logical path from the source of the data stream to its destination, may reduce the utilization of network resources and/or increase the blocking probability. For instance, in a dynamic traffic grooming scenario, using the non-bifurcated traffic grooming model, let the current logical topology of a WDM network be as shown in Figure 9.3[3]. This logical topology is currently supporting a number of data requests, based on the non-bifurcation model. Beside each logical edge, a label for the edge and the residual capacity of the edge are shown. For instance, the logical edge L_3 from E_1 to E_2 has a residual capacity of OC-6. At this point,

---

[3] This logical topology is the same as that shown in Figure 9.2.

let there be a request for data communication at OC-12 rate from E_1 to E_4. There are only two choices for the logical path from E_1 to E_4 as follows:

> Choice 1) the path $E_1 \Rightarrow E_2 \Rightarrow E_4$. The logical edge on this path with the minimum residual capacity is the edge $E_1 \Rightarrow E_2$ with a residual capacity of six units. This logical path can, therefore, service a request with a data communication rate of up to OC-6.
>
> Choice 2) the path $E_1 \Rightarrow E_3 \Rightarrow E_4$ which can service a request with a data communication rate of up to OC-9.

Since none of the logical paths can support a OC-12 request, the request will be rejected.

If this network used a bifurcation model, the request for data communication at OC-12 rate from E_1 to E_4 can be handled easily. For example, the data could be split into two components, one with a communication rate of OC-9, to be sent using the path $E_1 \Rightarrow E_3 \Rightarrow E_4$, the other with a rate of OC-3, to be sent using the path $E_1 \Rightarrow E2 \Rightarrow E_4$.

As discussed in Chapters 7 and 8, if a bifurcation model is used, routing on a logical topology requires only an LP. If the non-bifurcation model is used, the additional requirement that each request has to be routed on the logical topology using *exactly* 1 logical path means that a formulation for routing over a logical topology has to involve binary variables. In other words, solving an MILP formulation for the non-bifurcation model is more complex and time-consuming.

However, as pointed out in [390], bifurcation increases the complexity and the cost of traffic reassembly, and may also introduce delay jitter at the application layer. Many applications, especially real-time applications, require that their traffic be kept intact, i.e., without demultiplexing at the source, independent switching at intermediate nodes, and multiplexing at the destination.

The bifurcated model of traffic grooming has been discussed in detail in Chapter 7 and will not discussed in this chapter.

Non-bifurcated traffic grooming has exactly the same three subproblems as those discussed above for logical topology design problem, except for the routing strategy, and in some cases, the objective function for the problem formulation. The following aspects of non-bifurcated traffic grooming is considered, in detail, in this chapter:

> a) an ILP for static traffic grooming to
>   - minimize the cost of the network by minimizing the number of lightpaths,
>   - maximize the throughput of the network.
> b) a heuristic for static traffic grooming.
> c) a heuristic for dynamic traffic grooming.

9.1 Static Traffic Grooming

As mentioned earlier, in Chapter 7, one important problem is to consider situations where the traffic is relatively stable over long periods of time. In this case, it is worthwhile to spend time in determining the most cost-effective strategy for grooming. In this section, the issue of faults will not be considered. The idea is to consider the optimum strategy using multi-hops, whenever needed. Two ILP formulations presented in this section become computationally intractable for non-trivial networks. Simplifications have been proposed to solve these problems for practical networks.

9.1.1 Problem 1

The objective of formulation $FORM$ 9 is to design the optimal logical topology, the RWA, and the routing over the logical topology in such a way that *all* the requests can be satisfied. The discussion below is based on the formulation proposed in [201]. The RWA has also been simplified in the sense that the route of lightpaths though the physical topology, using, for instance, the shortest route, is determined ahead of time. The objective of the design is to minimize the number of transmitters and receivers by reducing the number of lightpaths used in the network.

Each end node (router node) in this formulation is assigned a unique identifying number $v, 1 \leq v \leq \mathcal{N}_E$ ($\mathcal{N}_E < v \leq \mathcal{N}_E + \mathcal{N}_R$), where $\mathcal{N}_E$ ($\mathcal{N}_R$) is the number of end nodes (router nodes) in the network. In the following discussions, end node i will denote the end node (router node) identified by number $i, 1 \leq i \leq \mathcal{N}_E$ ($\mathcal{N}_E < i \leq \mathcal{N}_E + \mathcal{N}_R$).

Notation used in $FORM$ 9

n_{ch} : the number of channels available on each fiber.

$\mathcal{N}_E$: the number of end nodes in the network.

R : the set of traffic requests.

s_r : the data communication rate of request $r \in R$, using the Optical Carrier level OC-n notation.

g : the data communication rate of a single lightpath, using the OC-n notation.

E : the set of all physical edges $x \to y$ in the network where x (y) is a node (router or end node) in the network.

$\mathcal{L}$: the set of potential logical edges in the network. This set consists of $\mathcal{N}_E(\mathcal{N}_E - 1)$ logical edges, one for each ordered pair of end nodes.

source(ℓ) (destination(ℓ)) : source (destination) end node for lightpath ℓ, $\ell \in \mathcal{L}$.

$\mathcal{V}_E$: $\{v : 1 \leq v \leq \mathcal{N}_E\}$, the set of all numbers identifying the end nodes in the network.

$\hat{A}$: the node, logical-edge incidence matrix with $\mathcal{N}_E$ rows and $|\mathcal{L}|$ columns corresponding to all the potential logical edges, where element $\hat{a}_{v,\ell}$ in row v and column ℓ of $\hat{A}$ is defined as follows:

$$\hat{a}_{v,\ell} = \begin{cases} 1 & \text{if } v = \text{source}(\ell), \\ -1 & \text{if } v = \text{destination}(\ell), \\ 0 & \text{otherwise.} \end{cases}$$

F : the fiber, logical-edge incidence matrix with $|E|$ rows and $|\mathcal{L}|$ columns, where the element $f_{e,\ell}$ in row e and column ℓ is defined as follows:

$$f_{e,\ell} = \begin{cases} 1 \text{ if fiber } e \text{ is used by lightpath } \ell, \\ 0 \text{ otherwise.} \end{cases}$$

$\overrightarrow{u_r}$: the source-destination column vector with $|\mathcal{V}_E|$ elements for request $r \in R$, where the element $u_{v,r}, v \in \mathcal{V}_E$ is defined as follows:

$$u_{v,r} = \begin{cases} 1 & \text{if } v \text{ is the source end node for request } r, \\ -1 & \text{if } v \text{ is the destination end node for request } r, \\ 0 & \text{otherwise.} \end{cases}$$

$\overrightarrow{x_r}$: the column vector with $|\mathcal{L}|$ elements containing lightpath routing variables for $r \in R$, where the element $x_{\ell,r}, \ell \in \mathcal{L}$ is defined as follows:

$$x_{\ell,r} = \begin{cases} 1 \text{ if request } r \text{ is allotted to lightpath } \ell, \\ 0 \text{ otherwise.} \end{cases}$$

$\overrightarrow{y_c}$: the column vector with $|\mathcal{L}|$ elements containing channel assignment variables $\forall c, 1 \leq c \leq n_{ch}$, where the element $y_{\ell,c}, \ell \in \mathcal{L}$ is defined as follows:

$$y_{\ell,c} = \begin{cases} 1 \text{ if channel } c \text{ is assigned to lightpath } \ell, \\ 0 \text{ otherwise.} \end{cases}$$

$\overrightarrow{1}$: $[1, 1, \ldots, 1]$, the unit column vector with $|E|$ elements.

The formulation for *FORM 9*

Formulation *FORM 9* uses matrix notation and minimizes the number of lightpaths used to set up a scheme to meet all the traffic requests in R.

Objective function:

$$\text{minimize} \sum_{\ell \in \mathcal{L}} \sum_{1 \le c \le n_{ch}} y_{\ell,c} \tag{9.1}$$

subject to:

1. The lightpaths selected, for request r, must define a logical path from the source end node to the destination end node of request r.

$$\hat{A}\vec{x_r} = \vec{u_r}, \quad \forall r \in R \tag{9.2}$$

2. If two or more lightpaths use the same fiber, they must be assigned distinct channels.

$$F\vec{y_c} \le \vec{1}, \quad \forall c, 1 \le c \le n_{ch} \tag{9.3}$$

3. The total weighted sum of all requests allotted to lightpath ℓ must not exceed the capacity of lightpath ℓ, where the weight of request r is its data communication rate s_r.

$$\sum_{r \in R} s_r x_{\ell,r} \le g \sum_{1 \le c \le n_{ch}} y_{\ell,c}, \quad \forall \ell \in \mathcal{L} \tag{9.4}$$

4. Variables $x_{\ell,r}$ and $y_{\ell,c}$, being binary variables, can only have values 0 and 1.

$$x_{\ell,r} \in \{0,1\}, \quad \forall \ell \in \mathcal{L}, \forall r \in R \tag{9.5}$$

$$y_{\ell,c} \in \{0,1\}, \quad \forall \ell \in \mathcal{L}, \forall c, 1 \le c \le n_{ch} \tag{9.6}$$

Justification of *FORM* 9

If a lightpath ℓ is needed for one or more of the requests $r \in R$, then ℓ has to be assigned a channel number. The idea here is that, if no request is allotted to a lightpath, the lightpath is not allotted a channel number and the lightpath is not established. Assuming, for the moment, that wavelength continuity is guaranteed by this formulation, this means that, only if a lightpath is needed for one or more of the requests $r \in R$, then $\sum_{1 \le c \le n_{ch}} y_{\ell,c} = 1$; otherwise, $\sum_{1 \le c \le n_{ch}} y_{\ell,c} = 0$. The objective function (9.1) is therefore to minimize the number of lightpaths assigned a channel number; in other words, to minimize the number of lightpaths.

Let the source (destination) end node of request $r \in R$ be the end node identified by x (y) and let this formulation determine a logical path $x = i_0 \Rightarrow$

$i_1 \Rightarrow \ldots \Rightarrow i_m = y$ for this request, where $i_0, i_1, \ldots, i_m$ are the numbers identifying the end nodes in the path from $x = i_0$ to $i_m = y$. Let the lightpath corresponding to the logical edge $i_p \Rightarrow i_{p+1}$ be ℓ_p, for all $p, 0 \le p < m$.

The column vector $\vec{u_r}$ for request r must be as follows:

- $u_{x,r} = 1$,
- $u_{y,r} = -1$,
- $u_{i,r} = 0$, otherwise.

Selected elements of matrix $\hat{A}$ must be as follows:

- $\hat{a}_{x,\ell_0} = 1$,
- $\hat{a}_{y,\ell_{m-1}} = -1$,
- $\hat{a}_{i_j \ell_{j-1}} = -1$, $\quad 1 \le j \le m$,
- $\hat{a}_{i_j \ell_j} = 1$, $\quad\quad 1 < j < m$.

The column vector $\vec{x_r}$ must be as follows:

- $x_{\ell_i} = 1$, for all $i = 0, 1, \ldots, m-1$,
- $x_{\ell_j} = 0$, otherwise.

(9.2) is therefore the flow balance equation[4] [4].

Row e of $F\vec{y_c}$ is $\sum\limits_{\ell=1}^{|\mathcal{L}|} f_{e,\ell} \times y_{\ell,c}$. This represents the number of lightpaths using channel c and fiber e. The purpose of (9.3) is to ensure that two lightpaths sharing fiber e cannot use channel c at the same time.

The maximum data communication rate of a lightpath is g. If request r is assigned lightpath ℓ to service this request, out of the total data rate of g of lightpath ℓ, a data rate of s_r must be dedicated to request r. Thus the left-hand side of (9.4) represents the sum of the data rates allotted to different requests that use lightpath ℓ. On the right-hand side of (9.4), $\sum\limits_{1 \le c \le n_{ch}} y_{\ell,c}$ is 1 only if lightpath ℓ is used to service one or more requests. The purpose of (9.4) is to ensure that, when ℓ is servicing a number of requests, the sum of all the requests using ℓ does not exceed g, the maximum data communication rate of lightpath ℓ.

It is important to note that the wavelength continuity constraint is not explicitly specified but is satisfied implicitly. If $y_{\ell,c} = 1$, it means that channel c is allotted to all fibers used by lightpath ℓ. Constraint (9.3) does allow more than one channel number for a lightpath ℓ, while ensuring that no two lightpaths sharing a fiber are allotted the same channel number. Since the objective function is to minimize the number of lightpaths, if more than one channel is allotted to a lightpath ℓ, this may be a solution only if the set of requests R cannot be satisfied with only one lightpath from source(ℓ) to

---

[4] Also reviewed in Appendix 3.

destination(ℓ). In other words, multiple lightpaths must exist from source(ℓ) to destination(ℓ).

The complexity of *FORM* 9

The number of binary variables is the crucial factor determining the time required to solve this formulation [294]. There are $|\mathcal{L}|$ column vectors $\overrightarrow{x_r}$, each column having $|R|$ elements. Since there are $\mathcal{N}_E \times (\mathcal{N}_E - 1)$ potential lightpaths to be considered, the number of variables $x_{\ell,r}$ is $\mathcal{N}_E(\mathcal{N}_E - 1)|R|$. Similarly there are $|\mathcal{L}|$ column vectors $\overrightarrow{y_c}$, each column having n_{ch} elements. The total number of (0/1) variables is therefore $\mathcal{N}_E(\mathcal{N}_E - 1)(|R| + n_{ch})$.

Simplification of *FORM* 9

Formulation *FORM* 9 can be simplified by considering the issue of RWA separately from the issues of grooming, logical topology design, and routing over the logical topology [201].

9.1.2 Problem 2

The objective of formulation *FORM* 10 is to design the optimal logical topol-ogy, the RWA, and the routing over the logical topology to maximize the *amount* of data communicated, corresponding to a specified set R of requests for data communication. Each request for data communication uses the OC-n notation to specify the data communication rate. The objective is to maximize the weighted sum of all the requests. If request r is for communication using a data communication rate of OC-p, the weight of request r is p. The discus-sion below is based on the formulation proposed in [444]. In this formulation, each node in the network (router or end node) is assigned a unique identifying number. Each end node is allotted a number $i, 1 \leq i \leq \mathcal{N}_E$, where $\mathcal{N}_E$ is the number of end nodes in the network and each router node is allotted a num-ber $i, \mathcal{N}_E < i \leq \mathcal{N}$, where $\mathcal{N}$ is the number of nodes in the network. In the following discussions, an end node (router node) identified by $i, 1 \leq i \leq \mathcal{N}_E$ ($i, \mathcal{N}_E < i \leq \mathcal{N}$), will be called end node (router node) i.

Notation used in *FORM* 10

n_{ch} : the number of channels available on each fiber.

$\mathcal{N}_E$: the number of end nodes in the network.

$\mathcal{N}$: the number of nodes (routers or end nodes) in the network.

R : the set of traffic requests.

s_r : the data communication rate of request for data communication $r \in R$, using the OC-n notation.

g : the data communication rate of a single lightpath, using the OC-n notation.

p_{mn} : the number of fibers from node m to node n. It is assumed that at most one fiber can exist from node m to node n so that
$$p_{mn} = \begin{cases} 1 \text{ if there exists a fiber link from node } m \text{ to } n, \\ 0 \text{ otherwise.} \end{cases}$$

TR_i (RR_i) : the number of transmitters (receivers) at end node i.

L_{ij} : the number of lightpaths from end node i to end node j in the logical topology.

L_{ij}^c : the number of lightpaths from end node i to end node j using channel c.

from(r) (to(r)) : source (destination) end node of request r.

$P_{mn}^{ij,c}$: number of lightpaths from end node i to end node j, routed through fiber $m \rightarrow n$ on channel c in the physical topology.

N_{max}^{TR} : the maximum value of the number of transmitters or receivers at any end node in the network.

$\lambda_{i,j}^r$: a binary variable where
$$\lambda_{i,j}^r = \begin{cases} 1 \text{ if the } r^{th} \text{ request uses the lightpath from } i \text{ to } j \text{ as an intermedi-} \\ \quad \text{ate logical edge (in other words, the logical path from the source} \\ \quad \text{to the destination of request } r \text{ uses the logical edge } i \Rightarrow j), \\ 0 \text{ otherwise.} \end{cases}$$

X_r : a binary variable where
$$X_r = \begin{cases} 1 \text{ if the } r^{th} \text{ request has been successfully allocated to a logical} \\ \quad \text{path,} \\ 0 \text{ otherwise.} \end{cases}$$

The formulation of FORM 10

Formulation $FORM\ 10$ sets up a scheme to design a logical topology and a routing scheme that maximizes the weighted sum of the requests in R. Objective function:

$$\text{maximize} \sum_{r \in R} s_r X_r \tag{9.7}$$

subject to:

1. The number of lightpaths starting from (terminating at) end node i cannot exceed the number of transmitters (receivers) at i.

$$\sum_{j=1}^{\mathcal{N}_E} L_{ij} \leq TR_i, \qquad \forall i, 1 \leq i \leq \mathcal{N}_E \qquad (9.8)$$

$$\sum_{i=1}^{\mathcal{N}_E} L_{ij} \leq RR_j, \qquad \forall j, 1 \leq j \leq \mathcal{N}_E \qquad (9.9)$$

2. The total number of lightpaths from end node i to end node j must be equal to the sum of the number of lightpaths from i to j for all possible channel numbers.

$$\sum_{c=1}^{n_{ch}} L_{ij}^c = L_{ij}, \qquad \forall i, j, 1 \leq i, j \leq \mathcal{N}_E \qquad (9.10)$$

3. If node k is neither the source nor the destination of a lightpath from i to j, the number of lightpaths, from i to j, entering node k must be the same as the number of lightpaths leaving k.

$$\sum_{m=1}^{N} P_{mk}^{ij,c} = \sum_{n=1}^{N} P_{kn}^{ij,c}, \qquad \forall i, j, c, k, 1 \leq i, j \leq \mathcal{N}_E, 1 \leq k \leq N, \qquad (9.11)$$
$$k \neq i, j, 1 \leq c \leq n_{ch}$$

4. The set of lightpaths from end node i to end node j cannot include any lightpath that uses a fiber to (from) i (j).

$$\sum_{m=1}^{N} P_{mi}^{ij,c} = 0, \qquad \forall i, j, c, 1 \leq i, j \leq \mathcal{N}_E, 1 \leq c \leq n_{ch} \qquad (9.12)$$

$$\sum_{n=1}^{N} P_{jn}^{ij,c} = 0, \qquad \forall i, j, c, 1 \leq i, j \leq \mathcal{N}_E, 1 \leq c \leq n_{ch} \qquad (9.13)$$

5. The total number of lightpaths from end node i to end node j must be the sum of the total number of lightpaths from i to j that use channel c on the fiber from (to) i (j), for all $c, 1 \leq c \leq n_{ch}$.

$$\sum_{n=1}^{N} P_{in}^{ij,c} = L_{ij}^c, \qquad \forall i, j, c, 1 \leq i, j \leq \mathcal{N}_E, 1 \leq c \leq n_{ch} \qquad (9.14)$$

$$\sum_{m=1}^{N} P_{mj}^{ij,c} = L_{ij}^c, \qquad \forall i, j, c, 1 \leq i, j \leq \mathcal{N}_E, 1 \leq c \leq n_{ch} \qquad (9.15)$$

6. The sum of all lightpaths that use the fiber $i \rightarrow j$ and channel c cannot exceed the number of fibers from i to j.

$$\sum_{i=1}^{\mathcal{N}_E} \sum_{j=1}^{\mathcal{N}_E} P_{mn}^{ij,c} \leq p_{mn}, \qquad \forall m, n, c, 1 \leq i, j \leq \mathcal{N}_E, 1 \leq c \leq n_{ch} \qquad (9.16)$$

7. If request r has been successfully allocated to a logical path (i.e., X_r is 1), there must be exactly one lightpath allocated to r which ends at (starts from) the destination (source) of request r.

$$\sum_i \lambda_{i,\text{to}(r)}^r = X_r, \qquad \forall r \in R \qquad (9.17)$$

$$\sum_j \lambda_{\text{from}(r),j}^r = X_r, \qquad \forall r \in R \qquad (9.18)$$

8. If end node k is neither the source nor the destination of request r, it can be the destination of a lightpath carrying request r only if there is a lightpath carrying request r whose source is node k.

$$\sum_i \lambda_{i,k}^r = \sum_j \lambda_{k,j}^r, \qquad \forall k, r, k \neq \text{from}(r), k \neq \text{to}(r), \qquad (9.19)$$
$$1 \leq k \leq \mathcal{N}_E, r \in R$$

9. Request r can not be allotted to a lightpath whose destination (source) is the source (destination) of r.

$$\sum_i \lambda_{i,\text{from}(r)}^r = 0, \qquad \forall r \in R \qquad (9.20)$$

$$\sum_i \lambda_{\text{to}(r),i}^r = 0, \qquad \forall r \in R \qquad (9.21)$$

10. The total amount of data on a logical edge $i \Rightarrow j$ cannot exceed the capacity g of that edge.

$$\sum_{r \in R} \mathfrak{s}_r \times \lambda_{i,j}^r \leq L_{ij} \times g, \qquad \forall i, j, 1 \leq i, j \leq \mathcal{N}_E \qquad (9.22)$$

Justification of *FORM* 10

The objective function of *FORM* 10 is to maximize the weighted sum of successfully routed connection requests. The weight of request r is its data communication rate $\mathfrak{s}_r$, and if it is successfully routed, $X_r = 1$ (9.7).

L_{ij} gives the number of lightpaths from i to j. $\sum_{j=1}^{\mathcal{N}_E} L_{ij}$ gives the total number of lightpaths whose source is in end node i. The purpose of (9.8) is to ensure that the total number of lightpaths from i does not exceed the number of transmitters at i. The explanation for (9.9) is similar.

(9.10), (9.12), (9.13), (9.14), (9.15), (9.20), and (9.21) are obvious. (9.11) ensures flow conservation to establish a logical path from i to j.

The constant p_{mn} gives the number of fibers from m to n. Since there may be at most one fiber from m to n, this means p_{mn} is either 1 or 0. The purpose of (9.16) is to ensure that the sum of all P_{mn}^{ijc}, representing all the lightpaths that use the edge $m \rightarrow n$ in the physical topology and channel number c, is 1 only if there exists a fiber from m to n. This also ensures that two lightpaths sharing a fiber are not allotted the same channel number.

(9.17), (9.18), and (9.19) are the flow conservation rules for requests since there must be a logical path from the source to the destination of a request if the request has been successfully allocated to a lightpath.

In (9.22), $\mathsf{s}_r \times \lambda_{ij}^r$ gives the data rate that has to be reserved for request r if the request is allocated to logical edge $i \Rightarrow j$. The left hand side of (9.22) gives the sum of the data rates of all requests allocated to logical edge $i \Rightarrow j$. This cannot exceed g, the capacity of a lightpath.

The complexity of *FORM* 10

The number of integer variables is the crucial factor determining the time required to solve this formulation. Table 9.2 shows the number of integer variables needed in this formulation. Clearly, a formulation that requires $O(\mathcal{N}_E^4)$ integer variables is not practical, even for a network with a moderate size.

Symbol	Permissible values	Total Number of variables
L_{ij}	$0, 1, \ldots, N_{max}^{TR}$	$\mathcal{N}_E(\mathcal{N}_E - 1)$
L_{ij}^c	$0, 1, \ldots, N_{max}^{TR}$	$\mathcal{N}_E(\mathcal{N}_E - 1)n_{ch}$
$P_{mn}^{ij,c}$	$0, 1, \ldots, N_{max}^{TR}$	$\mathcal{N}_E(\mathcal{N}_E - 1)\mathcal{N}(\mathcal{N} - 1)n_{ch}$
λ_{ij}^r	$0, 1$	$|R|\mathcal{N}_E(\mathcal{N}_E - 1)$
X_r	$0, 1$	$|R|$

Table 9.2. Integer variables needed in Problem 2

9.1.3 A Heuristic for Static Traffic Grooming

As mentioned in the introduction to this chapter, traffic grooming using the non-bifurcation model has three subproblems:

(i) logical topology design,
(ii) RWA for each logical edge in the logical topology and
(iii) allocation of communication requests to lightpaths.

Formulations *FORM* 9 and *FORM* 10 in Sections 9.1.1 and 9.1.2 are complex since the subproblems mentioned above are interrelated and have to be solved at the same time. To simplify this problem, the three subproblems have been decoupled completely and heuristics have been used to solve them independently. As observed in Chapter 7, there is no guarantee that this formulation gives an optimal solution. However, if a solution is obtained, it gives a feasible solution. In other words, the logical topology can be set up and the allocation of logical paths to individual requests can be achieved.

The objective of the following heuristic is to increase the throughput by allocating the requests with the higher data communication rates first. The heuristic maximizes, to the extent possible, the number of requests that use a single hop from the source to the destination of each request.

Techniques for solving the RWA problems have been reviewed in Chapters 4 and 5. The heuristic given below is taken from [444] and takes care of subproblems (i) and (iii). The heuristic has two parts, where the first part is similar to the greedy heuristic HLDA described in Section 7.2.

Part I. Finding a logical topology

The purpose of this heuristic is, given a set R of traffic requests, to set up a logical topology G_L, a directed graph where the label of each edge indicates the remaining capacity of the edge using the OC-n notation. The notation used in this heuristic is the same as that used in Section 9.1.2. In this description $label_{s \Rightarrow d}$ will denote the label on the logical edge $s \Rightarrow d$.

Step 1) Construct a traffic matrix $T = [t(i,j)]$ of size $\mathcal{N}_E \times \mathcal{N}_E$ where $\mathcal{N}_E$ is the number of end nodes in the network and $t(i,j) = \sum s_r$: $r \in R, \mathrm{from}(r) = i, \mathrm{to}(r) = j$.

Step 2) Let MaxTraffic $= t(s,d)$ be the maximum entry in T. If Max-Traffic $= 0$, stop.

Step 3) Using any of the RWA techniques discussed in Chapters 4 or 5 try to set up a lightpath from s to d. This may fail if
- there is no spare transmitter at end node s,
- there is no spare receiver at end node d,
- the strategy used for RWA could not succeed due to lack of available channels.

If the attempt fails, set $t(s,d) = 0$. If the attempt succeeds,
- set $t(s,d) = Max(t(s,d) - g, 0)$,
- if there is no directed edge $s \Rightarrow d$ in G_L add a directed edge $s \Rightarrow d$ with $label_{s \Rightarrow d} = g$,
- otherwise, change the existing value of $label_{s \Rightarrow d}$ to $label_{s \Rightarrow d} + g$.

Go to Step 2.

Part II. Finding a strategy for grooming

Step 1) Repeat Steps 2–3 for all (source-destination) pairs $(s,d), 1 \leq s, d \leq \mathcal{N}_E, s \neq d$.

Step 2) If there is a directed edge $s \Rightarrow d$ in G_L, repeat Step 3 until it fails.

Step 3) Find the request $r \in R$ having the largest value of $\mathfrak{s}_r$ such that
- from$(r) = s$,
- to$(r) = d$,
- $\mathfrak{s}_r \leq label_{s \Rightarrow d}$.

If the search for request r is successful,
- allocate r to the edge $s \Rightarrow d$,
- set $label_{s \Rightarrow d} = label_{s \Rightarrow d} - \mathfrak{s}_r$,
- remove request r from the set of requests R.

If the search for request r is not successful, Step 3 fails.

Step 4) Repeat Steps 5–8 until R is empty.

Step 5) Find the request $r \in R$ having the largest value of $\mathfrak{s}_r$.

Step 6) Check if there is a logical path $\wp$ in G_L from from(r) to to(r), such that every edge $x \Rightarrow y$ in $\wp$ has a label $label_{x \Rightarrow y} \geq \mathfrak{s}_r$.

Step 7) If $\wp$ exists, allocate r to each logical edge in $\wp$ and for every edge $x \Rightarrow y$ in $\wp$ set $label_{x \Rightarrow y} = label_{x \Rightarrow y} - \mathfrak{s}_r$.

Step 8) Remove request r from the set of requests R.

Informally, the intent of part I of the heuristic (finding the logical topology) is to set up a lightpath from s to d if the RWA for the lightpath is feasible and there is a "large" total traffic from s to d (Steps 2 and 3). Since single-hop communication uses only one channel, this will allow the grooming heuristic to accommodate, using a single hop, as many requests from s to d, having relatively high rates of data communication, as possible.

The intent of part II of the heuristic (find the grooming strategy) is to consider, first, all requests with "large" data communication rates from s to d, where there is a lightpath from s to d (Steps 2–3). In other words, a request r from s to d with a high value of $\mathfrak{s}_r$ is assigned to a lightpath from s to d, if such a lightpath exists. This allows the heuristic to maximize the number of requests with high data communication rates to be handled using a single-hop. Multi-hop paths are considered for the remaining requests in Steps 4–8.

Exercise 9.2. Consider the physical topology shown in Figure 9.1 where $n_{ch} = 2$. Using the heuristic described in this section, find a logical topology and a grooming to support all the requests in Table 9.1. □

9.2 Dynamic Traffic Grooming

In the case of dynamic traffic grooming, since the requests come one at a time and have to be processed as they are received, the only knowledge available is the current state of the network. One approach, proposed by [442], is based on finding the shortest path from the source to the destination of the request using a graph called the *auxiliary graph* [443]. The auxiliary graph is based on the physical topology of the network and represents the current state of the network, including the lightpaths in existence and the spare capacity on each of these lightpaths. Different grooming policies may be implemented using this auxiliary graph. In this graph model, it is possible to accommodate nodes having full or limited wavelength conversion capabilities. A simplified version of this technique using some terminology different from that used in [442, 443] is given below[5].

9.2.1 A Graph Model for a Network Supporting Dynamic Traffic Grooming

The physical topology of the network is represented by the graph $G = (V, E)$ where V is the set of all nodes (routers or end nodes) and E is the set of all edges, each representing a fiber. In the auxiliary graph model of G, there will be $2(n_{ch} + 2)$ nodes corresponding to each end node in V and $2n_{ch}$ nodes corresponding to each router node in V.

Each node in the physical topology corresponds to a subgraph in the auxiliary graph. The nodes in the subgraph of an auxiliary graph G_A corresponding to a node $N_i \in V$ are as follows:

- (only if N_i corresponds to an end node) *Access nodes* A_i^{in} and A_i^{out},
- (only if N_i corresponds to an end node) *Lightpath nodes* L_i^{in} and L_i^{out},
- *Channel nodes* C_{ij}^{in} and C_{ij}^{out}, for any channel $j, 1 \leq j \leq n_{ch}$.

Nodes which correspond to incoming (outgoing) optical signals use the superscript "*in*" ("*out*"). A low data rate communication request, whose source (destination) is end node N_i, is an input to (output from) access node A_i^{out} (A_i^{in}). Any lightpath that starts from (ends at) end node N_i is depicted by a directed edge from (to) lightpath node L_i^{out} (L_i^{in}). Access nodes and lightpath nodes do not appear in router nodes since lightpaths do not start or terminate at router nodes. Channel node C_{ij}^{in} (C_{ij}^{out}) are used to represent the fact that there is a fiber carrying incoming signals to (outgoing signals from) node N_i, and an edge to (from) C_{ij}^{in} (C_{ij}^{out}) means that there is a channel on a fiber to (from) N_i on which channel j is available (i.e., channel j has not been allocated to any lightpath that uses the fiber).

Each edge in an auxiliary graph has a length and, for convenience in describing algorithms on the graph, is denoted by a name. The length of an edge is the label for the edge. The edges of the graph and the lengths of the edges vary as new requests for communication are successfully processed and as existing requests for communication terminate. As in Section 9.1, $\mathfrak{s}_r$ will

---

[5] This description assumes that fixed-frequency transmitters and receivers are being used. Modifying this to handle tunable transmitters and receivers is simple.

be used to denote the data communication rate for request r. A description of the edges is given below:

1. The edge from channel node C_{ij}^{in} to channel node C_{ij}^{out} is called a *Channel Bypass Edge* $\mathcal{C}_{ij}$. This edge means that a lightpath using channel j entering node N_i may always be routed to another node using the same channel if other conditions are satisfied. This edge exists in all nodes N_i, for any channel j, and has length 0.

2. If N_i is an end node, the edge from access node A_i^{in} to A_i^{out} is called a *Grooming Edge* $\mathcal{G}_i$. This edge means that all low-rate communication, which uses N_i as an intermediate node in a multi-hop route to its destination, may form part of the payload of a lightpath that starts from node N_i. This edge exists in all end nodes and has length 0.

3. If N_i is an end node, the edge from the access node A_i^{out} to the lightpath node L_i^{out} is called a *mux edge* $\mathcal{M}_i$. This edge means that all low rate communication, which starts from end node N_i or uses N_i as an intermediate node in a multi-hop route to its destination, has access, through a multiplexer, to a lightpath that starts from node N_i. This edge exists in all end nodes and has a length 0.

4. If N_i is an end node, the edge from the lightpath node L_i^{in} to the access node A_i^{in} is called a *demux edge* $\mathcal{D}_i$. This edge means that all low-rate communication which ends at end node N_i or uses N_i as an intermediate node in a multi-hop route to its destination has access to the communication, through a demultiplexer, from a lightpath that ends at node N_i. This edge exists in all end nodes and has length 0.

5. If N_i is an end node, and there is at least one transmitter that is tuned to channel j, there is an edge from the access node A_i^{out} to the channel node C_{ij}^{out}. This edge is called a *Transmitter Edge* $\mathcal{T}_{ij}$. This edge means that it is possible to communicate data, from node N_i, by setting up a new lightpath which uses channel j. The length of transmitter edge $\mathcal{T}_{ij}$ gives the number of transmitters in node N_i that are tuned to the wavelength corresponding to channel j.

6. If N_i is an end node, and there is a receiver that is tuned to channel j, there is an edge from the channel node C_{ij}^{in} to the access node A_i^{in}. This edge is called a *Receiver Edge* $\mathcal{R}_{ij}$. This edge means that it is possible to receive, at node N_i, data carried by a new lightpath which uses channel j. The length of receiver edge $\mathcal{R}_{ij}$ gives the number of receivers in node N_i that are tuned to the wavelength corresponding to channel j.

7. If, at node N_i, there is a wavelength converter that can change the channel used by a lightpath passing through N_i from j to k, then there is an edge from channel node C_{ij}^{in} to C_{ik}^{out}. Such an edge is called a *Converter Edge* $\mathcal{X}_{ijk}$. This edge means that a lightpath using channel number j on an incoming fiber to node N_i may have its channel number converted to k before the lightpath is communicated to another node. The length of this edge is 0.

8. There is an edge from the channel node C_{ij}^{out} to the channel node C_{kj}^{in} if there is a fiber from N_i to N_k with an unused channel j. Such an edge is called a *Channel Availability Edge* C_{ikj}. This edge means that a new lightpath may use channel j on the fiber from node N_i to N_j. The length of each channel availability edge is $M \times g$ where M is a constant larger than the number of nodes $\mathcal{N}$ in the network and g is the capacity of a lightpath (using the Optical Carrier level OC-n notation) [6].

9. If there is a lightpath from end node N_i to end node N_j, in the auxiliary graph there is a *Lightpath Edge* $\mathcal{L}_{ij}$ from L_i^{out} to L_j^{in}. The length of a lightpath edge is the residual capacity of the corresponding lightpath. In other words, if $\sum \mathfrak{s}_r$ is the sum of all requests r that use the lightpath corresponding to $\mathcal{L}_{ij}$, then the length of $\mathcal{L}_{ij}$ will be $g - \sum \mathfrak{s}_r$.

Figure 9.4 shows the subgraph of the auxiliary graph that corresponds to end node N_i in a network with grooming facility. Since N_i is an end node, the corresponding subgraph has access nodes (A_i^{in} and A_i^{out}) and lightpath nodes (L_i^{in} and L_i^{out}). In this example, each fiber has two channels, having numbers 1 and 2, so that node N_i has channel nodes C_{i1}^{in}, C_{i1}^{out}, C_{i2}^{in}, and C_{i2}^{out}.

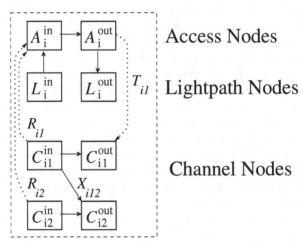

Fig. 9.4. The sub-graph of the auxiliary graph corresponding to end node N_i in the physical topology

There is a transmitter edge $\mathcal{T}_{i1}$ from A_i^{out} to C_{i1}^{out} with label 1, meaning that there is an available transmitter in node N_i, tuned at the wavelength corresponding to channel 1, that can be used to set up a lightpath that starts from N_i using channel 1. There are two receiver edges, both with labels 1. Receiver edge $\mathcal{R}_{i1}$ ($\mathcal{R}_{i2}$) is from C_{i1}^{in} (C_{i2}^{in}) to A_i^{in} meaning that there is an available receiver in node N_i, tuned at the wavelength corresponding to

---

[6] Later on, in this section, it will be clear why the label $M \times g$ is appropriate.

channel 1 (2), that can be used to set up a lightpath that terminates at N_i using channel 1 (2). There is a converter edge $\mathcal{X}_{i12}$ from C_{i1}^{in} to C_{i2}^{out}, meaning that there is a wavelength converter in node N_i which can convert the carrier wavelength corresponding to channel 1 to the wavelength corresponding to channel 2. In other words, a lightpath that uses channel 1 on an incoming fiber to N_i is allowed to change to channel 2 on an outgoing fiber from N_i.

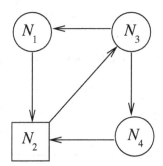

Fig. 9.5. The physical topology of a network

The physical topology of a network with end nodes N_1, N_3, N_4, and a router node N_2 is shown in Figure 9.5. In Figure 9.5, an end node (router node) is shown by a circle (square). Each end node has a transmitter and a receiver tuned to channel c_1 (c_2). Each fiber in the network supports two channels. Currently, there are two lightpaths L_1 and L_2 in the network. Lightpath L_1 (L_2) is from N_1 to N_3 (N_3 to N_4) using the path $N_1 \rightarrow N_2 \rightarrow N_3$ ($N_3 \rightarrow N_4$) and channel c_1 (c_2) as shown in Figure 9.6. Let each lightpath have a data transmission rate of OC-48. These lightpaths are sufficient to handle the following requests for communication:

1. An OC-24 request for communication from N_1 to N_3 using lightpath L_1,
2. An OC-6 request for communication from N_3 to N_4 using lightpath L_2,
3. An OC-12 request for communication from N_1 to N_4 using a multi-hop path consisting of lightpath L_1, followed by lightpath L_2.

Figure 9.7 shows the auxiliary graph corresponding to the network with two lightpaths shown in Figure 9.6.

For Figure 9.7, an explanation of the labels of selected edges, representing the lengths of those edges, is given below. Since the number of nodes $\mathcal{N} = 4$ and M must be greater than $\mathcal{N}$, in this discussion, M is chosen to be 5. The intent of the algorithm for servicing a new request (Given in Section 9.2.2) is that, when handling a new request, existing lightpaths must be used, if possible. The maximum value of the label of an existing lightpath is g, the

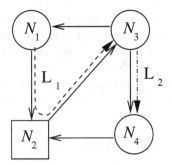

Fig. 9.6. Two lightpaths using the physical topology shown in Figure 9.5

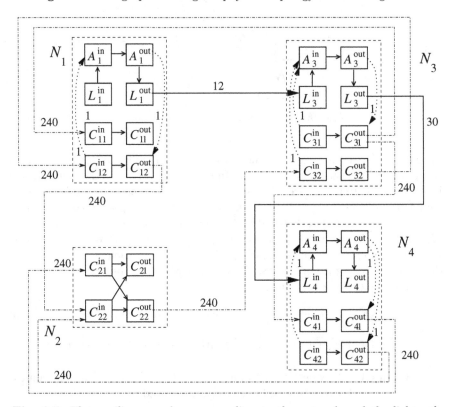

Fig. 9.7. The auxiliary graph corresponding to the network and the lightpaths shown in Figure 9.6

capacity of a lightpath. If a multi-hop logical path is used from the source to the destination of a request, the number of logical edges used in the multi-hop path cannot exceed $\mathcal{N}$, the number of nodes (end nodes or routers) in the network. Therefore the length of a multi-hop logical path from the source to the destination of a request cannot exceed $M \times g$. Since the algorithm for servicing a new request is based on a shortest-path algorithm, the label of $M \times g$ can be safely used on an unutilized channel on a fiber. In other words, if a new lightpath is set up, each channel used by the lightpath has a cost of $M \times g$. This explains why there is a label of $M \times g = 48 \times 5 = 240$ on the edge from C_{31}^{out} to C_{11}^{in}, an unutilized channel at the moment. Lightpath L_1 is used to support the OC-24 request from N_1 to N_3 and the OC-12 request from N_1 to N_4. The residual capacity of lightpath L_1 is $48 - 24 - 12 = 12$. This explains why there is a label of 12 on the edge from L_1^{out} to L_3^{in}. Since lightpath L_1 used channel c_1, the transmitter (receiver) in node N_1 (N_3) is used for this lightpath and there is no transmitter (receiver) edge from A_1^{out} to C_{11}^{out} (C_{31}^{in} to A_3^{in}). Since node N_1 still has a transmitter, tuned to channel c_2, there is an edge from A_1^{out} to C_{12}^{out}. The label on this edge is one since there is only one transmitter tuned to this channel.

Exercise 9.3. Show that all the other edges in Figure 9.7 are appropriate and the labels on the edges follow the scheme given above.   □

9.2.2 Algorithms for Supporting Dynamic Traffic Grooming

To support dynamic traffic grooming, it is necessary to

- create an initial auxiliary graph corresponding to the network,
- update the graph when a new request for communication at a specified rate arrives,
- modify the graph when an existing data communication request terminates.

The algorithms for handling a new request or the end of a request are "greedy" in the sense that the intent is to get the best possible scheme for handling the new request without perturbing any existing connection. In the following description, the objective is to create as few lightpaths as possible and each new lightpath is created using the shortest path. If possible, a request for communication will be handled without creating any new lightpaths. Algorithms for these operations are given below.

Creating the initial auxiliary graph

The construction of the initial auxiliary graph follows directly from the definition of different edges of the auxiliary graph described in Section 9.2.1.

The initial auxiliary graph corresponding to the physical topology shown in Figure 9.5 is shown in Figure 9.8. In this auxiliary graph, the edges without a specified length have length 0.

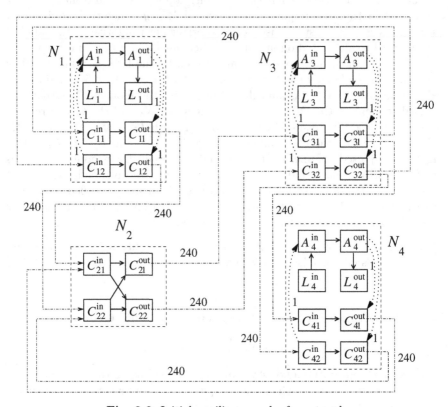

Fig. 9.8. Initial auxiliary graph of a network

Processing a new request for communication

The steps of the algorithm to process a request r for communication from $s =$ from(r) to $d =$ to(r) are given below. The data communication rate of request r is $\mathfrak{s}_r$, using the Optical Carrier level OC-n notation.

Step 1) Find the shortest path $\mathfrak{P}$ from the node A_s^{out} to the node A_d^{in}, in the auxiliary graph, if such a path exists. In this search, lightpath edges and the channel availability edges whose length is less than $\mathfrak{s}_r$ will be ignored. If such a path does not exist, the request r has to be blocked, possibly to be attempted some time later, with the expectation that one or more existing communications will be over by the time the new attempt is made and a path from s to d will be found.

Step 2) If path $\mathfrak{P}$ does not include any transmitter edge (or equivalently, any receiver edge), then go to Step 8, after setting $\mathfrak{P}_{new} = \mathfrak{P}$. In this situation the existing lightpaths are sufficient to handle the request r.

Step 3) For each subpath $\widehat{p} \in \mathfrak{P}$ in the form $\widehat{p} = A_i^{out} \rightarrow C_{ij}^{out} \rightarrow \ldots \rightarrow C_{qk}^{in} \rightarrow A_q^{in}$, from end node N_i to end node N_q, repeat Steps 4–6.

Step 4) Set up a lightpath from node N_i to N_q using channels as specified by the channel nodes in the subpath $\widehat{p}$. In the auxiliary graph, create a lightpath edge $\mathcal{L}_{iq}$ from lightpath node L_i^{out} to L_q^{in} having a length g.

Step 5) Delete, from the auxiliary graph, all the channel availability edges that appear in path $\widehat{p}$.

Step 6) Reduce, by 1, the length of the transmitter (receiver) edge $\mathcal{T}_{ij}$ ($\mathcal{R}_{qk}$) from access node A_i^{out} to channel node C_{ij}^{out} (from channel node C_{qk}^{in} to access node A_q^{in}). If the length of the edge $\mathcal{T}_{ij}$ ($\mathcal{R}_{ik}$) becomes 0, delete the edge from the auxiliary graph.

Step 7) Find the shortest path $\mathfrak{P}_{new}$ from the node A_s^{out} to the node A_d^{in}.

Step 8) Reduce the lengths of all the lightpath edges in $\mathfrak{P}_{new}$ by the data communication rate $\mathfrak{s}_r$ of request r.

Exercise 9.4. Show how a request for communication from N_1 to N_4 at OC-24 rate will be handled by the algorithm for dynamic traffic grooming. □

Exercise 9.5. Propose an algorithm to handle the situation when a communication terminates. (Hint: if there is no traffic on some of the lightpaths when the communication terminates, the lightpath may be dismantled.) □

Exercise 9.6. Propose an algorithm to handle requests in such a way that a dedicated protection scheme is used to take care of faults. □

Exercise 9.7. What modifications are needed in the heuristic described in 9.1.3 to handle the case of faults using shared-path protection? □

9.3 Bibliographic Notes

9.3.1 Books and Surveys on Traffic Grooming

A good introduction to traffic grooming, primarily for ring networks, appears in [279]. A comprehensive survey and classification of traffic grooming strategies is available in [118]. This includes possible cost functions, an ILP formulation and a review of techniques for traffic grooming, with an emphasis on ring structures. A survey of techniques for traffic grooming in ring and mesh networks has been covered in [397]. Traffic grooming for non-bifurcated traffic grooming appears in [202]. In [203], a useful categorization of important problems in dynamic grooming has been presented, along with some formulations for such problems. In [118, 203] traffic grooming with bifurcation has been considered. Detailed surveys of a number of traffic grooming strategies are available in [361, 449]. This includes static, dynamic and survivable traffic grooming.

9.3.2 Traffic Grooming on Rings

In [170] the authors have considered the problem of traffic grooming, considering a number of architectures based on the ring architecture. The maximum number of lightpaths on a fiber, the cost of the OLTs and the maximum number of hops are parameters to be minimized in [170]. The performances of different architectures for these cost metrics have been studied. In [54, 82, 86, 117, 438], traffic grooming in WDM/SONET ring networks, to minimize the cost of electronic add/drop multiplexers (ADMs), have been considered. It has been established that the traffic grooming problem, in general, is NP-complete, even for rings [86]. The need to specify a bound on the cost of individual nodes has been considered in [82]. It is not desirable to have a situation where some nodes have a relatively large number of OLTs. The objective of the formulation in [82] was to minimize the number of OLTs at the node where this cost is maximum. In [117] the logical topology design problem has been handled by a heuristic which achieves progressive tightening of the upper and lower bounds. The problem of traffic grooming for WDM ring networks has been studied in [398] and four different interconnection strategies for ring networks have been considered. Mathematical formulations of the traffic grooming problem for each of the four architectures have been presented and improvements to existing heuristics for traffic grooming have been outlined in [398]. The problem of minimizing the cost of OLTs in ring networks has been considered in [200], where the problem of non-bifurcated traffic grooming has been formulated as an integer linear program (ILP), and then converted into a simpler mixed ILP (MILP) by converting many of the integer variables to continuous variables, giving a form that is faster to solve.

Single-source multicast communication in WDM rings to minimize the number of ADMs, using traffic grooming, has been considered in [63]. It was shown that the problem is NP-complete, but a special case can be solved in polynomial time. Approximation algorithms and heuristics are provided for the general case.

9.3.3 Traffic Grooming Strategies

In [229] the problem of traffic grooming to reduce the number of transceivers in optical networks has been considered and a heuristic, based on the principle of duality, has been proposed. The objective was to determine an optimum logical topology, integrated with the problem of routing.

In [115] it was shown that the traffic grooming problem is NP-complete even for elementary structures such as the path and star networks. The paper considers both bifurcation and non-bifurcation models. The RWA in these two topologies is trivial, and thus it was shown that traffic grooming is an inherently difficult problem. The paper also shows how to compute lower and upper bounds for traffic grooming using heuristics.

An analysis of the upper bound on the variations of the fragmented capacities in STS-48 channels with the grooming capability of a network appears in [104]. Two upper bounds have been derived in [104], one of which is less tight but is easier to compute than the other. Different network scenarios have been considered and it was shown that the derived upper bounds are in agreement with results obtained using efficient routing and grooming algorithms.

The dynamic traffic grooming problem for WDM mesh networks, using the fixed-alternate-routing approach, has been studied in [421]. The paper proposes the fixed-order grooming (FOG) algorithm. A number of different grooming policies are introduced for route-selection, as well as for addressing wavelength and transceiver constraints. The results show that multi-hop traffic grooming outperforms single-hop traffic grooming and the performance of the FOG algorithm improves with the number of hops allowed in a route.

A Call Admission Control (CAC) mechanism for service differentiation in WDM networks with grooming capabilities and a heuristic with lower computational complexity has been proposed in [284]. A Markov decision process has been used to derive an optimal CAC and results indicate that the approach in [284] gives improvement over standard policies.

A queueing network model to calculate the blocking probabilities in traffic grooming optical networks, modeled as *multi-level overflow systems*, has been studied in [401]. A method that decomposes the queueing network into subsystems and analyzes each subsystem was shown to provide a good approximate analysis of large overflow systems supporting multiple traffic sources.

The capabilities of the nodes in an optical network may vary from one node to another, depending on grooming and wavelength conversion capabilities. When upgrading a network, it is important to determine where additional capabilities would have the maximum impact. In [368], an analytical model to evaluate the blocking performance of a WDM grooming network with heterogeneous grooming capability has been considered. The results of the analytical model has been compared in [368] to that obtained by simulation.

In [367] the dynamic routing of sub-wavelength traffic, based on request characteristics was discussed. A request-specific routing scheme, called *Available Shortest Path Routing* (ASPR), has been proposed. Simulation results indicate that ASPR improves the performance of the network, with respect to utilization and fairness metrics, as compared to destination-specific routing schemes.

Apart from the review of traffic grooming already mentioned, [397] also proposes an ILP formulation of the traffic grooming problem and a heuristic based on simulated annealing. It is shown that the multi-hop approach gives better utilization of transponders but requires more network bandwidth.

The analysis of traffic grooming in mesh WDM networks, using the client call blocking probability, appears in [413]. The paper describes a heuristic for single hop traffic grooming, followed by an exact and an approximation model for link blocking.

In [444], the problem of static traffic grooming using single hop and multi-hop communication has been studied. The paper gives an ILP formulation and two heuristics. The heuristic given in this chapter is based on this work.

In [437] the issue of changes in the traffic pattern over time and algorithms on how to respond to such changes by dynamic reconfiguration of the traffic grooming strategy has been considered. A heuristic, based on the Tabu search, and an ILP formulation have been proposed for the problem. Similar problems have also been considered in [397].

The characteristics and the performances of optical cross-connects for traffic grooming have been investigated in [447]. A number of grooming schemes have been proposed and the performances of different optical grooming cross-connects in dynamic environments have been studied in [447].

In [446] it was shown that every node in a network need not have grooming capability. It is possible to have "sparse grooming" — a few nodes with grooming facilities and yet have good utilization of the network. The dynamic traffic grooming problem for WDM mesh networks with sparse placement of grooming nodes has been studied in [198]. The problem of appropriately placing grooming nodes is formulated as a weighted minimum dominating set problem on graphs. Simulations on the NSFNET topology indicate that, with proper selection of grooming nodes, benefits of full grooming can be achieved with fewer nodes equipped with such grooming capability.

An ILP that determines the wavelength assignment and the grooming for a given set of connections, with specified routes and a given grooming factor is given in [246]. A heuristic based on binary search and LP relaxation of the ILP is also proposed.

A WDM backbone network is usually a multi-vendor, heterogeneous network since network equipment may come from different vendors with legacy equipment coexisting with new equipment. Some nodes may have optical cross-connects or wavelength conversion and traffic-grooming capabilities. In [448], the problem of efficiently provisioning connections of different bandwidth granularities in a heterogeneous WDM mesh network for dynamic traffic-grooming has been considered.

An analytical model for computing the blocking probability of WDM networks with grooming capabilities has been presented in [368]. The model uses an approximation for the distribution of the number of calls and is computationally less complex compared to exact models. Simulation results on different networks and with different link-load correlations are used to validate the model.

The static traffic grooming problem in IP over WDM framework has been considered in [141]. An ILP formulation, where the objective is to minimize the number of transceivers, has been presented for the problem. The ILP becomes computationally intractable for large networks. A heuristic, based on traffic aggregation, has been proposed for large networks. The heuristic is shown to improve significantly wavelength utilization, in both ring and mesh torus topologies.

The problem of joint reconfiguration and traffic grooming in WDM optical networks has been addressed in [265]. It was shown that splitting the overall problem into two subproblems and solving the two subproblems sequentially may be inadequate. An ILP and a heuristic for solving the integrated problem has been presented in [265].

In [411] a minimum edge-cost network flow approach for solving static and dynamic traffic grooming problems has been adopted. This has been achieved by appropriate graph transformation and assignment of edge weights, and then solved using an efficient heuristic algorithm.

Two new grooming concepts called *lightpath dropping* and *lightpath extension* have been considered in [142]. These concepts are based on the light-trail network architecture, discussed briefly in Section 3.5.2, in which incoming optical signals can be dropped at a node, while optically continuing to the next node [183]. Traffic grooming in light-trail networks has also been studied in [204, 392, 425].

9.3.4 Fault-Tolerant Traffic Grooming

The idea of fault tolerance in WDM networks has been described in Chapter 8 where path protection, using either dedicated path protection or shared-path protection has been discussed in detail. In Chapter 8, the idea was that, in the case of a fault, for each primary lightpath using the faulty component, the corresponding backup lightpath will be set up. All data carried by each primary lightpath affected by the fault will be switched over to the corresponding backup lightpath.

In shared-path protection, as described in Chapter 8, some significant saving in the overhead for the backup lightpaths was achieved by specifying that a backup lightpath B_1 can share both a channel and a fiber with another backup lightpath B_2, provided that the route of the primary lightpath P_1, corresponding to backup lightpath B_1, is edge-disjoint with respect to the route of the primary lightpath P_2, corresponding to backup lightpath B_2. In the case of traffic grooming, the interest is in guaranteeing that each low-rate data communication will continue when a fault occurs. Let a primary lightpath P_1 (P_2) carry a total payload of $P_1^{payload}$ ($P_2^{payload}$). Two cases are to be considered:

Case 1) If P_1 and P_2 are edge-disjoint, then the corresponding backup lightpaths B_1 and B_2 are allowed to share a channel and a fiber provided that the maximum possible payload of B_1 and B_2 is at least $max(P_1^{payload}, P_2^{payload})$.

Case 2) If P_1 and P_2 are not edge-disjoint, then the corresponding backup lightpaths B_1 and B_2 are allowed to share a channel and a fiber provided that the maximum possible payload of B_1 and B_2 is at least $(P_1^{payload} + P_2^{payload})$.

The logic behind the first situation is similar to that used in Chapter 8. If the routes of P_1 and P_2 are not edge-disjoint, and an edge used by both P_1 and P_2 fails, both P_1 and P_2 will become inoperative. If the spare payload of B_1 and B_2 is at least $(P_1^{payload} + P_2^{payload})$, then the payload of both disrupted lightpaths can be carried by the backup lightpath. This idea has been used in the formulation given in [140], for shared-path protection, where each node in the network (each node in this study is an end node and therefore has grooming facilities) and each communication is handled by a single hop. Two edge and node disjoint routes through the physical topology, ρ_{xy}^1 and ρ_{xy}^2, are pre-computed for each pair of end nodes (E_x, E_y). This means that the only choice is to determine the logical topology and to assign channel numbers to the primary and the backup lightpaths and, if a decision is taken to establish a logical edge from E_x to E_y, the routes for the primary and the backup lightpaths will be ρ_{xy}^1 and ρ_{xy}^2 or vice-versa.

In [297], survivable traffic grooming in the dynamic allocation model has been studied. The following approaches have been considered for grooming a connection request with shared-path protection:

   i) protection-at-lightpath (PAL) level,
  ii) mixed protection-at-connection (MPAC) level, and
 iii) separate protection-at-connection (SPAC) level.

Some heuristics for these three approaches have been proposed and it was observed that

   i) it is beneficial to groom working paths and backup paths separately, as done in PAL and SPAC,
  ii) SPAC yields the best performance when the number of grooming ports is sufficient, and
 iii) PAL achieves the best performance when the number of grooming ports is moderate or small.

The survivable traffic grooming (STG) problem using path protection has been studied in [422], using dedicated protection and shared-path protection. The paper looks at survivability under single failures (such as fiber cuts) as well as duct-cuts and some multiple failure scenarios. An ILP and some heuristics have been proposed in [422] to address these problems.

The problem of designing survivable traffic groomed optical WDM mesh networks has been considered in [139]. The paper introduces the concept of *quality of protection framework*, where the backup bandwidth for each connection can be less than or equal to the primary bandwidth. This leads to the concept of partial protection, which can reduce backup capacity requirements. An ILP has been presented in [139] for the static traffic grooming problem. The dynamic traffic grooming case has been analyzed using discrete event simulation models.

APPENDIX 1

Linear Programming in a Nutshell

Mathematical programming is a widely used family of techniques for making decisions under the restriction of limited resources. The first step in approaching a decision problem is to identify its two important components [377]:

1. *Objective*,
2. *Variables*.

The specification of a problem to be solved using mathematical programming is called the *formulation* of the problem. The objective is the desired end result, typically expressed as the maximization or minimization of some mathematical function. Such functions are often called *objective functions*. The variables are those parameters that are under the control of the designer and hence may be allowed to change (or vary) in order to reach the objective. The most popular type of problem formulation is called *linear program* where the objective function as well as the restrictions (often called *constraints*) are linear functions of the variables.

The standard revised simplex algorithm for processing a Linear Program (LP) to minimize an objective function subject to a number of linear constraints is given below. In this description of a linear program *FORM* 11, only the simplest form of the constraints has been considered. This is a concise description and more details, with proof of correctness, are available in standard textbooks such as [93, 377].

Notation used in *FORM* 11

$\widehat{n}_c$: the number of constraints in *FORM* 11.

$\widehat{n}_v$: the number of variables in *FORM* 11.

A : a matrix of size $\widehat{n}_c \times \widehat{n}_v$ where each element in the matrix is a constant, used to specify the constraints of *FORM* 11.

$[A|I]$: a matrix of size $\widehat{n}_c \times (\widehat{n}_v + \widehat{n}_c)$ such that the first $\widehat{n}_v$ columns of $[A|I]$ are taken from A and the remaining $\widehat{n}_c$ columns of $[A|I]$ are taken from identity matrix I.

$\overrightarrow{A^j}$: the jth column of $[A|I]$.

a_{ij} : the element in row i, column j of A — a constant.

$\overrightarrow{b}$: a column vector of $\widehat{n}_c$ nonnegative constants.

b^k : the kth element of vector $\overrightarrow{b}$.

$\overrightarrow{c}$: a row vector of $\widehat{n}_v$ constants.

$\overrightarrow{c_B}$: a row vector of $\widehat{n}_c$ constants.

$\overrightarrow{x}$: a column vector of $\widehat{n}_v$ variables.

$\overrightarrow{y}$: a row vector of $\widehat{n}_c$ variables.

$\overrightarrow{d}$: a column vector of $\widehat{n}_c$ variables.

x^i : the ith element of vector $\overrightarrow{x}$.

$\overrightarrow{x_B}$: a column vector of $\widehat{n}_c$ variables, called basic variables.

x_B^i : the ith element of $\overrightarrow{x_B}$.

B : a nonsingular matrix of size $\widehat{n}_c \times \widehat{n}_c$.

I : identity matrix of size $\widehat{n}_c \times \widehat{n}_c$.

$\overrightarrow{x_s}$: a column vector of $\widehat{n}_c$ variables.

x_s^i : the ith element of x_s, also called the ith *slack variable*.

$[\overrightarrow{x|x_s}]$: a column vector of $(\widehat{n}_v + \widehat{n}_c)$ elements, such that the first $\widehat{n}_v$ elements of $[\overrightarrow{x|x_s}]$ are taken from $\overrightarrow{x}$ and the remaining $\widehat{n}_c$ elements of $[\overrightarrow{x|x_s}]$ are taken from $\overrightarrow{x_s}$.

$\overrightarrow{0}$: a vector of all 0s.

The formulation for *FORM* 11

The input to the LP is a set of $\widehat{n}_c$ linear constraints and an objective function. The ith constraint is in the form $\sum_{j=0}^{\widehat{n}_v} a_{ij} x^j \leq b^i$, for all $i, 1 \leq i \leq \widehat{n}_c$. The cost function is in the form $\sum_{j=0}^{\widehat{n}_v} c_j x^j$.

Using matrix notation, the LP may be conveniently specified as follows:

Objective function:
$$\text{minimize} \quad \vec{c} \cdot \vec{x} \tag{A1.1}$$

subject to
$$A\vec{x} \leq b \tag{A1.2}$$

$$x^i \geq 0 \qquad \forall i, 1 \leq i \leq \widehat{n}_v \tag{A1.3}$$

The constraints in (A1.2) are called *linear constraints* because the left-hand side of each constraint (i.e., each row of $A\vec{x}$) is a linear function of the variables in $\vec{x}$. The first step is to add *slack* variable x_s^i in the ith constraint, for all $i, 1 \leq i \leq \widehat{n}_c$, in (A1.2) and replace the $\leq$ symbol by the equality symbol so that the LP now becomes

Objective function:
$$\text{minimize} \quad \vec{c} \cdot \vec{x} \tag{A1.4}$$

subject to
$$[A|I][\overrightarrow{x|x_s}] = \vec{b} \tag{A1.5}$$

$$x^i \geq 0 \tag{A1.6}$$

$$x_s^i \geq 0 \tag{A1.7}$$

It is well known that it is possible to obtain an $\widehat{n}_c \times \widehat{n}_c$ non-singular sub-matrix B of $[A|I]$ and a nonnegative vector $\overrightarrow{x_B}$ by choosing $\widehat{n}_c$ suitable variables from $[x|x_s]$ for inclusion in $\overrightarrow{x_B}$ and by selecting the corresponding columns of $[A|I]$ for inclusion in B [93], satisfying the following condition:

$$B\overrightarrow{x_B} = \vec{b} \tag{A1.8}$$

The matrix B is called the *basis*, the variables in $\overrightarrow{x_B}$ are called *basic variables*. (A1.8) can be solved to get a unique solution since B is non-singular. The variables in $\vec{x}$ or in $\vec{x}_s$ which are not included in $\overrightarrow{x_B}$ are called *nonbasic variables*. It is convenient to form a row vector $\overrightarrow{c_B}$ of all coefficients in vector $\vec{c}$ corresponding to the variables in $\overrightarrow{x_B}$. A solution of the LP may be found by setting all the nonbasic variables to 0. The value of the objective function is now $\overrightarrow{c_B} \cdot \overrightarrow{x_B}$. This process of determining the matrix B, the vector of variables $\overrightarrow{x_B}$, and the cost vector $\overrightarrow{c_B}$ before starting the iterations of the revised simplex method is called finding an *initial feasible solution (IFS)*. In the simple form considered here, an IFS is $B = I$ (the identity matrix), $\overrightarrow{x_B} = \vec{x}_s$ (the vector of slack variables), and $\overrightarrow{c_B} = \vec{0}$.

Once the initial feasible solution has been obtained, the idea is to improve the value of the objective function, if possible, in successive iterations of the revised simplex method until no further improvement in the value of the objective function is possible. In each iteration, one column of the basis is replaced by a column from $[A|I]$, currently not appearing in the basis. The column being replaced is called the *leaving column* and the column replacing it is called the *entering column*. The variable in $[x|x_s]$ corresponding to the leaving column is called the *leaving variable* and the variable corresponding to the entering column is called the *entering variable*. The steps in one iteration of the revised simplex method are as follows.

Step 1: Solve the equation $\vec{y} \cdot B = \vec{c_B}$.

Step 2: Find, if possible, a column $\vec{A^j}$ from $[A|I]$ such that the $\vec{y} \cdot \vec{A^j} > c_j$. If no such column can be found, stop. Otherwise, $\vec{A^j}$ is the entering column that has to be included in the basis.

Step 3: Solve the equation $B \cdot \vec{d} = \vec{A^j}$.

Step 4: Find the largest t such that $\vec{x_B} \geq t\vec{d}$. The component of $\vec{x_B} - t\vec{d}$ which is 0 is the leaving variable and the corresponding column in B is the leaving column.

Step 5: In the basis B, replace the leaving column by the entering column. Recompute $\vec{x_B}$ using the formula by $\vec{x_B} = B^{-1}\vec{b}$. Replace the cost of the leaving column in $\vec{c_B}$ by the cost of the entering column.

The elements in $\vec{y} = \vec{c_B}B^{-1}$, computed in Step 1, are called *simplex multipliers*. The ith simplex multiplier is the ith element in $\vec{y}, 1 \leq i \leq \hat{n}_c$.

In many problems, the variables must have integer values only. This situation occurs frequently in optical network design. Such problems are called *Integer Linear Programs* (ILPs). If a formulation contains some continuous variables and some integer variables, such formulations are called Mixed Integer Linear Programs (MILPs).

APPENDIX 2

The de Bruijn Graph

One interconnection architecture that has been widely studied is the de Bruijn graph [199, 240]. A very brief overview of the de Bruijn graph and related terms in graph theory is given below. It is convenient to model an interconnection architecture using a directed graph $G = (V, E)$ where each processor is a vertex in the set of vertices V in the graph G, and E is the set of edges in G. If two processors u and v are connected by a link allowing communication from u to v, there is a directed edge $e = u \rightarrow v$ in G. The edge e is an element in the set of edges E. Given two vertices $s, d \in V$, a *walk* W is a sequence of edges represented as $W = x_0 \rightarrow x_1 \rightarrow x_2 \ldots \rightarrow x_{k-1} \rightarrow x_k$, where $s = x_0$, $d = x_k$, and there is an edge $e = x_i \rightarrow x_{i+1} \in E$, for all $i, 0 \le i < k$. The length of this walk from s to d is k. A walk may involve a loop, meaning that the same vertex may appear more than once in a walk. A *path* is a walk where no vertex appears more than once. In general, there are many paths from any node s to any node d.

Using standard graph theoretic terminology [42], the *in-degree* (*out-degree*) of a vertex v will denote the number of edges to (from) v. Given two vertices s and d, a *routing algorithm* is used to construct a path from s to d. In most cases, the routing algorithm gives the shortest directed path from s to d. In a multiprocessor system, for example, such an algorithm is very useful to enable any processor to communicate with any other processor as quickly as possible. In characterizing graphs, the *diameter* is an important property and is defined as the largest value of the shortest path between any pair of vertices.

Topologies such as the de Bruijn graph are attractive because they have a low diameter ($O(\log_d N_{DB})$ where N_{DB} is the number of vertices in the graph and d is the in-degree and out-degree of every vertex). The de Bruijn graph is an example of a *regular* graph where the in-degree and out-degree of each vertex is the same. An attractive property of many regular graphs, including the de Bruijn graph, is that the interconnection between the vertices follows some simple rules so that the routing algorithm for communication between vertices is quite straightforward and does not involve the need for complex routing tables.

Interconnection rules and routing in a de Bruijn graph

A *de Bruijn* graph with in-degree and out-degree d and a diameter k, denoted by $B(d,k)$, is a directed graph with d^k vertices, where the vertices are represented by strings of length k, choosing the symbols of the string from the set $Z_d = \{0, 1, 2, \ldots, d-1\}$. Let vertex u (v) be a vertex in $B(d,k)$, represented by the string of k digits $u_k u_{k-1} \ldots u_1$ $(v_k v_{k-1} \ldots v_1)$, where $u_i \in Z_d$ $(v_i \in Z_d)$, for all $i, 1 \leq i \leq k$. There will be an edge $u \to v$ in $B(d,k)$ iff $v_i = u_{i-1}, 1 < i \leq k$. Here, the value of v_1 can be any digit i in Z_d. The edge $u_k u_{k-1} \ldots u_1 \to u_{k-1} u_{k-2} \ldots u_1 i$ will be called the ith edge of vertex $u_k u_{k-1} \ldots u_1$.

$B(d,k)$ is regular, since the rule for the existence of edge $u \to v$ is the same for all pairs of vertices (u, v). Each vertex has in-degree and out-degree d. There are d vertices with self-loops.[7] It may be readily verified that if a vertex in $B(d,k)$ is represented by a string $iii \ldots i, 0 \leq i < d$, then the vertex has a self-loop.

It is convenient to associate numbers $0, 1, 2, \ldots, d^k - 1$ with the vertices of $B(d,k)$. If n is the number associated with a vertex u represented by the string $u_k u_{k-1} \ldots u_1$, then $n = \sum_{i=1}^{k} u_i \cdot d^{i-1}$.

Example A2.1. The diagram of a de Bruijn graph $B(d,k)$ with $d = 2$ and $k = 3$ is shown in Figure A2.1. Each vertex has a vertex number (shown on the first line inside the square representing the vertex) and a string representing the vertex (shown on the second line inside the square representing the vertex). Vertex number 5 has the string representation 101 since $5 = 1 \times 2^2 + 0 \times 2^1 + 1 \times 2^0$. The vertex with the string representation 010 has directed edges to two vertices with string representations 100 and 101, following the rules of interconnection given above. The vertices with string representations 000 and 111 have one edge to themselves. These are the self-loops.   □

A somewhat more complex de Bruijn graph, $B(3, 2)$, is shown in Figure A2.2.

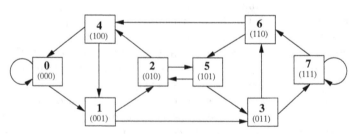

Fig. A2.1. The de Bruijn graph $B(2,3)$ with eight vertices

---

[7] A *self-loop* is any edge $u \to u$ for some $u \in V$.

Since the string representation of a vertex in a de Bruijn graph $B(d, k)$ identifies the vertex, it is convenient to refer to the string representation as the *address* to identify a vertex. Let any two vertices u and v have addresses $u_k u_{k-1} \ldots u_1$ and $v_k v_{k-1} \ldots v_1$. Clearly, there is a walk in $B(d, k)$ of length k given by $u_k u_{k-1} \ldots u_1 \rightarrow u_{k-1} u_{k-2} \ldots u_1 v_k \rightarrow u_{k-2} u_{k-3} \ldots u_1 v_k v_{k-1} \rightarrow \ldots \rightarrow v_k v_{k-1} \ldots v_1$.

Example A2.2. In the de Bruijn graph $B(2, 3)$ shown in Figure A2.1, there is a walk $100 \rightarrow 000 \rightarrow 000 \rightarrow 001$ from vertex number 4 to vertex number 1. This has a loop $000 \rightarrow 000$.

There is a path $100 \rightarrow 001 \rightarrow 011 \rightarrow 111$ from vertex number 4 to vertex number 7.   □

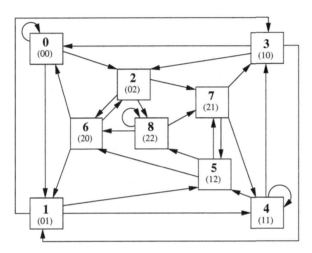

Fig. A2.2. The de Bruijn graph $B(3, 2)$ with nine vertices

Clearly, in a $B(d, k)$, for any two vertices u and v, a walk of length k from u to v is always possible. For $u = 000 \ldots 0$ and $v = 111 \ldots 1$, this walk is a path. This illustrates that the diameter of $B(d, k)$ is k.

Definition A2.1. *The suffix(x, p) (prefix(x, p)) of a vertex x in a $B(d, k)$ having an address $x_k x_{k-1} \ldots x_1$ and $p < k$, is the string $x_p x_{p-1} \ldots x_1$ $(x_k x_{k-1} \ldots x_{k-p+1})$.*

The shortest path from a vertex u to a vertex v in a $B(d, k)$ is unique and is of length $k - t$ if t is the length of the longest suffix of the string $u_k u_{k-1} \ldots u_1$ that appears as the prefix of $v_k v_{k-1} \ldots v_1$. The shortest path from u to v is $u_k u_{k-1} \ldots u_1 \rightarrow u_{k-1} u_{k-2} \ldots u_1 v_{k-t} \rightarrow u_{k-2} u_{k-3} \ldots u_1 v_{k-t} v_{k-t-1} \rightarrow \ldots \rightarrow u_t u_{t-1} \ldots u_1 v_{k-t} v_{k-t-1} \ldots v_1$.

Since suffix$(u, t) = $ prefix(v, t), $u_t u_{t-1} \ldots u_1 v_{k-t} v_{k-t-1} \ldots v_1 = v_k v_{k-1} \ldots v_1$.

Example A2.3. In the shortest path from $u = 100$ to $v = 011$ in $B(3,2)$, the longest suffix of u which matches the prefix of v has length $t = 1$, since suffix$(100, 1) = $ prefix$(011, 1) = 0$, suffix$(100, 2) = 10 \neq $ prefix$(011, 2) = 01$. Therefore, the shortest path from 100 to 011 is $100 \to 001 \to 011$. □

APPENDIX 3

Network Flow Programming

This appendix is a very concise overview of network flow programming. Standard textbooks, such as [4], are available for more information. Network flow programming is used to solve problems where one or more distinct *commodities* have to be shipped over a transportation network (e. g., a roadway system or a set of railway tracks). Each commodity is something (e.g., grains or minerals) that has to be transported from one or more sources for the commodity to one or more destinations. A transportation network is represented by a directed graph $G = (V, E)$. Each vertex in V represents a potential source or destination of a commodity (e.g., a rail station) and each edge $i \rightarrow j$ represents a link on the transportation network (e.g., the rail line between two stations). An edge $i \rightarrow j$ is an element in the form (i, j) in the set of edges E. Each edge $i \rightarrow j$ has a capacity and a cost/unit flow (u_{ij}, c_{ij}) associated with it. Such graphs are called *capacitated networks* because each edge has a capacity associated with it. Figure A3.1 shows a directed network with a pair (capacity, cost) associated with every edge.

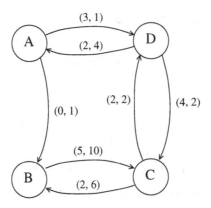

Fig. A3.1. A capacitated network.

A problem involving a network where there is only one commodity flowing from one or more sources to one or more destinations is called a *single commodity network flow problem*. If the network supports the flow of more than one commodity, it is called a *multi-commodity network flow problem*. The following formulations of network flow problems are of interest.

Problem 1

Given a capacitated network G, one source s, and one destination d of a single commodity flowing on network G, find the scheme to transport the maximum amount of the commodity.

This is called the *max-flow* problem and can be handled using linear program formulation *FORM* 12 given below.

Notation used in *FORM* 12

G : a directed graph (V, E) with V representing the set of vertices and E representing the set of edges so that if $(i, j) \in E$ there is an edge $i \rightarrow j$ from i to j. Each vertex is assigned a unique number $i, 1 \leq i \leq N$.

s (d) : the source (destination) of the commodity.

c_{ij} : cost/unit flow of using the edge $(i, j) \in E$.

u_{ij} : capacity of the edge $(i, j) \in E$.

$\widehat{v}$: amount of commodity transported from s to d.

X_{ij} : continuous variable denoting the amount of flow on the edge $(i, j) \in E$.

N : number of vertices in V.

The Formulation for *FORM* 12

Objective function:
$$\text{maximize } \widehat{v} \tag{A3.1}$$

subject to

1. Satisfy flow constraints.

$$\sum_{j:(i,j)\in E} X_{ij} - \sum_{j:(j,i)\in E} X_{ji} = \begin{cases} v & \text{if } i = s, \\ -v & \text{if } i = d, \quad \forall i, 1 \leq i \leq N \\ 0 & \text{otherwise.} \end{cases} \tag{A3.2}$$

2. Ensure that the capacities of edges are not exceeded.

$$X_{ij} \leq u_{ij}, \quad \forall i, j, (i, j) \in E \tag{A3.3}$$

3. Flows cannot be negative.

$$X_{ij} \geq 0, \quad \forall i, j, (i,j) \in E \qquad \text{(A3.4)}$$

Justification of *FORM* 12

The objective function is to maximize the amount $\hat{v}$ of a commodity transported from s to d.

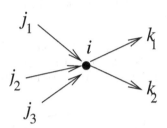

Fig. A3.2. Node i in a network

Figure A3.2 shows a typical node i with three incoming edges ($j_1 \rightarrow i$, $j_2 \rightarrow i$, $j_3 \rightarrow i$) to i and two outgoing edges ($i \rightarrow k_1$, $i \rightarrow k_2$) from i. The total *inflow* (i.e., the sum of flows into node i on all incoming edges $j \rightarrow i$) is the sum of the flows on edges $j_1 \rightarrow i$, $j_2 \rightarrow i$, and $j_3 \rightarrow i$. The total *outflow* (i.e., the sum of flows from node i, on all outgoing edges $i \rightarrow j$ for node i) is the sum of the flows on edges $i \rightarrow k_1$ and $i \rightarrow k_2$.

The commodity flowing from s to d may involve node i. If so, the following cases have to be considered:

Case I) ($i = s$) No flow can use an edge to i but one or more flows will start from i, using one or more of the outgoing edges from i.

Case II) ($i = d$) No flow can start from i but one or more flows will terminate at i, using one or more of the incoming edges to i.

Case III) ($i \neq s$ and $i \neq d$) There will be one or more flows into i, using one or more incoming edges to i, so that the total inflow to i is matched by the total outflow from i, using one or more outgoing edges from i.

Since the objective is to maximize $\hat{v}$, any solution where the total inflow to (outflow from) node s (d) is nonzero decreases the value of $\hat{v}$ and hence is not optimal.

The total inflow for node i is $\displaystyle\sum_{j:(j,i)\in E} X_{ji}$. Similarly the total outflow is

$$\sum_{j:(i,j)\in E} X_{ij}.$$

(A3.2) should be interpreted as follows:

Case I) $(i = s)$ Since i is the source of the commodity, the total outflow from i is $\widehat{v}$ and the total inflow is 0. Thus the difference between the total outflow and the total inflow must be $\widehat{v}$.

Case II) $(i = d)$ Since i is the destination of the commodity, the total inflow to i is $\widehat{v}$ and the total outflow is 0. Thus the difference between the total outflow and the total inflow must be $-v$.

Case III) $(i \neq s$ and $i \neq d)$ Since i is neither the source nor the destination of the commodity, the total inflow must be matched by the total outflow so that the difference between the total outflow and the total inflow must be 0.

(A3.2) is called *flow balance equation* (also called *flow conservation equation*). The purpose of (A3.3) is to state that the flow on edge $i \to j$ cannot exceed the capacity u_{ij} of the edge. All flows must be positive (A3.4).

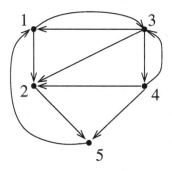

Fig. A3.3. A network

It is very easy to generate constraints corresponding to formulations such as this one. For instance, in the network shown in Figure A3.3, if node 3 is the source and 5 is the destination, the constraints corresponding to (A3.2) for selected nodes are as follows:

$$X_{31} + X_{32} + X_{34} - X_{13} - X_{43} = \widehat{v} \tag{A3.5}$$

$$X_{12} + X_{32} + X_{42} - X_{25} = 0 \tag{A3.6}$$

$$X_{51} - X_{25} - X_{45} = -v \tag{A3.7}$$

(A3.5), (A3.6), and (A3.7) arise from (A3.2) applied to nodes 3, 2, and 5, respectively.

Problem 2

Given a network $G = (V, E)$, one source s, and one destination d, find the minimum cost path from s to d, where each edge $(i, j) \in E$ has a cost c_{ij} associated with it.

This problem may be solved using integer linear program formulation $FORM$ 13 given below.

Notation used in $FORM$ 13

G, s, d, c_{ij} : same as in Problem 1.

x_{ij} : a binary variable (i.e., a variable which is either 0 or 1) denoting the amount of flow on the edge $(i, j) \in E$.

The Formulation for $FORM$ 13

Objective function:
$$\text{minimize} \quad \sum_{(i,j) \in E} c_{ij} \cdot x_{ij} \tag{A3.8}$$

subject to

1. Satisfy flow constraints.

$$\sum_{j:(i,j) \in E} x_{ij} - \sum_{j:(j,i) \in E} x_{ji} = \begin{cases} 1 & \text{if } i = s, \\ -1 & \text{if } i = d, \\ 0 & \text{otherwise.} \end{cases} \quad \forall i, 1 \le i \le N \tag{A3.9}$$

2. Ensure that the variable x_{ij} is either 0 or 1.

$$x_{ij} \in \{0, 1\} \quad \forall i, j, (i, j) \in E \tag{A3.10}$$

Justification of $FORM$ 13

It will be shown below that the above constraints generate a path $\mathbb{P}$ from s to d by ensuring that $x_{ij} = 1$ if and only if the edge $i \to j$ appears in the path $\mathbb{P}$. Figure A3.4 shows a typical path $\mathbb{P} = s \to i \to j \to k \to l \to m \to d$. Since $x_{ij} = 1$ if and only if $i \to j$ is in path $\mathbb{P}$, $\sum_{(i,j) \in E} c_{ij} \cdot x_{ij}$ gives the cost of the path from s to d. The objective function in (A3.8) is to find the minimum value of $\sum_{(i,j) \in E} c_{ij} \cdot x_{ij}$ and hence the minimum cost path.

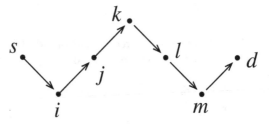

Fig. A3.4. A path from s to d

The purpose of (A3.10) is to ensure that x_{ij} is either 0 or 1. (A3.9) is similar to (A3.2). Here, the traffic is 1 so that the outflow from (inflow to) the source (destination) s (d) is 1. When $i = s$, (A3.9) becomes $\sum_{j:(s,j)\in E} x_{sj} = 1$. Since $x_{ij} \in \{0, 1\}$, this means that the value of $x_{sj} = 1$ for exactly one value of j and there is only one edge from s carrying the outflow from s. The same argument shows that there is exactly one edge to d carrying the inflow to s. For node $i, i \neq s, i \neq d$, (A3.9) states that $\sum_{j:(i,j)\in E} x_{ij} - \sum_{j:(j,i)\in E} x_{ji} = 0$. If there is exactly one edge, say $j_1 \to i$ carrying an inflow of 1 to node i, (A3.9) becomes $\sum_{j:(i,j)\in E} x_{ij} = x_{j_1 i} = 1$ so that there must be exactly one outgoing edge $i \to k_1$ to some k_1 such that $x_{ik_1} = 1$. In summary, (A3.9) ensures that the formulation finds a path from s to d.

Problem 3

The idea of the Multi-Commodity Network Flow (MCNF) problem is to optimize some parameter associated with the flow of an arbitrary number of commodities. Two popular problems, for instance, are the maximum multi-commodity network flow problem [152] and the minimum cost network flow problem [4]. The maximum multi-commodity network flow problem is to maximize the sum of flows for all commodities between their respective source/destination pairs. The minimum cost network flow problem is to determine the flow satisfying the demands of all commodities at a minimum cost without violating the capacity constraints of all the edges. An example similar to the single commodity problem (Problem 1) discussed above illustrates the MCNF problem involving a minimum cost network.

On a capacitated network G, q commodities are flowing; each commodity identified by a commodity number $k, 1 \leq k \leq q$. Each commodity has one source, one destination, and a specified amount of required traffic. The flows of different commodities may share edges. In general, an edge $i \to j$ has a number of commodities flowing on the edge. It must be ensured that the total traffic on edge $i \to j$, taking into account all the commodities flowing on the edge, does not exceed the capacity of the edge. The problem is to determine flows on each edge such that the specified traffic, for all commodities, is transported at a minimum cost. The problem may be solved using linear program formulation $FORM$ 14 given below.

Notation used in $FORM$ 14

G, c_{ij}, u_{ij} : same as in Problem 1.

s^k (d^k) : the source (destination) of commodity number k.

q : number of commodities to be transported.

τ^k : amount of commodity number k to be transported from s^k to d^k.

z_{ij}^k : continuous variable denoting the amount of flow of commodity number k on the edge $(i,j) \in E$.

The Formulation for $FORM$ 14

Objective function:

$$\text{minimize } \sum_{k=1}^{q} \sum_{(i,j)\in E} z_{ij}^k \times c_{ij} \tag{A3.11}$$

subject to

1. Satisfy flow constraints for each commodity k, $1 \leq k \leq q$.

$$\sum_{j:(i,j)\in E} z_{ij}^k - \sum_{j:(j,i)\in E} z_{ji}^k = \begin{cases} \tau^k & \text{if } i = s, \quad \forall k, 1 \leq k \leq q \\ -\tau^k & \text{if } i = d, \quad \forall i, 1 \leq i \leq N \\ 0 & \text{otherwise.} \end{cases} \tag{A3.12}$$

2. Ensure that the capacities of edges are not exceeded.

$$\sum_{k=1}^{q} z_{ij}^k \leq u_{ij}, \quad \forall i, j, (i,j) \in E \tag{A3.13}$$

3. Flows cannot be negative.

$$z_{ij}^k \geq 0, \quad \forall i, j, (i,j) \in E, \forall k, 1 \leq k \leq q \tag{A3.14}$$

Justification of *FORM* 14

The flow of commodity k on edge $i \to j$ is z_{ij}^k (which may very well be 0, meaning that there is no flow of commodity k on edge $i \to j$). Since the cost per unit flow on edge $i \to j$ is c_{ij}, the cost of flow z_{ij}^k is $z_{ij}^k \times c_{ij}$. The objective function in (A3.11) is to find the minimum value of $\sum_{k=1}^{q} \sum_{(i,j) \in E} z_{ij}^k \times c_{ij}$, the minimum cost for transporting τ^k quantity from s^k to d^k, for all commodities $k, 1 \leq k \leq q$.

(A3.12) is similar to (A3.3), except that there is one set of equations for each commodity, and should be interpreted as follows:

Case I) $(i = s^k)$ Since i is the source of the commodity k, the outflow for commodity k from i is τ^k and the inflow is 0. Thus the difference between the outflow and the inflow must be τ^k.

Case II) $(i = d^k)$ Since i is the destination of the commodity k, the inflow to i for commodity k is τ^k and the outflow is 0. Thus the difference between the outflow and the inflow must be $-\tau^k$.

Case III) $(i \neq s^k$ and $i \neq d^k)$ Since i is neither the source nor the destination of commodity k, the total inflow for commodity k must be matched by the total outflow for commodity k, so that the difference between the outflow and the inflow must be 0.

(A3.13) is like (A3.2) except that the sum of the flows for all commodities cannot exceed the capacity u_{ij} of the edge.

References

1. A. S. Acampora and S. I. A. Shah. Multihop lightwave networks: a comparison of store-and-forward and hot-potato routing. *IEEE Transactions on Communications*, 40(6):1082–1090, June 1992.
2. A. Aggarwal, A. Bar-Noy, D. Coppersmith, R. Ramaswami, B. Schieber, and M. Sudan. Efficient routing in optical networks. *Journal of the ACM*, 43(6):973–1001, 1996.
3. G. P. Agrawal. *Fiber-Optic Communication Systems*. Wiley, 2002.
4. R. K. Ahuja, T. L. Magnanti, and J. B. Orlin. *Network Flows: Theory, Algorithms, and Applications*. Prentice Hall, 1993.
5. M. Alanyali and E. Ayanoglu. Provisioning algorithms for WDM optical networks. *IEEE/ACM Transactions on Networking*, 7(5):767–778, October 1999.
6. M. Ali and J. Deogun. Power-efficient design of multicast wavelength-routed networks. *IEEE Journal on Selected Areas in Communications*, 18:1852–1862, October 2000.
7. D. Amar, A. Raspaud, and O. Togni. All-to-all wavelength-routing in all-optical compound networks. *Discrete Mathematics*, 235(1-3):353–363, 2001.
8. R. Andersen, F. Chung, A. Sen, and G. Xue. On disjoint path pairs with wavelength continuity constraint in WDM networks. In *IEEE International Conference on Computer Communications (INFOCOM)*, pages 524–535, March 2004.
9. Y. Aneja, A. Jaekel, and S. Bandyopadhyay. An improved ILP formulation for path protection in WDM networks. In *7th IEEE International Conference on High Speed Networks and Multimedia Communications (HSNMC)*, volume 3079, pages 903–911, June/July 2004.
10. Y. Aneja, A. Jaekel, and S. Bandyopadhyay. Approximation algorithms for traffic routing in wavelength routed WDM networks. In *International Conference on Broadband Networks (BroadNets)*, volume 1, pages 23–30, October 2005.
11. Y. P. Aneja, A. Jaekel, and S. Bandyopadhyay. Logical topology design for WDM networks using survivable routing. In *IEEE International Conference on Communications (ICC)*, volume 6, pages 2471–2476, June 2006.
12. A. Arora, S. Subramaniam, and H.-A. Choi. Logical topology design for linear and ring optical networks. *IEEE Journal on Selected Areas in Communications*, 20(1):62–74, January 2002.

13. A. S. Arora and S. Subramaniam. Converter placement in wavelength routing mesh topologies. In *IEEE International Conference on Communications (ICC)*, pages 1282–1288, June 2000.

14. V. Auletta, I. Caragiannis, L. Gargano, C. Kaklamanis, and P. Persiano. Sparse and limited wavelength conversion in all-optical tree networks. *Theoretical Computer Science*, 266(1-2):887–934, 2001.

15. V. Auletta, I. Caragiannis, C. Kaklamanis, and P. Persiano. Bandwidth allocation algorithms on tree-shaped all-optical networks with wavelength converters. In D. Krizanc and P. Widmayer, editors, *International Colloquium on Structural Information and Communication Complexity (SIROCCO)*, pages 24–39. Carleton Scientific, 1997.

16. V. Auletta, I. Caragiannis, C. Kaklamanis, and P. Persiano. Efficient wavelength routing in trees with low–degree converters. In *Multichannel Optical Networks: Theory and Practice, DIMACS Series in Discrete Mathematics and Computer Science, (AMS)*, volume 46, pages 1–14, 1998.

17. V. Auletta, I. Caragiannis, C. Kaklamanis, and P. Persiano. On the complexity of wavelength converters. In L. Brim, J. Gruska, and J. Zlatuska, editors, *International Symposium on Mathematical Foundations(MFCS)*, volume 1450 of *Lecture Notes in Computer Science*, pages 771–779. Springer, 1998.

18. V. Auletta, I. Caragiannis, C. Kaklamanis, and P. Persiano. Randomized path coloring on binary trees. In K. Jansen and S. Khuller, editors, *Workshop on Approximation Algorithms for Combinatorial Optimization Problems (APPROX)*, volume 1913 of *Lecture Notes in Computer Science*, pages 60–71. Springer, 2000.

19. V. Auletta, I. Caragiannis, C. Kaklamanis, and P. Persiano. Randomized path coloring on binary trees. *Theoretical Computer Science*, 289(1):355–399, 2002.

20. B. Awerbuch, Y. Azar, A. Fiat, S. Leonardi, and A. Rosén. On-line competitive algorithms for call admission in optical networks. *Algorithmica*, 31(1):29–43, 2001.

21. S. Balasubramanian, W. He, and A. K. Somani. Light-trail networks: Design and survivability. In *IEEE Conference on Local Computer Networks (LCN)*, pages 174–181, November 2005.

22. I. Baldine and G. N. Rouskas. Traffic adaptive WDM networks: a study of reconfiguration issues. *IEEE/OSA Journal of Lightwave Technology*, 19(4): 433–455, April 2001.

23. S. Bandyopadhyay, A. Jaekel, Y. P. Aneja, and S. Saha. A comparative study of schemes for robust logical topology design in wavelength routed WDM networks. In *IASTED International Conference on Optical Communication Systems and Networks (OCSN)*, pages 230–237, 2006.

24. S. Bandyopadhyay, A. Jaekel, A. Sengupta, and W. Lang. A virtual wavelength translation scheme for routing in all-optical networks. *Photonic Networks Communications*, 4(3-4):391–407, July 2002.

25. D. Banerjee and B. Mukherjee. A practical approach for routing and wavelength assignment in large wavelength-routed optical networks. *IEEE Journal on Selected Areas in Communications*, 14(5):903–908, June 1996.

26. D. Banerjee and B. Mukherjee. Wavelength-routed optical networks: linear formulation, resource budgeting tradeoffs, and a reconfiguration study. *IEEE/ACM Transactions on Networking*, 8(5):598–607, October 2000.

27. S. Baroni and P. Bayvel. Wavelength requirements in arbitrarily connected wavelength-routed optical networks. *IEEE/OSA Journal of Lightwave Technology*, 15(2):242–251, February 1997.

28. S. Baroni, R. J. Gibbens, and P. Bayvel. On the number of wavelengths in arbitrarily-connected wavelength-routed optical networks. In *Optical Networks and Their Applications (ONA)*, January 1998.

29. J. R. Barry and E. A. Lee. Performance of coherent optical receivers. In *Proceedings of the IEEE*, volume 78, pages 1369–1394, August 1990.

30. R. A. Barry. Will optical networks see the light of day? *IEEE Network*, 10(6):6–6, November 1996.

31. R. A. Barry and P. A. Humblet. On the number of wavelengths and switches in all-optical networks. *IEEE Transactions on Communications*, 42(2/3/4): 583–591, 1994.

32. R. A. Barry and P. A. Humblet. Models of blocking probability in all-optical networks with and without wavelength changers. *IEEE Journal on Selected Areas in Communications*, 14(5):858–867, June 1996.

33. Y. Bartal, A. Fiat, and S. Leonardi. Lower bounds for on-line graph problems with application to on-line circuit and optical routing. In *ACM Symp. on Theory of Computing (STOC)*, pages 531–540, May 1996.

34. Y. Bartal and S. Leonardi. On-line routing in all-optical networks. *Theoretical Computer Science*, 221(1-2):19–39, 1999.

35. M. S. Bazaraa, J. J. Jarvis, and H. D. Sherali. *Linear Programming and Network Flows*. Wiley, 1990.

36. B. Beauquier. Broadcasting in WDM optical rings and tori. In *DIMACS Workshop:Multichannel Optical Networks:Theory and Practice*, volume 46, pages 63–73, March 1998.

37. B. Beauquier. All-to-all communication for some wavelength-routed all-optical networks. *Networks*, 33(3):179–187, May 1999.

38. B. Beauquier, J.-C. Bermond, L. Gargano, P. Hell, S. Pérennes, and U. Vaccaro. Graph problems arising from wavelength-routing in all-optical networks. In *Workshop on Optics and Computer Science (WOCS)*, April 1997.

39. B. Beauquier, P. Hell, and S. Perennes. Optimal wavelength-routed multicasting. *Discrete Applied Mathematics*, 84(1-3):15–20, 1998.

40. B. Beauquier, S. Prennes, and M. Syska. Efficient access to optical bandwidth, routing and grooming in WDM networks: State-of-the-art survey. Technical report, Projet MASCOTTE (CNRS/INRIA/UNSA), Sophia Antipolis, July 2002.

41. V. E. Benes. *Mathematical Theory of Connecting Networks and Telephone Traffic*. Academic Press, 1965.

42. C. Berge. *Graphs and Hypergraphs*. American Elsevier Pub. Co, 1973.

43. J.-C. Bermond, L. Braud, and D. Coudert. Traffic grooming on the path. In *12th International Colloquium on Structural Information and Communication Complexity (SIROCCO)*, pages 34–48, Le Mont Saint-Michel, France, May 24-26 2005. LNCS 3499.

44. J.-C. Bermond, L. Braud, and D. Coudert. Traffic grooming on the path. *Theoretical Computer Science*, 2007, to appear.

45. J.-C. Bermond and S. Ceroi. Minimizing SONET ADMs in unidirectional WDM rings with grooming ratio 3. *Networks*, 41(2):83–86, 2003.

46. J.-C. Bermond, C. Colbourn, D. Coudert, G. Ge, A. Ling, and X. Muñoz. Traffic grooming in unidirectional WDM rings with grooming ratio C=6. *SIAM Journal on Discrete Mathematics*, 19(2):523–542, 2005.

47. J.-C. Bermond, M. Cosnard, D. Coudert, and S. Perennes. Optimal solution of the maximum all request path grooming problem. In *Advanced International Conference on Telecommunications (AICT)*. IEEE, 2006.

48. J.-C. Bermond and D. Coudert. Traffic grooming in unidirectional WDM ring networks using design theory. In *IEEE International Conference on Communications (ICC)*, volume 2, pages 1402–1406, Anchorage, Alaska, May 2003.

49. J.-C. Bermond and D. Coudert. *The CRC Handbook of Combinatorial Designs (2nd edition)*, volume 42 of *Discrete Mathematics and Its Applications*, chapter VI.27, Grooming, pages 493–496. CRC Press, C.J. Colbourn and J.H. Dinitz edition, Nov. 2006.

50. J.-C. Bermond, D. Coudert, and X. Muñoz. Traffic grooming in unidirectional WDM ring networks: The all-to-all unitary case. In *7th IFIP Working Conference on Optical Network Design & Modelling (ONDM)*, pages 1135–1153, Budapest, Hungary, 2003.

51. J.-C. Bermond, D. Coudert, X. Muñoz, and I. Sau Valls. Traffic grooming in bidirectional WDM ring networks. In *IEEE-LEOS International Conference on Transparent Optical Networks (ICTON/COST 293 GRAAL)*, volume 3, pages 19–22, June 2006.

52. J.-C. Bermond, L. Gargano, S. Perennes, A. A. Rescigno, and U. Vaccaro. Efficient collective communication in optical networks. In *International Colloquium on Automata, Languages and Programming (ICALP), year = 1996, pages = 574-585, crossref = DBLP:conf/icalp/1996, bibsource = DBLP, http://dblp.uni-trier.de*.

53. J.-C. Bermond, L. Gargano, S. Perennes, A. A. Rescigno, and U. Vaccaro. Efficient collective communication in optical networks. *Theoretical Computer Science*, 233(1-2):165–189, 2000.

54. R. Berry and E. Modiano. Reducing electronic multiplexing costs in SONET/WDM rings with dynamically changing traffic. *IEEE Journal on Selected Areas in Communications*, 18(10):1961–1971, October 2000.

55. G. Bhati and G. Saraph. ADM reduction in SONET-WDM rings for stochastically varying traffic. In *International Conference on Broadband Networks (BroadNets)*, pages 333–335. IEEE, 2005.

56. L. Bhuyan and D. P. Agarwal. Generalized hypercube and hyperbus structure for a computer network. *IEEE Transactions on Computers*, C-33:323–333, April 1984.

57. Z. Bian, Q.-P. Gu, and X. Zhou. Wavelength assignment on bounded degree trees of rings. In *International Conference on Parallel and Distributed Systems (ICPADS)*, pages 73–80. IEEE Computer Society, 2004.

58. Z. Bian, Q.-P. Gu, and X. Zhou. Tight bounds for wavelength assignment on trees of rings. In *IEEE International Parallel and Distributed Processing Symposium (IPDPS)*. IEEE Computer Society, 2005.

59. V. Bilò, M. Flammini, and L. Moscardelli. On Nash equilibria in non-cooperative all-optical networks. In V. Diekert and B. Durand, editors, *International Symposium on Theoretical Aspects of Computer Science (STACS)*, volume 3404 of *Lecture Notes in Computer Science*, pages 448–459. Springer, 2005.

60. A. Birman. Computing approximate blocking probabilities for a class of all-optical networks. *IEEE Journal on Selected Areas in Communications*, 14(5):852–857, June 1996.

61. E. Bouillet, J.-F. Labourdette, G. Ellinas, R. Ramamurthy, and S. Chaudhuri. Stochastic approaches to compute shared mesh restored lightpaths in optical network architectures. In *IEEE International Conference on Computer Communications (INFOCOM)*, volume 2, pages 801–807, June 2002.

62. M. Bouklit, D. Coudert, J.-F. Lalande, C. Paul, and H. Rivano. Approximate multicommodity flow for WDM networks design. In J. Sibeyn, editor, *International Colloquium on Structural Information and Communication Complexity (SIROCCO)*, number 17, in Proceedings in Informatics, pages 43–56, Umeå, Sweden, 2003. Carleton Scientific.

63. D. Buchfuhrer, T. Carnes, B. Tagiku, E. Celis, and R. Libeskind-Hadas. Traffic grooming for single-source multicast communication in WDM rings. In *IEEE International Conference on Communications (ICC)*, volume 3, pages 1613–1619, 2005.

64. B. V. Caenegem, W. V. Parys, F. D. Turck, and P. M. Demeester. Dimensioning of survivable WDM networks. *IEEE Journal on Selected Areas in Communications*, 16(7):1146–1157, September 1998.

65. G. Călinescu and P.-J. Wan. Traffic partition in WDM/SONET rings to minimize SONET ADMs. In *IEEE International Parallel and Distributed Processing Symposium (IPDPS)*, page 201. IEEE Computer Society, 2001.

66. G. Călinescu and P.-J. Wan. Splittable traffic partition in WDM/SONET rings to minimize SONET ADMs. *Theoretical Computer Science*, 276(1-2):33–50, 2002.

67. G. Călinescu and P.-J. Wan. Traffic partition in WDM/SONET rings to minimize SONET ADMs. *Journal of Combinatorial Optimization*, 6(4):425–453, 2002.

68. I. Caragiannis. Wavelength management in WDM rings to maximize the number of connections. In W. Thomas and P. Weil, editors, *International Symposium on Theoretical Aspects of Computer Science (STACS)*, volume 4393 of *Lecture Notes in Computer Science*, pages 61–72. Springer, 2007.

69. I. Caragiannis, A. Ferreira, C. Kaklamanis, S. Perennes, P. Persiano, and H. Rivano. Approximate constrained bipartite edge coloring. *Discrete Applied Mathematics*, 143(1-3):54–61, 2004.

70. I. Caragiannis, A. Ferreira, C. Kaklamanis, S. Perennes, and H. Rivano. Fractional path coloring with applications to WDM networks. In F. Orejas, P. G. Spirakis, and J. van Leeuwen, editors, *International Colloquium on Automata, Languages and Programming (ICALP)*, volume 2076 of *Lecture Notes in Computer Science*, pages 732–743. Springer, 2001.

71. I. Caragiannis and C. Kaklamanis. Approximate path coloring with applications to wavelength assignment in WDM optical networks. In V. Diekert and M. Habib, editors, *International Symposium on Theoretical Aspects of Computer Science (STACS)*, volume 2996 of *Lecture Notes in Computer Science*, pages 258–269. Springer, 2004.

72. I. Caragiannis, C. Kaklamanis, and P. Persiano. Bounds on optical bandwidth allocation on directed fiber tree topologies. In *International Parallel Processing Symposium, 2nd Workshop on Optics and Computer Science (WOCS'97)*, Geneva, Switzerland, April 1997. IEEE Press.

73. I. Caragiannis, C. Kaklamanis, and P. Persiano. Wavelength routing of symmetric communication requests in directed fiber trees. In L. Gargano and D. Peleg, editors, *International Colloquium on Structural Information and Communication Complexity (SIROCCO)*, pages 10–19. Carleton Scientific, 1998.

74. I. Caragiannis, C. Kaklamanis, and P. Persiano. Edge coloring of bipartite graphs with constraints. In M. Kutylowski, L. Pacholski, and T. Wierzbicki, editors, *International Symposium on Mathematical Foundations(MFCS)*, volume 1672 of *Lecture Notes in Computer Science*, pages 376–386. Springer, 1999.

75. I. Caragiannis, C. Kaklamanis, and P. Persiano. Symmetric communication in all-optical tree networks. *Parallel Processing Letters*, 10(4):305–314, 2000.

76. I. Caragiannis, C. Kaklamanis, and P. Persiano. Wavelength routing in all-optical tree networks: A survey. *Computing and Informatics (formerly Computers and Artificial Intelligence)*, 20(2):95–120, 2001.

77. I. Caragiannis, C. Kaklamanis, and P. Persiano. Edge coloring of bipartite graphs with constraints. *Theoretical Computer Science*, 270(1-2):361–399, 2002.

78. I. Caragiannis, C. Kaklamanis, and P. Persiano. Approximation algorithms for path coloring in trees. In *Efficient Approximation and Online Algorithms*, LNCS 3484, pages 74–96. Springer, 2006.

79. I. Caragiannis, C. Kaklamanis, P. Persiano, and A. Sidiropoulos. Fractional and integral coloring of locally-symmetric sets of paths on binary trees. In K. Jansen and R. Solis-Oba, editors, *Workshop on Approximation and Online Algorithms (WAOA)*, volume 2909 of *Lecture Notes in Computer Science*, pages 81–94. Springer, 2003.

80. G.-K. Chang, G. Ellinas, J. K. Gamelin, M. Z. Iqbal, and C. A. Brackett. Multi-wavelength reconfigurable WDM/ATM/SONET network testbed. *IEEE/OSA Journal of Lightwave Technology*, 14(6):1320–1340, June 1996.

81. M. W. Chbat, E. Grard, L. Berthelon, A. Jourdan, P. A. Perrier, A. Leclert, B. Landousies, A. Ramdane, N. Parnis, E. V. Jones, E. Limal, H. N. Poulsen, R. J. S. Pedersen, N. Flaarønning, D. Vercauteren, M. Puleo, E. Ciaramella, G. Marone, R. Hess, H. Melchior, W. V. Parys, P. M. Demeester, P. J. Gødsvang, T. Olsen, and D. R. Hjelme. Toward wide-scale all-optical transparent networking: the ACTS Optical Pan-European Network (OPEN) project. *IEEE Journal on Selected Areas in Communications*, 16(7):1226–1244, September 1998.

82. B. Chen, G. N. Rouskas, and R. Dutta. Traffic grooming in WDM ring networks with the min-max objective. In *LNCS NETWORKING*, volume 3042, pages 174–185, May 2004.

83. B. Chen and J. Wang. Efficient routing and wavelength assignment for multicast in WDM networks. *IEEE Journal on Selected Areas in Communications*, 20(1):97–109, January 2002.

84. L. Chen and E. Modiano. Efficient routing and wavelength assignment for reconfigurable WDM ring networks with wavelength converters. *IEEE/ACM Transactions on Networking*, 13(1):173–186, February 2005.

85. Y. Chen, C. Qaio, and X. Yu. Optical burst switching: a new area in optical networking research. *IEEE Network*, 18(3):16–23, May-June 2004.

86. A. L. Chiu and E. H. Modiano. Traffic grooming algorithms for reducing electronic multiplexing costs in WDM ring networks. *IEEE/OSA Journal of Lightwave Technology*, 18(1):2–12, January 2000.

87. I. Chlamtac, A. Farago, and T. Zang. Lightpath (wavelength) routing in large WDM networks. *IEEE Journal on Selected Areas in Communications*, 14(5):909–913, June 1996.

88. I. Chlamtac, A. Ganz, and G. Karmi. Lightpath communications: an approach to high bandwidth optical WAN's. *IEEE Transactions on Communications*, 40(7):1171–1182, July 1992.

89. I. Chlamtac, A. Ganz, and G. Karmi. Lightnets: topologies for high-speed optical networks. *IEEE/OSA Journal of Lightwave Technology*, 11(5):951–961, May-June 1993.

90. H. Choi, S. Subramaniam, and H. Choi. Loopback recovery from double-link failures in optical mesh networks. *IEEE/ACM Transactions on Networking*, 12(6):1119–1130, December 2004.

91. T. Y. Chow, F. Chudak, and A. M. Ffrench. Fast optical layer mesh protection using pre-cross-connected trails. *IEEE/ACM Transactions on Networking*, 12(3):539–548, June 2004.

92. X. Chu and B. Li. Dynamic routing and wavelength assignment in the presence of wavelength conversion for all-optical networks. *IEEE/ACM Transactions on Networking*, 13(3):704–715, June 2005.

93. V. Chvatal. *Linear Programming*. W. H. Freeman, 1983.

94. C. Clos. A study of nonblocking switching networks. *Bell System Technical Journal*, 32:406–424, March 1953.

95. C. J. Colbourn and A. C. H. Ling. Graph decompositions with application to wavelength add-drop multiplexing for minimizing SONET ADMs. *Discrete Mathematics*, 261(1-3):141–156, 2003.

96. D. Coudert, P. Datta, S. Perennes, H. Rivano, and M.-E. Voge. Shared risk resource group: Complexity and approximability issues. *Parallel Processing Letters*, 2007, to appear.

97. D. Coudert and X. Muñoz. *Recent Research Developments in Optics, 3*, chapter 37, Graph Theory and Traffic Grooming in WDM Rings, pages 759–778. Research Signpost. Kerala, India, 2003. ISBN: 81-271-0028-5.

98. D. Coudert, S. Perennes, H. Rivano, and M.-E. Voge. Shared risk resource groups and survivability in multilayer networks. In *IEEE-LEOS International Conference on Transparent Optical Networks (ICTON/COST 293 GRAAL)*, volume 3, pages 235–238, 2006. Invited Paper.

99. D. Coudert and H. Rivano. Lightpath assignment for multifibers WDM optical networks with wavelength translators. In *IEEE Global Telecommunications Conference (GLOBECOM)*, volume 3, pages 2686–2690, Taiwan, Nov. 2002. OPNT-01-5.

100. O. Crochat, J.-Y. L. Boudec, and O. Gerstel. Protection interoperability for WDM optical networks. *IEEE/ACM Transactions on Networking*, 8(3): 384–395, June 2000.

101. G. Dantzig and R. V. Slyke. Generalized upper bounding technique. *Journal of Computer and System Sciences*, 1(3):213–226, 1967.

102. D. Das, M. De, and B. Sinha. A new network topology with multiple meshes. *IEEE Transactions on Computers*, 48(5):536–551, May 1999.

103. S. Datta, S. Sengupta, S. Biswas, and S. Datta. Efficient channel reservation for backup paths in optical mesh networks. In *IEEE Global Telecommunications Conference (GLOBECOM)*, volume 4, pages 2104–2108, November 2001.

104. S. Datta, S. Sengupta, S. Biswas, D. Saha, and H. Kobayashi. Analysis of sub-wavelength traffic grooming efficiency in optical mesh networks. In *IEEE International Conference on Communications*, volume 3, pages 1811–1815, 2004.

105. P. Degano, R. Gorrieri, and A. Marchetti-Spaccamela, editors. *Automata, Languages and Programming, 24th International Colloquium, ICALP'97, Bologna, Italy, 7-11 July 1997, Proceedings*, volume 1256 of *Lecture Notes in Computer Science*. Springer, 1997.

106. P. Demeester, M. Gryseels, A. Autenrieth, C. Brianza, L. Castagna, G. Signorelli, R. Clemenfe, M. Ravera, A. Jajszczyk, D. Janukowicz, K. V. Doorselaere, and Y. Harada. Resilience in multilayer networks. *IEEE Communications Magazine*, 37(8):70–76, August 1999.

107. P. Demeester, T. H. Wu, and N. Yoshikai. Survivable communication networks. *IEEE Communications Magazine*, 37(8), Aug 1999. Special Issue.

108. E. W. Dijkstra. A note on two problems in connexion with graphs. *Numerische Mathematik*, 1:269–271, 1959.

109. A. Ding and G. S. Poo. A survey of optical multicast over WDM networks. *Computer Communications*, 26(2):193–200, February 2003.

110. B. T. Doshi, S. Dravida, P. Harshavardhana, O. Hauser, and Y. Wang. Optical network design and restoration. *Bell Labs Technical Journal*, 4(1):58–84, January 1999.

111. P. W. Dowd. Wavelength division multiple access channel hypercube processor interconnection. *IEEE Transactions on Computers*, 41(10):1223–1241, October 1992.

112. F. J. Duarte. *Tunable Laser Optics*. Academic Press, 2003.

113. A. K. Dutta, N. K. Dutta, and M. Fujiwara, editors. *WDM Technologies: Optical Networks*. Elsevier, 2004.

114. N. K. Dutta, A. K. Dutta, and M. Fujiwara, editors. *WDM Technologies: Passive Optical Components*. Elsevier, 2003.

115. R. Dutta, S. Huang, and G. N. Rouskas. Traffic grooming in path, star, and tree networks: complexity, bounds, and algorithms. In *2003 ACM SIGMETRICS International Conference on Measurement and Modeling of Computer Systems*, pages 298–299, New York, NY, USA, 2003. ACM Press.

116. R. Dutta and G. N. Rouskas. A survey of virtual topology design algorithms for wavelength routed optical networks. *SPIE Optical Networks Magazine*, 1(1):73–89, January 2000.

117. R. Dutta and G. N. Rouskas. On optimal traffic grooming in WDM rings. *IEEE Journal on Selected Areas in Communications*, 20(1):110–121, January 2002.

118. R. Dutta and G. N. Rouskas. Traffic grooming in WDM networks: past and future. *IEEE Network*, 16(6):46–56, November 2002.

119. T. Eilam, S. Moran, and S. Zaks. Approximation algorithms for survivable optical networks. In *International Symposium on Distributed Computing (ISDC)*, pages 104–118, 2000.

120. T. Eilam, S. Moran, and S. Zaks. Lightpath arrangement in survivable rings to minimize the switching cost. *IEEE Journal on Selected Areas in Communications*, 20(1):172–182, January 2002.

121. G. Ellinas, K. Bala, and G. K. Chang. Scalability of a novel wavelength assignment algorithm for WDM shared protection rings. In *IEEE/OSA Optical Fiber Communications Conference (OFC)*, February 1998.

122. J. M. H. Elmirghani and H. T. Mouftah. All-optical wavelength conversion: technologies and applications in DWDM networks. *IEEE Communications Magazine*, 38(3):86–92, March 2000.

123. L. Epstein and A. Levin. Better bounds for minimizing SONET ADMs. In G. Persiano and R. Solis-Oba, editors, *Workshop on Approximation and Online Algorithms (WAOA)*, volume 3351 of *Lecture Notes in Computer Science*, pages 281–294. Springer, 2004.

124. L. Epstein and A. Levin. The chord version for SONET ADMs minimization. *Theoretical Computer Science*, 349(3):337–346, 2005.

125. L. Epstein and A. Levin. SONET ADMs minimization with divisible paths. In T. Erlebach and G. Persiano, editors, *Workshop on Approximation and Online Algorithms (WAOA)*, volume 3879 of *Lecture Notes in Computer Science*, pages 119–132. Springer, 2005.

126. T. Erlebach and K. Jansen. Scheduling of virtual connections in fast networks. In *Proc. of 4th Workshop on Parallel Systems and Algorithms (PASA'96)*, pages 13–32. World Scientific Publishing, 1996.

127. T. Erlebach and K. Jansen. Efficient implementation of an optimal greedy algorithm for wavelength allocation in directed tree networks. In K. Mehlhorn, editor, *Algorithm Engineering*, pages 13–24. Max-Planck-Institut für Informatik, 1998.

128. T. Erlebach and K. Jansen. Maximizing the number of connections in optical tree networks. In K.-Y. Chwa and O. H. Ibarra, editors, *International Symposium on Algorithms & Computation (ISAAC)*, volume 1533 of *Lecture Notes in Computer Science*, pages 179–188. Springer, 1998.

129. T. Erlebach and K. Jansen. Efficient implementation of an optimal greedy algorithm for wavelength assignment in directed tree networks. *ACM Journal of Experimental Algorithms*, 4:article 4, 1999.

130. T. Erlebach and K. Jansen. The complexity of path coloring and call scheduling. *Theoretical Computer Science*, 255(1-2):33–50, March 2001.

131. T. Erlebach and K. Jansen. The maximum edge-disjoint paths problem in bidirected trees. *SIAM J. Discrete Math.*, 14(3):326–355, 2001.

132. T. Erlebach, K. Jansen, C. Kaklamanis, M. Mihail, and P. Persiano. Optimal wavelength routing on directed fiber trees. *Theoretical Computer Science*, 221(1-2):119–137, 1999.

133. T. Erlebach, K. Jansen, C. Kaklamanis, and P. Persiano. An optimal greedy algorithm for wavelength allocation in directed tree networks. In *DIMACS Workshop on Network Design: Connectivity and Facilities Location (April 28-30, 1997)*, volume 40 of *DIMACS Series in Discrete Mathematics and Theoretical Computer Science*, pages 117–129, Rutgers University, NJ, 1998. American Mathematical Society.

134. T. Erlebach, K. Jansen, C. Kaklamanis, and P. Persiano. Directed tree networks. In *Encyclopedia of Optimization, Volume I*, pages 444–453. Kluwer Academic Publishers, June 2001.

135. T. Erlebach, A. Pagourtzis, K. Potika, and S. Stefanakos. Resource allocation problems in multifiber WDM tree networks. In H. L. Bodlaender, editor, *Workshop on Graph Theoretical Concepts in Computer Science (WG)*, volume 2880 of *Lecture Notes in Computer Science*, pages 218–229. Springer, 2003.

136. T. Erlebach and S. Stefanakos. On shortest-path all-optical networks without wavelength conversion requirements. In H. Alt and M. Habib, editors, *International Symposium on Theoretical Aspects of Computer Science (STACS)*,

volume 2607 of *Lecture Notes in Computer Science*, pages 133–144. Springer, 2003.

137. T. Erlebach and S. Stefanakos. Wavelength conversion in shortest-path all-optical networks. In T. Ibaraki, N. Katoh, and H. Ono, editors, *International Symposium on Algorithms & Computation (ISAAC)*, volume 2906 of *Lecture Notes in Computer Science*, pages 595–604. Springer, 2003.

138. T. Erlebach and S. Stefanakos. Wavelength conversion in all-optical networks with shortest-path routing. *Algorithmica*, 43(1-2):43–61, 2005.

139. J. Fang, M. Sivakumar, A. K. Somani, and K. M. Sivalingam. On partial protection in groomed optical WDM mesh networks. In *IEEE International Conference on Dependable Systems and Networks (DSN) — Dependable Computing and Communications Symposium (DCCS)*, pages 228–237, June 2005.

140. J. Fang and A. K. Somani. Enabling subwavelength level traffic grooming in survivable WDM optical network design. In *IEEE Global Telecommunications Conference (GLOBECOM)*, volume 5, pages 2761–2766, December 2003.

141. J. Fang and A. K. Somani. IP traffic grooming over WDM optical networks. In *9th IFIP/IEEE conference on Optical Networks Design and Modeling (ONDM)*, pages 393–402, February 2005.

142. F. Farahmand, X. Huang, and J. Jue. Efficient online traffic grooming algorithms in WDM mesh networks with drop-and-continue node architecture. In *International Conference on Broadband Networks (BroadNets)*, pages 180–189, 2004.

143. A. Ferreira, S. Perennes, H. Rivano, A. W. Richa, and N. E. S. Moses. Models, complexity and algorithms for the design of multi-fiber WDM networks. *Journal of Telecommunication Systems*, 24(2-4):123–138, October 2003.

144. A. Ferreira, S. Prennes, A. W. Richa, H. Rivano, and N. Stier. On the design of multifiber WDM networks. In *10th IEEE International Conference on Telecommunications – ICT 2003*, volume I, pages 12–18, Tahiti, 2003.

145. M. Flammini, G. Monaco, L. Moscardelli, M. Shalom, and S. Zaks. Approximating the traffic grooming problem in tree and star networks. In F. V. Fomin, editor, *Workshop on Graph Theoretical Concepts in Computer Science (WG)*, volume 4271 of *Lecture Notes in Computer Science*, pages 147–158. Springer, 2006.

146. M. Flammini, L. Moscardelli, M. Shalom, and S. Zaks. Approximating the traffic grooming problem. In X. Deng and D.-Z. Du, editors, *International Symposium on Algorithms & Computation (ISAAC)*, volume 3827 of *Lecture Notes in Computer Science*, pages 915–924. Springer, 2005.

147. M. Flammini, A. Navarra, and A. Proskurowski. On routing of wavebands for gossiping in all-optical paths and cycles. In J. F. Sibeyn, editor, *International Colloquium on Structural Information and Communication Complexity (SIROCCO)*, volume 17 of *Proceedings in Informatics*, pages 133–145. Carleton Scientific, 2003.

148. M. Flammini, A. Navarra, and A. Proskurowski. On routing of wavebands for all-to-all communications in all-optical paths and cycles. *Theoretical Computer Science*, 333(3):401–413, March 2005.

149. M. Flammini and C. Scheideler. Simple, efficient routing schemes for all-optical networks. *Theory of Computing Systems*, 32(3):387–420, June 1999.

150. M. Flammini, M. Shalom, and S. Zaks. On minimizing the number of ADMs — tight bounds for an algorithm without preprocessing. In T. Erlebach, editor,

CAAN, volume 4235 of *Lecture Notes in Computer Science*, pages 72–85. Springer, 2006.

151. M. Flammini, M. Shalom, and S. Zaks. On minimizing the number of ADMs in a general topology optical network. In S. Dolev, editor, *International Symposium on Distributed Computing (DISC)*, volume 4167 of *Lecture Notes in Computer Science*, pages 459–473. Springer, 2006.

152. L. Ford and D. Fulkerson. A suggested computation for maximal multicommodity network flows. *Management Science*, 5(1):97–101, November 1958.

153. A. Frank. Edge-disjoint paths in planar graphs. *Journal of Combinatorial Theory, Series B*, 39(2):164–178, 1985.

154. A. Frank, T. Nishizeki, N. Saito, and H. Suzuki. Algorithms for routing around a rectangle. *Discrete Applied Mathematics*, 40(3):363–378, 1992.

155. A. Fumagalli, I. Cerutti, M. Tacca, F. Masetti, R. Jagannathan, and S. Alagar. Survivable networks based on optimal routing and WDM self-healing rings. In *IEEE International Conference on Computer Communications (INFOCOM)*, volume 2, pages 726–733, March 1999.

156. A. Fumagalli and L. Valcarenghi. IP restoration vs. WDM protection: is there an optimal choice? *IEEE Network*, 14(6):34–41, November/December 2000.

157. A. Ganz and X. Wang. Efficient algorithm for virtual topology design in multihop lightwave networks. *IEEE/ACM Transactions on Networking*, 2(3): 217–225, June 1994.

158. M. R. Garey, D. S. Johnson, G. L. Miller, and C. H. Papadimitriou. The complexity of coloring circular arcs and chords. *SIAM J. Algebraic Discrete Methods*, 1:216–227, 1980.

159. L. Gargano. Limited wavelength conversion in all-optical tree networks. In K. G. Larsen, S. Skyum, and G. Winskel, editors, *International Colloquium on Automata, Languages and Programming (ICALP)*, volume 1443 of *Lecture Notes in Computer Science*, pages 544–555. Springer, 1998.

160. L. Gargano. Multicasting in optical networks. In R. Freivalds, editor, *Fundamentals of Computation Theory (FCT)*, volume 2138 of *Lecture Notes in Computer Science*, pages 459–460. Springer, 2001.

161. L. Gargano, P. Hell, and S. Perennes. Colouring paths in directed symmetric trees with applications to WDM routing. In Degano et al. [105], pages 505–515.

162. L. Gargano, P. Hell, and S. Pérennes. Coloring all directed paths in a symmetric tree, with an application to optical networks. *Journal of Graph Theory*, 38(4):183–196, 2001.

163. L. Gargano and A. Rescigno. Coloring circular arcs with applications to WDM routing. In *Workshop on Approximation and Randomization Algorithms in Communication Networks (ARACNE'00)*, pages 155–166, July 2000.

164. L. Gargano, A. Rescigno, and U. Vaccaro. Multicasting to groups in optical networks and related combinatorial optimization problems. In *IEEE International Parallel and Distributed Processing Symposium (IPDPS)*, page 223a. IEEE Computer Society, 2003.

165. L. Gargano and V. Vaccaro. *Numbers, Information and Complexity, I. Althofer et al. (Eds.)*, chapter Routing in All-Optical Networks: Algorithmic and Graph-Theoretic Problems, Tutorial, pages 555–578. Kluwer Academic Publisher, February 2000.

166. O. Gerstel and S. Kutten. Dynamic wavelength allocation in all-optical ring networks. In *IEEE International Conference on Communications (ICC)*, volume 1, pages 432–436, June 1997.

167. O. Gerstel and R. Ramaswami. Optical layer survivability – an implementation perspective. *IEEE Journal on Selected Areas in Communications*, 18(10): 1885–1899, October 2000.

168. O. Gerstel and R. Ramaswami. Optical layer survivability: a services perspective. *IEEE Communications Magazine*, 38(3):104–113, March 2000.

169. O. Gerstel, R. Ramaswami, and G. Sasaki. Benefits of limited wavelength conversion in WDM ring networks. In *IEEE/OSA Optical Fiber Communications Conference (OFC)*, pages 119–120, 16-21 February 1997.

170. O. Gerstel, R. Ramaswami, and G. Sasaki. Cost-effective traffic grooming in WDM rings. *IEEE/ACM Transactions on Networking*, 8(5):618–630, October 2000.

171. A. Ghatak and K. Thyagarajan. *An Introduction to Fiber Optics*. Cambridge University Press, 1998.

172. A. M. Glass, D. J. DiGiovanni, T. A. Strasser, A. J. Stentz, R. E. Slusher, A. E. White, A. R. Kortan, and B. J. Eggleton. Advances in fiber optics. *Bell Labs Technical Journal*, 5(1):168–187, 2000.

173. D. R. Goff. *Fiber Optic Reference Guide*. Elsevier Science, 2002.

174. J.-P. Goure and I. Verrier. *Optical Fibre Devices*. CRC Press, 2002.

175. P. Green. Progress in optical networking. *IEEE Communications Magazine*, 39(1):54–61, January 2001.

176. P. E. Green. *Fiber Optic Networks*. Prentice Hall, 1992.

177. A. Grosso, E. Leonardi, M. Mellia, and A. Nucci. Logical topologies design over WDM wavelength routed networks robust to traffic uncertainties. *IEEE Communications Letters*, 5(4):172–174, April 2001.

178. J. Gruber and R. Ramaswami. Moving toward all-optical networks. *Lightwave*, pages 60–68, December 2000.

179. Q.-P. Gu. On-line permutation routing on WDM all-optical networks. In *IEEE International Conference on Parallel Processing (ICPP)*, pages 419–428. IEEE Computer Society, August 2002.

180. Q.-P. Gu and S. Peng. Efficient protocols for permutation routing on all-optical multistage interconnection networks. In *IEEE International Conference on Parallel Processing (ICPP)*, pages 513–520, 2000.

181. Q.-P. Gu and S. Peng. Wavelengths requirement for permutation routing in all-optical multistage interconnection networks. In *IEEE International Parallel and Distributed Processing Symposium (IPDPS)*, pages 761–768. IEEE Computer Society, 2000.

182. Q.-P. Gu and S. Peng. Multihop all-to-all broadcast on WDM optical networks. *IEEE Transactions on Parallel and Distributed Systems*, 14(5):477–486, May 2003.

183. A. Gumaste and I. Chlamtac. Light-trails: an optical solution for IP transport. *Journal of Optical Networking*, 3(5):261–281, May 2004.

184. K. P. Gummadi, M. J. Pradeep, and C. S. R. Murthy. An efficient primary-segmented backup scheme for dependable real-time communication in multihop networks. *IEEE/ACM Transactions on Networking*, 11(1), 2003.

185. A. M. Hamad and A. E. Kamal. A survey of multicasting protocols for broadcast-and-select single-hop networks. *IEEE Network*, 16(4):36–48, July/August 2002.

186. H. Harai, M. Murata, and H. Miyahara. Performance of alternate routing methods in all-optical switching networks. In *IEEE International Conference*

on Computer Communications (INFOCOM), volume 2, pages 516–525, April 1997.

187. H. Harai, M. Murata, and H. Miyahara. Heuristic algorithms of allocation of wavelength convertible nodes and routing coordination in all-optical networks. *Journal of Lightwave Technology*, 17(4):535–545, April 1999.

188. J. R. K. Hartline, R. Libeskind-Hadas, K. M. Dresner, E. W. Drucker, and K. J. Ray. Optimal virtual topologies for one-to-many communication in WDM paths and rings. *IEEE/ACM Transactions on Networking*, 12(2):375–383, April 2004.

189. O. Hauser, M. Kodialam, and T. V. Lakshman. Capacity design of fast path restorable optical networks. In *IEEE International Conference on Computer Communications (INFOCOM)*, volume 2, pages 817–826, 2002.

190. J. Hecht. *City of Light: The Story of Fiber Optics*. Oxford University Press, New York, US, 1999.

191. J. Hecht. *Understanding Fiber Optics*. Prentice Hall, 2005.

192. P. Hell and F. S. Roberts. Analogues of Shannon capacity of a graph. *Annals of Discrete Mathematics*, pages 155–168, 1982.

193. M.-C. Heydemann, J.-C. Meyer, and D. Sotteau. On forwarding indices of networks. *Discrete Applied Mathematics*, 23:103–123, 1989.

194. M. G. Hluchyj and M. J. Karol. Shuffle net: an application of generalized perfect shuffles to multihop lightwave networks. *Journal of Lightwave Technology*, 9(10):1386–1397, October 1991.

195. P. Ho and H. Mouftah. Routing and wavelength assignment with multi-granularity traffic in optical networks. *Journal of Lightwave Technology*, 20: 1292–1303, August 2002.

196. P.-H. Ho and H. Mouftah. A framework for service-guaranteed shared protection in WDM mesh networks. *IEEE Communications Magazine*, 40(2):97–103, February 2002.

197. P.-H. Ho and H. T. Mouftah. Allocation of protection domains in dynamic WDM mesh networks. In *International Conference on Network Protocols (ICNP)*, pages 188–189, November 2002.

198. M. E. Houmaidi, M. Bassiouni, and G. Li. Optimal traffic grooming in WDM mesh networks under dynamic traffic. In *IEEE/OSA Optical Fiber Communications Conference (OFC)*, volume 2, pages 23–27, February 2004.

199. J. Hromkovič, R. Klasing, A. Pelc, P. Ružička, and W. Unger. *Dissemination of Information in Communication Networks: Part I. Broadcasting, Gossiping, Leader Election, and Fault-Tolerance*. Springer-Verlag, 2005.

200. J. Hu. Traffic grooming in wavelength-division-multiplexing ring networks: a linear programming solution. *Journal of Optical Networking*, 1(11):397–408, November 2002.

201. J. Q. Hu and B. Leida. Traffic grooming, routing, and wavelength assignment in optical WDM mesh networks. In *IEEE International Conference on Computer Communications (INFOCOM)*, volume 1, pages 495–501, March 2004.

202. J.-Q. Hu and E. Modiano. *Optical WDM Networks: Principles and Practice*, volume II, chapter Traffic Grooming in WDM Networks. Kluwer Academic Publishers, 2004.

203. S. Huang and R. Dutta. Research problems in dynamic traffic grooming in optical networks. In *International Conference on Broadband Networks (Broad-Nets)*, October 2004.

204. X. Huang, F. Farahmand, and J. P. Jue. An algorithm for traffic grooming in WDM mesh networks with dynamically changing light-trees. In *IEEE Global Telecommunications Conference (GLOBECOM)*, volume 3, pages 1813–1817, November 2004.

205. K. Hwang and F. A. Briggs. *Computer Architecture and Parallel Processing*. McGraw-Hill Companies, 1984.

206. ILOG CPLEX 9.1. Documentation available at *http://www.iitb.ac.in/ ieor/resources/cplex91/refcallablelibrary/index.html.*

207. J. Iness, S. Banerjee, and B. Mukherjee. Gemnet: A generalized shuffle-exchange-based regular, scalable and modular multihop network based on WDM lightwave network. *IEEE/ACM Transactions on Networking*, August 1995.

208. R. R. Iraschko and W. D. Grover. A highly efficient path-restoration protocol for management of optical network transport integrity. *IEEE Journal on Selected Areas in Communications*, 18(5):779–794, May 2000.

209. A. Jaekel, S. Bandyopadhyay, S. Roychoudhury, and A. Sengupta. A scalable logical topology for optical networks. *Journal of High Speed Networks*, 11(2):79–87, 2002.

210. S. Janardhanan, A. Mahanti, and D. Saha. A heuristic search based optimal wavelength assignment algorithm to minimize the number of SONET ADMs in WDM rings. In M. Y. Sanadidi, editor, *Communications and Computer Networks*, pages 152–157. IASTED/ACTA Press, 2005.

211. S. Janardhanan, A. Mahanti, D. Saha, and S. K. Sadhukhan. A routing and wavelength assignment (RWA) technique to minimize the number of SONET ADMs in WDM rings. In *Hawaii International Conference on System Sciences (HICSS)*. IEEE Computer Society, 2006.

212. K. Jansen. Approximation results for wavelength routing in directed trees. In *International Parallel Processing Symposium, 2nd Workshop on Optics and Computer Science (WOCS'97)*, Geneva, Switzerland, 1997. IEEE Press.

213. K. Jansen. Approximate strong separation with application in fractional graph coloring and preemptive scheduling. *Theoretical Computer Science*, 1-3(302):239–256, 2003.

214. X.-H. Jia, D. Du, X. Hu, M. Lee, and J. Gu. Optimization of wavelength assignment for QoS multicast in WDM networks. *IEEE Transactions on Communications*, 49(2):341–350, February 2001.

215. A. Jukan and G. Franzl. Path selection methods with multiple constraints in service-guaranteed WDM networks. *IEEE/ACM Transactions on Networking*, 12(1):59–72, February 2004.

216. J.Wang, L. Sahasrabuddhe, and B. Mukherjee. Path vs. subpath vs. link restoration for fault management in IP-over-WDM networks: Performance comparisons using GMPLS control signaling. *IEEE Communication Magazine*, 40(11):80–87, November 2002.

217. C. Kaklamanis and G. Persiano. Efficient wavelength routing on directed fiber trees. In J. Díaz and M. J. Serna, editors, *European Symposium on Algorithms (ESA)*, volume 1136 of *Lecture Notes in Computer Science*, pages 460–470. Springer, 1996.

218. C. Kaklamanis, P. Persiano, T. Erlebach, and K. Jansen. Constrained bipartite edge coloring with applications to wavelength routing. In Degano et al. [105], pages 493–504.

219. I. A. Karapetyan. On coloring of arc graphs. *Doklady Akad. Nauk Armianskoi CCP*, 70(5):306–311, 1980. In Russian.

220. E. Karasan and E. Ayanoglu. Effects of wavelength routing and selection algorithms on wavelength conversion gain in WDM optical networks. *IEEE/ACM Transactions on Networking*, 6(2):186–196, April 1998.

221. G. Keiser. *Optical Communications Essentials*. McGraw-Hill, 2003.

222. F. W. Kerfoot and W. C. Marra. Undersea fiber optic networks: past, present, and future. *IEEE Journal on Selected Areas in Communications*, 16(7): 1220–1225, September 1998.

223. D. Kettler, H. Kafka, and D. Spears. Driving fiber to the home. *IEEE Communications Magazine*, 38(11):106–110, November 2000.

224. S. Kim and S. S. Lumetta. Capacity-efficient protection with fast recovery in optically transparent mesh networks. In *International Conference on Broadband Networks (BroadNets)*, pages 290–299, 2004.

225. T. Kitani, M. Yonedu, N. Funabiki, T. Nakanishi, K. Okayama, and T. Higashino. A two-stage hierarchical algorithm for wavelength assignment in WDM-based bidirectional Manhattan Street networks. In *IEEE International Conference on Networks*, pages 419–424, Sept-Oct 2003.

226. R. Klasing. Methods and problems of wavelength-routing in all-optical networks. In *International Symposium on Mathematical Foundations(MFCS) Workshop on Communication*, August 1998.

227. M. Kodialam and T. V. Lakshman. Integrated dynamic IP and wavelength routing in IP over WDM networks. In *IEEE International Conference on Computer Communications (INFOCOM)*, volume 1, pages 258–366, April 2001.

228. M. Kodialam and T. V. Lakshman. Dynamic routing of restorable bandwidth-guaranteed tunnels using aggregated network resource usage information. *IEEE/ACM Transactions on Networking*, 11(3):399–410, June 2003.

229. V. R. Konda and T. Y. Chow. Algorithm for traffic grooming in optical networks to minimize the number of transceivers. In *IEEE Workshop on High Performance Switching and Routing (HPSR)*, pages 218–221, May 2001.

230. M. Kovacevic and A. Acampora. Benefits of wavelength translation in all-optical clear-channel networks. *IEEE Journal on Selected Areas in Communications*, 14(5):868–880, June 1996.

231. R. M. Krishnaswamy and K. N. Sivarajan. Algorithms for routing and wavelength assignment based on solutions of the LP-relaxations. *IEEE Communications Letters*, 5(10):435–437, October 2001.

232. R. M. Krishnaswamy and K. N. Sivarajan. Design of logical topologies: A linear formulation for wavelength-routed optical networks with no wavelength changers. *IEEE/ACM Transactions on Networking*, 9(2):186–198, April 2001.

233. V. Kumar and E. J. Schwabe. Improved access to optimal bandwidth in trees. In *ACM-SIAM Symposium on Discrete Algorithms (SODA)*, pages 437–444, 1997.

234. J. Kuri, N. Puech, M. Gagnaire, and E. Dotaro. Routing foreseeable lightpath demands using a tabu search meta-heuristic. In *IEEE Global Telecommunications Conference (GLOBECOM)*, pages 2803–2807, November 2002.

235. J. Labourdette, E. Bouillet, R. Ramamurthy, and A. A. Akyama. Fast approximate dimensioning and performance analysis of mesh optical networks. *IEEE/ACM Transactions on Networking*, 13(4):906–917, August 2005.

236. J.-F. P. Labourdette and A. S. Acampora. Logically rearrangeable multihop lightwave networks. *IEEE Transactions on Communications*, 39(8):1223–1230, August 1991.

237. C.-C. Lam, C.-H. Huang, and P. Sadayappan. Optimal algorithms for all-to-all personalized communication on rings and two dimensional tori. *Journal of Parallel and Distributed Computing*, 43(1):3–13, 1997.

238. J. P. Lang, V. Sharma, and E. A. Varvarigos. An analysis of oblivious and adaptive routing in optical networks with wavelength translation. *IEEE/ACM Transactions on Networking*, 9(4):503–517, August 2001.

239. K.-C. Lee and V. O. K. Li. A wavelength-convertible optical network. *Journal of Lightwaveve Technology*, 11(5):962–970, May/June 1993.

240. F. Leighton. *Introduction to Parallel Algorithms and Architectures: Array, Trees, Hypercubes*. Morgan Kaufmann Publishers, September 1992.

241. E. Leonardi, M. Mellia, and M. A. Marsan. Algorithms for the logical topology design in WDM all-optical networks. *Optical Network Magazine*, 1(1):35–46, January 2000.

242. S. Leonardi. On-line network routing. In A. Fiat and G. J. Woeginger, editors, *Online Algorithms*, volume 1442 of *Lecture Notes in Computer Science*, pages 242–267. Springer, 1996.

243. S. Leonardi, A. Marchetti-Spaccamela, A. Presciutti, and A. Rosén. On-line randomized call control revisited . *SIAM Journal on Computing*, 31(1):86–112, 2001.

244. S. Leonardi and A. Vitaletti. Randomized lower bounds for online path coloring. In M. Luby, J. D. P. Rolim, and M. J. Serna, editors, *Randomization and Approximation Techniques in Computer Science(RANDOM)*, volume 1518 of *Lecture Notes in Computer Science*, pages 232–247. Springer, 1998.

245. B. Li, X. Chu, and K. Sohraby. Routing and wavelength assignment vs. wavelength converter placement in all-optical networks. *IEEE Communications Magazine*, 41(8):S22–S28, August 2003.

246. D. Li, Z. Sun, X. Jia, and K. Makki. Traffic grooming on general topology WDM networks. *IEE Proceedings Communications*, 150:197–201, 2003.

247. G. Li and R. Simha. On the wavelength assignment problem in multifiber WDM star and ring networks. In *IEEE International Conference on Computer Communications (INFOCOM)*, volume 3, pages 1771–1780, March 2000.

248. G. Li, D. Wang, C. Kalmanek, and R. Doverspike. Efficient distributed path selection for shared restoration connections. In *IEEE International Conference on Computer Communications (INFOCOM)*, volume 1, pages 140–149, June 2002.

249. L. Li and A. K. Somani. Dynamic wavelength routing using congestion and neighborhood information. *IEEE/ACM Transactions on Networking*, 7(5): 779–786, October 1999.

250. L. Li and A. K. Somani. A new analytical model for multifiber WDM networks. *IEEE Journal on Selected Areas in Communications*, 18(10):2138–2145, October 2000.

251. R. Libeskind-Hadas and R. Melhem. Multicast routing and wavelength assignment in multihop optical networks. *IEEE/ACM Transactions on Networking*, 10(5):621–629, October 2002.

252. E. Limal, S. L. Danielsen, and K. E. Stubkjaer. Capacity utilization in resilient wavelength-routed optical networks using link restoration. In *IEEE/OSA*

Optical Fiber Communications Conference (OFC), volume 2, pages 297–298, February 1998.

253. C. F. Lin. *Optical Components for Communications: Principles and Applications*. Springer, 2004.

254. H. C. Lin and C. H. Wang. A hybrid multicast scheduling algorithm for single-hop WDM networks. In *IEEE International Conference on Computer Communications (INFOCOM)*, volume 1, pages 169–178, April 2001.

255. K. Liu and J. Ryan. All the animals in the zoo: the expanding menagerie of optical components. *IEEE Communication Magazine*, 39(7):110–115, July 2001.

256. L. Liu, X.-Y. Li, P.-J. Wan, and O. Frieder. Wavelength assignment in WDM rings to minimize SONET ADMs. In *IEEE International Conference on Computer Communications (INFOCOM)*, pages 1020–1025, 2000.

257. X. Liu and Q.-P. Gu. Multicasts on WDM all-optical multistage interconnection networks. In *International Conference on Parallel and Distributed Systems (ICPADS)*, pages 601–608, 2001.

258. X. Liu and Q.-P. Gu. Multicasts on WDM all-optical butterfly networks. *Journal of Information Science and Engineering*, 18(6):1049–1058, November 2002.

259. Y. Liu, D. Tipper, and P. Siripongwutikorn. Approximating optimal spare capacity allocation by successive survivable routing. *IEEE/ACM Transactions on Networking*, 13(1):198–211, February 2005.

260. C. Looi, W. Liao, and D. Yang. Service differentiation in optical burst switched networks. In *IEEE Global Telecommunications Conference (GLOBECOM)*, volume 3, pages 2313–2317, November 2002.

261. K. Lu, G. Xiao, and I. Chlamtac. Behavior of distributed wavelength provisioning in wavelength-routed networks with partial wavelength conversion. In *IEEE International Conference on Computer Communications (INFOCOM)*, volume 3, pages 1816–1825, April 2003.

262. S. S. Lumetta and M. Médard. Towards a deeper understanding of link restoration algorithms for mesh networks. In *IEEE International Conference on Computer Communications (INFOCOM)*, volume 1, pages 367–375, April 2001.

263. S. S. Lumetta, M. Médard, and Y.-C. Tseng. Capacity versus robustness: a tradeoff for link restoration in mesh networks. *Journal of Lightwave Technology*, 18(12):1765–1775, December 2000.

264. M. W. Maeda. Management and control of transparent optical networks. *IEEE Journal on Selected Areas in Communications*, 16(7):1008–1023, September 1998.

265. R. Mahalati and R. Dutta. Reconfiguration of traffic grooming optical networks. In *International Conference on Broadband Networks (BroadNets)*, pages 170–179, 2004.

266. J. Manchester, P. Bonenfant, and C. Newton. The evolution of transport network survivability. *IEEE Communications Magazine*, 37(8):44–51, August 1999.

267. Y. Manoussakis and Z. Tuza. The forwarding index of directed networks. *Discrete Applied Mathematics*, 68(3):279–291, July 1996.

268. M. A. Marsan, A. Bianco, E. Leonardi, and F. Neri. Topologies for wavelength-routing all-optical networks. *IEEE/ACM Transactions on Networking*, 1(5):534–546, October 1993.

269. C. Mas and P. Thiran. An efficient algorithm for locating soft and hard failures in WDM networks. *IEEE Journal on Selected Areas in Communications*, 18(10):1900–1911, October 2000.

270. M. F. Maxemchuk. Routing in Manhattan Street networks. *IEEE Transactions on Communications*, COM-35(5):503–512, May 1987.

271. M. Médard, R. A. Barry, S. G. Finn, W. He, and S. Lumetta. Generalized loop-back recovery in optical mesh networks. *IEEE/ACM Transactions on Networking*, 10(1):153–164, February 2002.

272. M. Médard, S. G. Finn, and R. A. Barry. WDM loop-back recovery in mesh networks. In *IEEE International Conference on Computer Communications (INFOCOM)*, volume 2, pages 752–759, March 1999.

273. E. Medova. Network flow algorithms for routing in networks with wavelength division multiplexing. *IEE Proceedings Communications*, 142(4): 238–242, August 1995.

274. Y. Mei and C. Qiao. Efficient distributed control protocols for WDM all-optical networks. In *International Conference on Computer Communications and Networks (ICCCN)*, pages 150–153, September 1997.

275. M. Mihail, C. Kaklamanis, and S. Rao. Efficient access to optical bandwidth — wavelength routing on directed fiber trees, rings, and trees of rings. In *Symposium on Foundations of Computer Science (FOCS)*, pages 548–557, October 1995.

276. K. Miliotis, G. I. Papadimitriou, and A. S. Pomportsis. Adaptive weight functions for wavelength-continuous WDM multi-fiber networks. In *IEEE International Conference on Networks*, pages 413–418, Sept-Oct 2003.

277. D. Mitra and J. B. Seery. Comparative evaluations of randomized and dynamic routing strategies for circuit-switched networks. *IEEE Transactions on Communications*, 39(1):102–116, January 1991.

278. E. Modiano. Random algorithms for scheduling multicast traffic in WDM broadcast-and-select networks. *IEEE/ACM Transactions on Networking*, 7(3):425–434, June 1999.

279. E. Modiano and P. J. Lin. Traffic grooming in WDM networks. *IEEE Communications Magazine*, 39(7):124–129, July 2001.

280. E. Modiano and A. Narula-Tam. Survivable lightpath routing: A new approach to the design of WDM-based networks. *IEEE Journal on Selected Areas in Communications*, 20(4):800–809, May 2002.

281. G. Mohan and C. S. R. Murthy. Lightpath restoration in WDM optical networks. *IEEE Network Magazine*, 14(6):24–32, November 2000.

282. G. Mohan, C. S. R. Murthy, and A. K. Somani. Efficient algorithms for routing dependable connections in WDM optical networks. *IEEE/ACM Transactions on Networking*, 9(5):553–566, October 2001.

283. A. Mokhtar and M. Azizoglu. Adaptive wavelength routing in all-optical networks. *IEEE/ACM Transactions on Networking*, 6(2):197–206, April 1998.

284. K. Mosharaf, J. Talim, L. Lambadaris, and L. Marmorkos. Service differentiation and fairness control in WDM grooming networks. In *IEEE Global Telecommunications Conference (GLOBECOM)*, volume 3, pages 1968–1973, November 2004.

285. H. T. Mouftah and J. Zheng. *Optical WDM Networks: Concepts and Design Principles*. John Wiley & Sons, 2004.

286. B. Mukherjee. *Optical Communication Networks*. McGraw-Hill, 1997.

287. B. Mukherjee. WDM optical communication networks: progress and challenges. *IEEE Journal on Selected Areas in Communications*, 18(10):1810–1824, October 2000.

288. B. Mukherjee. *Optical WDM Networks*. Springer, 2006.

289. B. Mukherjee, D. Banerjee, S. Ramamurthy, and A. Mukherjee. Some principles for designing a wide-area WDM optical network. *IEEE/ACM Transactions on Networking*, 4(5):684–696, October 1996.

290. M. Murata, H. Harai, and H. Miyahara. Performance analysis of wavelength assignment policies in all optical networks with limited-range wavelength conversion. *IEEE Journal on Selected Areas in Communications*, 16(7):1051–1060, September 1998.

291. L. Narayanan, J. Opatrny, and D. Sotteau. All-to-all optical routing in chordal rings of degree 4. *Algorithmica*, 31(2):155–178, 2001.

292. A. Narula-Tam, P. J. Lin, and E. Modiano. Efficient routing and wavelength assignment for reconfigurable WDM networks. *IEEE Journal on Selected Areas in Communication*, 20(1):75–88, January 2002.

293. T. K. Nayak and K. N. Sivarajan. A new approach to dimensioning optical networks. *IEEE Journal on Selected Areas in Communications*, 20(1):134–148, January 2002.

294. G. L. Nemhauser and L. A. Wolsey. *Integer and Combinatorial Optimization*. John Wiley and Sons, 1988.

295. C. Ou, H. Zang, N. K. Singhal, K. Zhu, L. H. Sahasrabuddhe, R. A. MacDonald, and B. Mukherjee. Subpath protection for scalability and fast recovery in optical WDM mesh networks. *IEEE Journal on Selected Areas in Communications*, 22(9):1859–1875, November 2004.

296. C. Ou, J. Zhang, H. Zang, L. H. Sahasrabuddhe, and B. Mukherjee. New and improved approaches for shared-path protection in WDM mesh networks. *Journal of Lightwave Technology*, 22(5):1223–1234, May 2004.

297. C. Ou, K. Zhu, H. Zang, L. H. Sahasrabuddhe, and B. Mukherjee. Traffic grooming for survivable WDM networks — shared protection. *IEEE Journal on Selected Areas in Communications*, 21(9):1367–1383, November 2003.

298. A. Ozdaglar and D. P. Bertsekas. Routing and wavelength assignment in optical networks. *IEEE/ACM Transactions on Networking*, 11(2):259–272, April 2003.

299. J. C. Palais. *Fiber Optic Communications*. Prentice Hall, 1992.

300. G. Panchapakesan and A. Sengupta. On a lightwave network topology using Kautz digraphs. *IEEE Transactions on Computers*, 48(10):1131–1137, October 1999.

301. P. K. Pankaj. Wavelength requirements for multicasting in all-optical networks. *IEEE/ACM Transactions on Networking*, 7(3):414–424, June 1999.

302. R. K. Pankaj and R. G. Gallager. Wavelength requirements of all-optical networks. *IEEE/ACM Transactions on Networking*, 3(3):269–280, June 1995.

303. G. I. Papadimitriou, C. Papazoglou, and A. S. Pomportsis. Optical switching: switch fabrics, techniques, and architectures. *IEEE/OSA Journal of Lightwave Technology*, 21(2):384–405, February 2003.

304. M. Paterson, H. Schröder, O. Sýkora, and I. Vrto. Permutation communication in all-optical rings. *Parallel Processing Letters*, 12(1):23–29, 2002.

305. D. K. Pradhan and S. M. Reddy. A fault tolerant communication architecture for distributed systems. *IEEE Transactions on Computers*, C-31(9):863–870, September 1982.

306. C. Qiao and D. Xu. Distributed partial information management (DPIM) schemes for survivable networks. Part I. In *IEEE International Conference on Computer Communications (INFOCOM)*, volume 1, pages 302–311, June 2002.

307. C. Qiao and M. Yoo. Optical burst switching (OBS) – a new paradigm for an optical internet. *Journal of High Speed Networks*, 8(1):69–84, 1999.

308. X. Qin and Y. Yang. Multicast connection capacity of WDM switching networks with limited wavelength conversion. *IEEE/ACM Transactions on Networking*, 12(3):526–538, June 2004.

309. P. Raghavan and E. Upfal. Efficient routing in all-optical networks. In *ACM Symp. on Theory of Computing (STOC)*, pages 134–143, 1994.

310. B. Ramamurthy and B. Mukherjee. Wavelength conversion in WDM networking. *IEEE Journal on Selected Areas in Communications*, 16(7):1061–1073, September 1998.

311. B. Ramamurthy and A. Ramakrishnan. Virtual topology reconfiguration of wavelength routed optical networks. In *IEEE Global Telecommunications Conference (GLOBECOM)*, volume 2, pages 1269–1275, November 2000.

312. R. Ramamurthy, A. Akyamac, J.-F. Labourdette, and S. Chaudhuri. Preemptive reprovisioning in mesh optical networks. In *IEEE/OSA Optical Fiber Communications Conference (OFC)*, volume 2, pages 785–787, March 2003.

313. R. Ramamurthy, Z. Bogdanowicz, S. Samieian, D. Saha, B. Rajagopalan, S. Sengupta, S. Chaudhuri, and K. Bala. Capacity performance of dynamic provisioning in optical networks. *IEEE/OSA Journal of Lightwave Technology*, 19(1):40–48, January 2001.

314. R. Ramamurthy and B. Mukherjee. Fixed-alternate routing and wavelength conversion in wavelength-routed optical networks. *IEEE/ACM Transactions on Networking*, 10(3):351–367, June 2002.

315. R. Ramamurthy, S. Sengupta, and S. Chaudhuri. Comparison of centralized and distributed provisioning of lightpaths in optical networks. In *IEEE/OSA Optical Fiber Communications Conference (OFC)*, volume 1, pages MH4–1–MH4–3, 2001.

316. S. Ramamurthy and B. Mukherjee. Survivable WDM mesh networks. Part I-Protection. In *IEEE International Conference on Computer Communications (INFOCOM)*, volume 2, pages 744–751, March 1999.

317. S. Ramamurthy and B. Mukherjee. Survivable WDM mesh networks. Part II-Restoration. In *IEEE International Conference on Communications (ICC)*, volume 3, pages 2023–2030, June 1999.

318. S. Ramamurthy, L. Sahasrabuddhe, and B. Mukherjee. Survivable WDM mesh networks. *IEEE/OSA Journal of Lightwave Technology*, 21(4):870–883, April 2003.

319. R. Ramaswami and G. Sasaki. Multiwavelength optical networks with limited wavelength conversion. *IEEE/ACM Transactions on Networking*, 6(6):744–754, December 1998.

320. R. Ramaswami and A. Segall. Distributed network control for wavelength routed optical networks. In *IEEE International Conference on Computer Communications (INFOCOM)*, volume 1, pages 138–147, March 1996.

321. R. Ramaswami and A. Segall. Distributed network control for optical networks. *IEEE/ACM Transactions on Networking*, 5(6):936–943, December 1997.

322. R. Ramaswami and K. N. Sivarajan. Routing and wavelength assignment in all-optical networks. *IEEE/ACM Transactions on Networking*, 3(5):489–500, October 1995.

323. R. Ramaswami and K. N. Sivarajan. Design of logical topologies for wavelength-routed optical networks. *IEEE Journal on Selected Areas in Communications*, 14(5):840–851, June 1996.

324. R. Ramaswami and K. N. Sivarajan. *Optical Networks:A Practical Perspective*. Morgan Kaufmann Publishers, 2002.

325. S. Ramesh, G. N. Rouskas, and H. G. Perros. Computing blocking probabilities in multiclass wavelength-routing networks with multicast calls. *IEEE Journal on Selected Areas in Communications*, 20(1):89–96, January 2002.

326. A. J. Rogers. *Understanding Optical Fiber Communications*. Artech House, 2001.

327. G. N. Rouskas and M. H. Ammar. Multidestination communication over tunable-receiver single-hop WDM networks. *IEEE Journal on Selected Areas in Communications*, 15(3):501–511, 1997.

328. G. N. Rouskas and V. Sivaraman. Packet scheduling in broadcast WDM networks with arbitrary transceiver tuning latencies. *IEEE/ACM Transactions on Networking*, 5(3):359–370, June 1997.

329. I. Rubin and J. Ling. Failure protection methods for optical meshed-ring communications networks. *IEEE Journal on Selected Areas in Communications*, 18(10):1950–1960, October 2000.

330. P. Saengudomlert, E. Modiano, and R. G. Gallager. Dynamic wavelength assignment for WDM all-optical tree networks. *IEEE/ACM Transactions on Networking*, 13(4):895–905, August 2005.

331. P. Saengudomlert, E. Modiano, and R. G. Gallager. On-line routing and wavelength assignment for dynamic traffic in WDM ring and torus networks. *IEEE/ACM Transactions on Networking*, 14(2):330–340, April 2006.

332. L. Sahasrabuddhe, S. Ramamurthy, and B. Mukherjee. Fault management in IP-over-WDM networks: WDM protection versus IP restoration. *IEEE Journal on Selected Areas in Communications*, 20(1):21–33, January 2002.

333. L. H. Sahasrabuddhe and B. Mukherjee. Light trees: Optical multicasting for improved performance in wavelength-routed networks. *IEEE Communication Magazine*, 37(2):67–73, February 1999.

334. L. H. Sahasrabuddhe and B. Mukherjee. Multicast routing algorithms and protocols: a tutorial. *IEEE Network*, 14(1):90–102, Jan/Feb 2000.

335. G. Sahin and M. Azizoglu. Optical layer survivability for single and multiple service classes. *Journal on High Speed Networks*, 10(2):91–108, 2001.

336. H. Schröder, O. Sýkora, and I. Vrto. Optical all-to-all communication for some product graphs. In *Annual Conference on Current Trends in Theory and Practice of Informatics (SOFSEM)*, pages 555–562, 1997.

337. A. Sen, S. Bandyopadhyay, and B. P. Sinha. A new architecture and a new metric for lightwave networks. *IEEE/OSA Journal of Lightwave Technology*, 19(7):913–925, July 2001.

338. A. Sen, B. Hao, and B. H. Shen. Survivable routing in WDM networks. In *Proceedings IEEE International Symposium Computers and Communications*, pages 726–731, July 2002.

339. A. Sen, B. Hao, B. H. Shen, and G. Lin. Survivable routing in WDM networks — logical ring in arbitrary physical topology. In *IEEE International Conference on Communications (ICC)*, pages 2771–2775, April 2002.

340. A. Sen, S. Murthy, and S. Bandyopadyay. gStreams: A new technique for fast recovery with capacity efficient protection in WDM mesh networks. In *IEEE International Conference on Communications (ICC)*, 2007.

341. A. Sen, B. H. Shen, and S. Bandyopadhyay. Survivability of lightwave networks – path lengths in WDM protection scheme. *Journal of High Speed Networks*, 10(4):303–315, 2001.

342. A. Sen, B. H. Shen, B. Hao, H. Jayakumar, and S. Bandyopadhyay. On a preemptive multi-class routing scheme with protection paths for WDM networks. In *IEEE International Conference on Communications (ICC)*, volume 2, pages 1417–1422, May 2003.

343. A. Sengupta, S. Bandyopadhyay, A. Balla, and A. Jaekel. Algorithms for dynamic routing in all-optical networks. *Photonic Networks Communications*, 2(2):163–184, May 2000.

344. A. Sengupta, S. Bandyopadhyay, and A. Jaekel. A distributed fault management scheme for wavelength routed all-optical networks. *Photonic Networks Communications*, 2(4):369–382, November 2000.

345. A. Sengupta, A. Sen, and S. Bandyopadhyay. Fault-tolerant distributed system design. *IEEE Transactions on Circuits and Systems*, 35(2):168–172, February 1988.

346. S. Sengupta and R. Ramamurthy. Capacity efficient distributed routing of mesh-restored lightpaths in optical networks. In *IEEE Global Telecommunications Conference (GLOBECOM)*, volume 4, pages 2129–2133, November 2001.

347. M. Shalom and S. Zaks. A $10/7 + \varepsilon$ approximation for minimizing the number of ADMs in SONET rings. In *International Conference on Broadband Networks (BroadNets)*, pages 254–262. IEEE Computer Society, 2004.

348. M. Shalom and S. Zaks. Minimizing the number of ADMs in SONET rings with maximum throughput. In A. Pelc and M. Raynal, editors, *International Colloquium on Structural Information and Communication Complexity (SIROCCO)*, volume 3499 of *Lecture Notes in Computer Science*, pages 277–291. Springer, 2005.

349. V. Sharma and E. A. Varvarigos. Limited wavelength translation in all-optical WDM mesh networks. In *IEEE International Conference on Computer Communications (INFOCOM)*, volume 2, pages 893–901, April 1998.

350. V. Sharma and E. A. Varvarigos. An analysis of limited wavelength translation in regular all-optical WDM networks. *Journal of Lightwave Technology*, 18(12):1606–1619, December 2000.

351. B. H. Shen, B. Hao, and A. Sen. Minimum cost ring survivability in WDM networks. In *IEEE Workshop on High Performance Switching and Routing (HPSR)*, pages 183–188, 2003.

352. G. Shen, S. K. Bose, and T. H. Cheng. Operation of WDM networks with different wavelength conversion capability. *IEEE Communication Letters*, 4(7): 239–241, July 2000.

353. L. Shen, X. Yang, and B. Ramamurthy. Shared risk link group (SRLG) - diverse path provisioning under hybrid service level agreements in wavelength-routed optical mesh networks. *IEEE/ACM Transactions on Networking*, 13(4): 918–931, August 2005.

354. X. Shen, Q. Hu, H. Dai, and X. Wang. Optimal routing of permutations on rings. In *International Symposium on Algorithms & Computation (ISAAC)*, pages 360–368, 1994.

355. W. T. Silfvast. *Laser Fundamentals*. Cambridge University Press, 2003.

356. D. Sima, T. Fountain, and P. Karsuk. *Advanced Computer Architectures*. Addison-Wesley Longman Publishing, 1997.

357. S. Sinha and C. S. R. Murthy. Information theoretic approach to traffic adaptive WDM networks. *IEEE/ACM Transactions on Networking*, 13(4):881–894, August 2005.

358. K. M. Sivalingam and S. Subramaniam. *Optical WDM Networks: Principles and Practice*. Springer, 2000.

359. K. N. Sivarajan and R. Ramaswami. Lightwave networks based on de Bruijn graphs. *IEEE/ACM Transactions on Networking*, 2(1):70–79, February 1994.

360. P. Solé. Expanding and forwarding. *Discrete Applied Mathematics*, 58(1): 67–78, 1995.

361. A. K. Somani. *Survivability and Traffic Grooming in WDM Optical Networks*. Cambridge University Press, February 2006.

362. A. K. Somani, M. Mina, and L. Li. On trading wavelengths with fibers: a cost-performance based study. *IEEE/ACM Transactions on Networking*, 12(5): 944–951, October 2004.

363. R. Spanke. Architectures for guided-wave optical space switching systems. *IEEE Communications Magazine*, 25(5):42–48, May 1987.

364. M. A. Sridhar and C. S. Raghavendra. Fault-tolerant networks based on the de Bruijn graph. *IEEE Transactions on Computers*, 40(10):1167–1174, 1991.

365. M. Sridharan, A. K. Somani, and M. V. Salapaka. Approaches for capacity and revenue optimization in survivable WDM networks. *Journal of High Speed Networks*, 10(2):109–125, 2001.

366. N. Srinath, C. S. R. Murthy, B. H. Gurucharan, and G. Mohan. A two-stage approach for virtual topology reconfiguration of WDM optical networks. *Optical Network Magazine*, 2(3):58–71, May-June 2001.

367. R. Srinivasan and A. K. Somani. Request-specific routing in WDM grooming networks. In *IEEE International Conference on Communications*, volume 5, pages 2876–2880, April 2002.

368. R. Srinivasan and A. K. Somani. Analysis of optical networks with heterogeneous grooming architectures. *IEEE/ACM Transactions on Networking*, 12(5):931–943, October 2004.

369. S. Stefanakos and T. Erlebach. Routing in all-optical ring networks revisited. In *IEEE Symposium on Computers and Communications (ISCC)*, volume 1, pages 288–293. IEEE Computer Society, 2004.

370. T. Stern and K. Bala. *Multiwavelength Optical Networks: A Layered Approach*. Addison-Wesley Longman Publishing, 1999.

371. H. S. Stone and J. Cocke. Computer architecture in the 1990s. *Computer*, 24(9):30–38, September 1991.

372. J. Strand, A. Chiu, and R. Tkach. Issues for routing in the optical layer. *IEEE Communications Magazine*, 39(2):81–87, February 2001.

373. S. Subramaniam, M. Azizoglu, and A. K. Somani. All-optical networks with sparse wavelength conversion. *IEEE/ACM Transactions on Networking*, 4(4):544–557, August 1996.

374. S. Subramaniam, M. Azizoglu, and A. K. Somani. Connectivity and sparse wavelength conversion in wavelength-routing networks. In *IEEE International Conference on Computer Communications (INFOCOM)*, volume 1, pages 148–155, March 1996.

375. S. Subramaniam, M. Azizoglu, and A. K. Somani. On optimal converter placement in wavelength-routed networks. *IEEE/ACM Transactions on Networking*, 7(5):754–766, October 1999.

376. S. Subramaniam, A. K. Somani, M. Azizoglu, and R. A. Barry. A performance model for wavelength conversion with non-Poisson traffic. In *IEEE International Conference on Computer Communications (INFOCOM)*, volume 2, pages 499–506, April 1997.

377. H. Taha. *Operations Research: An Introduction*. Macmillan, 1982.

378. A. N. Tam and E. Modiano. Dynamic load balancing in WDM based packet networks with and without wavelength constraints. *IEEE Journal on Selected Areas in Communications*, 18(10):1972–1979, October 2000.

379. K. Tang. Cayleynet: A multihop WDM-based lightwave network. In *IEEE International Conference on Computer Communications (INFOCOM)*, volume 3, pages 1260–1267, 1994.

380. D. Thaker and G. N. Rouskas. Multi-destination communication in broadcast WDM networks, a survey. *Optical Networks*, 3(1):34–44, January-February 2002.

381. S. Thiagarajan and A. K. Somani. An efficient algorithm for optimal wavelength converter placement on wavelength-routed networks with arbitrary topologies. In *IEEE International Conference on Computer Communications (INFOCOM)*, volume 2, pages 916–923, March 1999.

382. B. J. Thompson and D. Malacara. *Handbook of Optical Engineering*. Marcel Dekker Inc., 2001.

383. O. Togni. Optical all-to-all communication in inflated networks. In *Workshop on Graph Theoretical Concepts in Computer Science (WG)*, pages 78–87, 1998.

384. O. Togni. Optical index of circulant graphs. Manuscript, LABRI, Université Bordeaux I, 1999.

385. J. Tomlin. Minimum-cost multicommodity network flows. *Operations Research*, 14(1):45–51, January-February 1966.

386. M. Tornatore, G. Maier, and A. Pattavina. WDM network optimization by ILP based on source formulation. In *IEEE International Conference on Computer Communications (INFOCOM)*, volume 3, pages 1813–1821, 2002.

387. S. Tridandapani and B. Mukherjee. Channel sharing in multi-hop WDM lightwave networks: Realization and performance of multicast traffic. *IEEE Journal on Selected Areas in Communications*, 15(3):488–500, April 1997.

388. T. Tripathi and K. N. Sivarajan. Computing approximate blocking probabilities in wavelength routed all-optical networks with limited-range wavelength conversion. In *IEEE International Conference on Computer Communications (INFOCOM)*, volume 1, pages 329–336, March 1999.

389. A. Tucker. Coloring a family of circular arcs. *SIAM Journal of Applied Mathematics*, 29(3):493–502, November 1975.

390. R. Ul-Mustafa and A. E. Kamal. Design and provisioning of WDM networks with multicast traffic grooming. *IEEE Journal on Selected Areas in Communications*, 24(4):37–53, April 2006.

391. E. W. van Stryland (Editor) and S. Chapman. *Fiber Optics Handbook: Fiber, Devices, and Systems for Optical Communications*. Optical Society of America, 2001.

392. N. Vanderhorn, S. Balasubramanian, M. Mina, R. J. Weber, and A. K. Somani. Light-trail testbed for metro optical networks. In *IEEE Testbeds and Research*

Infrastructures for the Development of Networks and Communities (Trident-com), pages 310–315, 2006.

393. A. Venkateswaran and A. Sengupta. On a scalable topology for lightwave networks. In *IEEE International Conference on Computer Communications (INFOCOM)*, volume 2, pages 427–434, March 1996.

394. K. R. Venugopal, M. Shivakumar, and P. S. Kumar. A heuristic for placement of limited range wavelength converters in all-optical networks. In *IEEE International Conference on Computer Communications (INFOCOM)*, volume 2, pages 908–915, March 1999.

395. H. Wang, E. Modiano, and M. Médard. Partial path protection for WDM networks: End-to-end recovery using local failure information. In *Proceedings IEEE International Symposium Computers and Communications (ISCC)*, pages 719–725, July 2002.

396. J. Wang, B. Chen, and R. N. Uma. Dynamic wavelength assignment for multicast in all-optical WDM networks to maximize the network capacity. *IEEE Journal on Selected Areas in Communications*, 21(8):1274–1284, October 2003.

397. J. Wang, W. Cho, V. Vemuri, and B. Mukherjee. Improved approaches for cost-effective traffic grooming in WDM ring networks: ILP formulations and single-hop and multihop connections. *IEEE/OSA Journal of Lightwave Technology*, 19(11):1645–1653, November 2001.

398. J. Wang and B. Mukherjee. Interconnected WDM ring networks: Strategies for interconnection and traffic grooming. *Optical Networks Magazine*, 3(5):10–20, Sept/Oct 2002.

399. J. Wang, X. Qi, and B. Chen. Wavelength assignment for multicast in all-optical WDM networks with splitting constraints. *IEEE/ACM Transactions on Networking*, 14(1):169–182, February 2006.

400. Y. Wang and Q.-P. Gu. Efficient algorithms for traffic grooming in SONET/WDM networks. In *IEEE International Conference on Parallel Processing (ICPP)*, pages 355–364. IEEE Computer Society, 2006.

401. A. N. Washington and H. G. Perros. Analysis of a traffic-groomed optical network with alternate routing. In *International Conference on Information Technology:New Generations (ITNG'07)*, pages 854–862, 2007.

402. J. Y. Wei. Advances in the management and control of optical internet. *IEEE Journal on Selected Areas in Communications*, 20(4):768–785, May 2002.

403. J. Y. Wei and R. I. McFarland. Just-in-time signaling for WDM optical burst switching networks. *Journal of lightwave technology*, 18(12):2019–2037, December 2000.

404. Y. Wen and V. W. S. Chan. Ultra-reliable communication over unreliable optical networks via lightpath diversity: system characterization and optimization. In *IEEE Global Telecommunications Conference (GLOBECOM)*, volume 5, pages 2529–2535, December 2003.

405. G. T. Wilfong. Minimum wavelength in an all-optical ring network. In T. Asano, Y. Igarashi, H. Nagamochi, S. Miyano, and S. Suri, editors, *International Symposium on Algorithms & Computation (ISAAC)*, volume 1178 of *Lecture Notes in Computer Science*, pages 346–355. Springer, 1996.

406. G. T. Wilfong and P. Winkler. Ring routing and wavelength translation. In *ACM-SIAM Symposium on Discrete Algorithms (SODA)*, pages 333–341, 1998.

407. T. Wu and A. K. Somani. Cross-talk attack monitoring and localization in all-optical networks. *IEEE/ACM Transactions on Networking*, 13(6):1390–1401, December 2005.

408. T. H. Wu. *Fiber Network Service Survivability*. Artech House Publishers, 1992.
409. T. H. Wu. Emerging technologies for fiber network survivability. *IEEE Communications Magazine*, 33(2):58–59, 62–74, February 1995.
410. G. Xiao and Y. W. Leung. Algorithms for allocating wavelength converters in all-optical networks. *IEEE/ACM Transactions on Networking*, 7(4):545–557, August 1999.
411. S. Xiao, G. Xiao, and Y.-W. Leung. A network flow approach for static and dynamic traffic grooming in WDM networks. *Computer Networks*, 50(17):3400–3415, 2006.
412. C. Xiaowen, L. Bo, and I. Chlamtac. Wavelength converter placement under different RWA algorithms in wavelength-routed all-optical networks. *IEEE Transactions on Communications*, 51(4):607–617, April 2003.
413. C. Xin, C. Qiao, and S. Dixit. Traffic grooming in mesh WDM optical networks – performance analysis. *IEEE Journal on Selected Areas in Communications*, 22(9):1658–1669, November 2004.
414. C. Xin, Y. Ye, S. Dixit, and C. Qiao. A joint working and protection path selection approach in WDM optical networks. In *IEEE Global Telecommunications Conference (GLOBECOM)*, volume 4, pages 2165–2168, 2001.
415. Y. Xiong, D. Xu, and C. Qiao. Achieving fast and bandwidth-efficient shared-path protection. *Journal of Lightwave Technology*, 21(2):365–371, February 2003.
416. D. Xu, Y. Xiong, and C. Qiao. Novel algorithms for shared segment protection. *IEEE Journal on Selected Areas in Communication*, 21(8):1320–1331, October 2003.
417. D. Xu, Y. Xiong, C. Qiao, and G. Li. Trap avoidance and protection schemes in networks with shared risk link groups. *Journal of Lightwave Technology*, 21(11):2683–2693, November 2003.
418. S. Xu, L. Li, and S. Wang. Dynamic routing and assignment of wavelength algorithms in multifiber wavelength division multiplexing networks. *IEEE Journal on Selected Areas in Communications*, 18(10):2130–2137, October 2000.
419. S. Xu, L. Li, S. Wang, and C. Chen. Wavelength assignment for dynamic traffic in WDM networks. In *IEEE International Conference on Networks*, pages 375–379, 2000.
420. X. Yang and B. Ramamurthy. An analytical model for virtual topology reconfiguration in optical networks and a case study. In *International Conference on Computer Communications and Networks (ICCCN)*, pages 302–308, October 2002.
421. W. Yao and B. Ramamurthy. Dynamic traffic grooming using fixed-alternate routing in WDM mesh optical networks. In *First Workshop on Traffic Grooming in WDM Networks, Co-located with BroadNets 2004*, 2004.
422. W. Yao and B. Ramamurthy. Survivable traffic grooming with path protection at the connection level in WDM mesh networks. In *International Conference on Broadband Networks (BroadNets)*, pages 310–319, October 2004.
423. J. Yates, J. Lacey, D. Everitt, and M. Summerfield. Limited-range wavelength translation in all-optical networks. In *IEEE International Conference on Computer Communications (INFOCOM)*, volume 3, pages 954–961, March 1996.
424. Y. Ye, S. Dixit, and M. Ali. On joint protection/restoration in IP-centric DWDM-based optical transport networks. *IEEE Communication Magazine*, 38(6):174–183, June 2000.

425. Y. Ye, H. Woesner, R. Grasso, T. Chen, and I. Chlamtac. Traffic grooming in light trail networks. In *IEEE Global Telecommunications Conference (GLOBECOM)*, volume 4, pages 1957–1962, November 2005.

426. J. Y. Yoo and S. Banerjee. Design, analysis, and implementation of wavelength-routed all-optical networks: Routing and wavelength assignment approach. *IEEE Communication Survey, Broadband Networks area*, 1997.

427. M. Yoo, C. Qiao, and S. Dixit. QoS performance of optical burst switching in IP-over-WDM networks. *IEEE Journal of Selected Areas in Communications*, 18(10):2062–2071, October 2000.

428. M. Yoo, C. Qiao, and S. Dixit. Optical burst switching for service differentiation in the next-generation optical internet. *IEEE Communications Magazine*, 39(2):98–104, February 2001.

429. S. J. B. Yoo. Wavelength conversion technologies for WDM network applications. *Journal of Lightwave Technology*, 14(6):955–966, June 1996.

430. X. Yuan and A. Fulay. A wavelength assignment heuristic to minimize SONET ADMs in WDM rings. In *IEEE International Conference on Parallel Processing (ICPP) Workshops*, pages 257–262. IEEE Computer Society, 2001.

431. X. Yuan, R. Melhem, R. Gupta, Y. Mei, and C. Qiao. Distributed control protocols for wavelength reservation and their performance evaluation. *Photonic Networks Communications*, 1(3):207–218, November 1999.

432. H. Zang, J. Jue, L. Sahasrabuddhe, R. Ramamurthy, and B. Mukherjee. Dynamic lightpath establishment in wavelength-routed WDM networks. *IEEE Communications Magazine*, 39(9):100–108, September 2001.

433. H. Zang, J. P. Jue, and B. Mukherjee. A review of routing and wavelength assignment approaches for wavelength-routed optical WDM networks. *SPIE Optical Networks Magazine*, 1(1), January 2000.

434. H. Zang, C. Ou, and B. Mukherjee. Path-protection routing and wavelength assignment (RWA) in WDM mesh networks under duct-layer constraints. *IEEE/ACM Transactions on Networking*, 11(2):248–258, April 2003.

435. J. Zhang and B. Mukherjee. A review of fault management in WDM mesh networks: Basic concepts and research challenges. *IEEE Network*, 18(2):41–48, March/April 2004.

436. J. Zhang, K. Zhu, and B. Mukherjee. A comprehensive study on backup reprovisioning to remedy the effect of multiple-link failures in WDM mesh networks. In *IEEE International Conference on Communications (ICC)*, pages 1654–1658, 2004.

437. S. Zhang and B. Ramamurthy. Dynamic traffic grooming algorithms for reconfigurable SONET over WDM networks. *IEEE Journal on Selected Areas in Communications*, 21(7):1165–1172, September 2003.

438. X. Zhang and C. Qiao. An effective and comprehensive approach for traffic grooming and wavelength assignment in SONET/WDM rings. *IEEE/ACM Transactions on Networking*, 8(5):608–617, October 2000.

439. X. Zhang, J. Y. Wei, and C. Qiao. Constrained multicast routing in WDM networks with sparse light splitting. *Journal of Lightwave Technology*, 18(12):1917–1927, December 2000.

440. Z. Zhang and A. S. Acampora. A heuristic wavelength assignment algorithm for multihop WDM networks with wavelength routing and wavelength re-use. *IEEE/ACM Transactions on Networking*, 3(3):281–288, June 1995.

441. F. Zhou, G. Chen, Y. Xu, and J. Gu. Minimizing ADMs on WDM directed fiber trees. *Journal of Computer Science and Technology*, 18(6):725–731, 2003.

442. H. Zhu, H. Zang, K. Zhu, and B. Mukherjee. Dynamic traffic grooming in WDM mesh networks using a novel graph model. In *IEEE Global Telecommunications Conference (GLOBECOM)*, volume 3, pages 2681–2685, November 2002.

443. H. Zhu, H. Zang, K. Zhu, and B. Mukherjee. A novel generic graph model for traffic grooming in heterogeneous WDM mesh networks. *IEEE/ACM Transactions on Networking*, 11(2):285–299, April 2003.

444. K. Zhu and B. Mukherjee. Traffic grooming in an optical WDM mesh network. *IEEE Journal on Selected Areas in Communications*, 20(1):122–133, January 2002.

445. K. Zhu and B. Mukherjee. A review of traffic grooming in WDM optical networks: Architectures and challenges. *Optical Networks Magazine*, 4(2):55–64, March 2003.

446. K. Zhu, H. Zang, and B. Mukherjee. Design of WDM mesh networks with sparse grooming capability. In *IEEE Global Telecommunications Conference (GLOBECOM)*, pages 2696–2700, November 2002.

447. K. Zhu, H. Zang, and B. Mukherjee. A comprehensive study on next-generation optical grooming switches. *IEEE Journal on Selected Areas in Communications*, 21(7):1173–1186, September 2003.

448. K. Zhu, H. Zhu, and B. Mukherjee. Traffic engineering in multigranularity heterogeneous optical WDM mesh networks through dynamic traffic grooming. *IEEE Networks*, 17:8–15, March/April 2003.

449. K. Zhu, H. Zhu, and B. Mukherjee. *Traffic Grooming in Optical WDM Mesh Networks*. Springer, 2005.

450. Y. Zhu, G. N. Rouskas, and H. G. Perros. A comparison of allocation policies in wavelength routing networks. *Photonic Networks Communications*, 2(3):267–295, August 2000.

451. Y. Zhu, G. N. Rouskas, and H. G. Perros. A path decomposition approach for computing blocking probabilities in wavelength-routing networks. *IEEE/ACM Transactions on Networking*, 8(6):747–762, December 2000.

List of symbols used

$\overrightarrow{1}$: the unit column vector $[1, 1, \ldots, 1]$ with $|E|$ elements.

$\overrightarrow{0}$: a vector of all 0s.

$x \rightarrow y$: denotes a fiber in the network from node x to node y (i.e., an edge in physical topology G).

$x \Rightarrow y$: denotes a lightpath in the network from end node x to end node y (i.e., an edge in the logical topology G_L).

α : a string in σ_{sd}^j.

α_1 : a z-conjugate of α.

α_{oc} : coupling ratio of an optical coupler.

α_{ke} : continuous variable which is only permitted to have a value of either 0 or 1.

A : constraints matrix.

$\overrightarrow{A^p}$: column p of constraints matrix A.

a_{ij} : the element in row i, column j of constraints matrix A.

$\tilde{A}, \tilde{B}, \tilde{T}$: edge-path incidence matrices where each matrix has one row for every physical edge $i \rightarrow j$ and one column for every logical edge in the network.

A_i^{in}, A_i^{out} : Access nodes in a network with traffic grooming capabilities.

$\hat{A}$: node, logical-edge incidence matrix of graph G_L.

AC : arc-chain incidence matrix of size $m \times \hat{n}$ where m is the number of logical edges and $\hat{n}$ is the total number of chains in the logical network.

AC^k : arc-chain incidence sub-matrix of size $m \times n^k$ where m is the number of logical edges and n^k is the total number of chains in the logical network for commodity k.

$\overrightarrow{AC_j^k}$: the column vector of AC corresponding to the jth chain of commodity K^k.

$\tilde{a}_{ij}^k, \tilde{b}_{ij}^k, \tilde{t}_{ij}^k$: the entry in the row corresponding to the physical edge $i \to j$ and the column corresponding to the kth logical edge of the network for matrix $\tilde{A}, \tilde{B}$, or $\tilde{T}$ respectively.

ac_{ij}^k : the ith element of the jth chain of the kth commodity.

$\aleph$: number of paths in the logical topology from s to d.

$B(d, k)$: de Bruijn graph of degree d and diameter k.

$\overrightarrow{b}$: a column vector of $\widehat{n_c}$ nonnegative constants.

b^k : the kth element of vector $\overrightarrow{b}$.

B : basis matrix.

β : a string in σ_{sd}^j.

b_{ij} : a binary variable such that
$$b_{ij} = \begin{cases} 1 \text{ if a lightpath exists from end node } E_i \text{ to end node } E_j, \\ 0 \text{ otherwise.} \end{cases}$$

b_p : a binary variable relevant when lightpath ℓ_p is selected.

$\overrightarrow{c}$: a vector of cost coefficients where the pth element c_p represents the cost of the pth column of constraints matrix A.

$\overrightarrow{c_B}$: a vector of cost coefficients, each coefficient corresponding to a variable in the basis.

c_i : the ith channel.

c_{ij} : cost/unit flow associated with edge $i \to j$ in a capacitated network.

C_i : set of channels.

C_{ij}^{in} : channel node.

C_{ij}^{out} : channel node.

C_{ij} : channel bypass edge.

χ : list of channel numbers.

c_P^k : the channel assigned to the kth primary path, for all $k, 1 \le k \le M$.

c_B^k : the channel assigned to the kth backup path, for all $k, 1 \le k \le M$.

$c_{lightpath}$: the capacity of a lightpath (varies from 2.5 to 40.0 Gbps now).

$\overrightarrow{C_j^k}$: the jth chain of the kth commodity.

C_{ikj} : channel availability edge.

C_m^d : the d-dimensional torus.

$c(G_{\Re}^{\mathcal{I}})$: congestion (load) of a network G having instance $\mathcal{I}$ and using the routing $\Re$.

$\mathbf{Color}(i, j)$: color assigned to the request (i, j).

d : degree of the de Bruijn graph $B(d, k)$.

$\deg(G)$: maximum degree of graph G.

d_i : the destination of the ith logical edge, $1 \le i \le q$.

d_i : the destination of the ith communication request, $1 \le i \le q$.

d^k : destination of commodity k.

d_{pr}^e : a precomputed binary coefficient for each pair of end nodes (i, j).

Δ_{in}^i : number of receivers at end node i.

Δ_{in} : a constant denoting the number of receivers at every end node.

Δ_{out} : a constant denoting the number of transmitters at every end node.

Δ_{out}^i : number of transmitters at end node i.

Δ_G : maximum degree of graph G.

δ_k : a continuous variable to characterize whether the backup path of the new connection may share an edge and a channel with the kth backup path, $1 \le k \le M$.

δ_e^{kp} : continuous variable which is only permitted to have a value of either 0 or 1.

δ_{ij}^k : a binary variable to denote whether the backup path of the new connection may share a channel with the backup path for the kth connection on the physical edge $i \to j$.

$\mathbf{dest}(k)$: the destination of the kth (source, destination) pair, $1 \le k \le \hat{q}$, having a nonzero value in T.

$\mathbf{destination}(i)$: the destination of logical edge i.

$\mathcal{D}_i$: demux edge.

E : the set of all pairs of nodes (i, j) such that $i \to j$ is a physical link (i.e., a fiber connects nodes i and j) in the network.

E : the set of edges in a graph G.

$|E|$: number of edges in a graph G.

$E(G)$: the set of edges of a graph G.

E_i : the ith end node in the network.

E_i^{sd} : an end node for multi-hop communication from E_s to E_d.

$\overrightarrow{e^k}$: a vector $[1, 1, \ldots, 1]$ of $\hat{n}^k$ 1s.

E_L : the set of logical edges.

ϵ : a very small constant.

η_e^{kp} : a continuous variable which is only permitted to have a value of either 0 or 1.

$F = [f_{e,\ell}]$: the fiber, logical-edge incidence matrix of size $|E| \times |\mathcal{L}|$, where
$$f_{e,\ell} = \begin{cases} 1 \text{ if fiber } e \text{ is used by lightpath } \ell, \\ 0 \text{ otherwise.} \end{cases}$$

f_p^{sd} : a continuous variable denoting the traffic on logical edge ℓ_p from source end node s to destination end node d.

from(r) : source of request r.

g : the data communication capacity of a single lightpath, using the OC-n notation.

$G = (V, E)$: the (directed) graph model of the physical topology of a network where V is the set of nodes in the network and E is the set of edges.

$G_{\mathfrak{R}}^{\mathcal{I}} = (V_{\mathfrak{R}}, E_{\mathfrak{R}})$: conflict graph (path interference graph).

$\mathcal{G}_i$: grooming edge.

G_L : the graph model of the logical topology of a network where V is the set of end nodes in the network and E_L is the set of logical edges.

$G_S = (V_s, E_s)$: the graph model of a scalable topology where V_s is the set of vertices in the network and E_s is the set of edges.

γ_e^{kp} : continuous variable which is permitted to have a value of either 0 or 1.

γ_{ij} : a binary variable to denote whether the backup path of the new connection may share a channel with the backup path for any existing connection on the physical edge $i \to j$.

H : number of hops in a route from E_x to E_y.

H_d : the d-dimensional hypercube.

I_{SD} : collection of source-destination pairs for which route and wavelength assignment needs to be carried out.

I : identity matrix.

$\mathcal{I}$: an instance of the RWA problem.

I_{one} : one-to-all instance on G with source v.

I_1, I_2 : inputs to an optical coupler.

$(i, i+1, \ldots, j)$: left-to-right dipath from i to j for a request (i,j) in a bidirectional path.

$(i, i-1, \ldots, j)$: right-to-left dipath from i to j for a request (i,j) in a bidirectional path.

(i, j) : communication request from i to j.

(i, j) : dipath from i to j.

k : diameter of the de Bruijn graph $B(d, k)$.

k_B : number of transmitters and receivers in each end node of a broadcast and select network.

K^k : commodity corresponding to the traffic from source s_k to destination d_k.

k_{OC} : number of inputs to an OXC.

ℓ_{sd} : the length of the longest suffix of the string $s = s_k \ldots s_2 s_1$ that is also a prefix of $d = d_k \ldots d_2 d_1$.

$\ell(s, d)$: length of the shortest path from s to d.

λ_i : wavelength of optical carrier.

λ_{pq} : wavelength to communicate from end node E_p to end node E_q.

Λ_{max} : the $\mathcal{L}$ congestion of a logical topology.

$\overrightarrow{\Lambda}$: a column vector $[\Lambda_{max}, \Lambda_{max}, \ldots, \Lambda_{max}]$ with m occurrences of the variable Λ_{max}.

$\lambda^r_{i,j}$: 1 if the rth request uses the lightpath from i to j as an intermediate logical edge.

λ_{sd} : wavelength to communicate from s to d.

$\mathcal{L}$: set of all potential logical edges in a network.

ℓ_p : the pth element of $\mathcal{L}$.

L_i : the lightpath corresponding to the ith logical edge.

L_{ij} : the number of lightpaths from i to j.

$\mathcal{L}_{ij}$: lightpath edge.

L : the set of all pairs of end nodes (i, j) such that $i \Rightarrow j$ is a logical edge (i.e., a lightpath exists from end node i end node j).

L^c_{ij} : the number of lightpaths from i to j using channel c.

L^{new}_i : a lightpath to replace lightpath L_i if L_i becomes inoperative.

L^P_i : the primary lightpath corresponding to the ith logical edge.

L_i^B : the backup lightpath corresponding to the ith logical edge.

L_i^{in}, L_i^{out} : lightpath nodes.

m : number of logical edges.

m_{DM} : number of optical carriers carried by the input to a demultiplexer.

M : number of connections already established in the network.

$\mathcal{M}_i$: mux edge.

$M_{r,s}$: the $r \times s$ mesh.

μ_i : refractive index of an optical medium.

$\hat{M}$: the number characterizing the set $S_{\hat{M}}$ containing the largest address in a graph G_S.

$\aleph$: the number of routes from s_i to d_i in G.

n : a number uniquely identifying an end node in a de Bruijn graph.

n_{oc} : number of inputs to (outputs from) an optical coupler.

N : number of nodes in a capacitated network.

$\mathcal{N}$: number of nodes (router nodes and end nodes) in a WDM network.

$\mathcal{N}_E$: number of end nodes in the network.

$\mathcal{N}_R$: number of router nodes in the network.

N_S : number of vertices in a scalable graph G_S.

N_i : node i (either a router node or an end node) in a network.

N_{max}^{TR} : the maximum value of the number of transmitters or receivers at any end node in the network.

n_{ch} : maximum number of channels that each edge/fiber in this network can accommodate.

n_c : number of input/output port of a switch.

n_{DB} : number of nodes in a de Bruijn graph.

$\widehat{n_c}$: number of constraints.

$\widehat{n_v}$: number of variables.

$\hat{n}$: total number of chains considering all the commodities.

$\hat{n}^k$: number of chains for the kth commodity.

n_{ij}^{light} : number of lightpaths from end node E_i to E_j.

O_1, O_2 : outputs from an optical coupler.

Ω_{max} : the maximum number of lightpaths that share the same fiber.

$\mathfrak{p}$: probability that a channel is used on a fiber.

$\widehat{p}$: number of extra edges added to a path in G_S.

P_n : the bidirectional path of n nodes.

P : set of left-to-right (or right-to-left) dipaths on a bidirectional path.

P' : set of transformed dipaths on a bidirectional path/ring.

$\mathbb{P}, \mathbb{P}_1, \mathbb{P}_2, \mathbb{P}_3$: a path in G_S.

ζ : number of MNH routes from s_i to d_i.

$\wp(s, d)$: the shortest path from s to d.

$\widehat{p}$: a subpath in the physical topology.

$\widetilde{p}$: number of extra edges required for a path in G_S.

$\wp$: number of logical paths from a specified source to a specified destination.

P_b : blocking probability when trying to establish a lightpath.

$\mathcal{P}_1, \mathcal{P}_2$: power levels of signals to an optical coupler.

P : number of binary variables in a MILP.

$\mathfrak{P}$: shortest path from a source to a destination in the auxiliary graph for traffic grooming.

P^1 : list of primary paths already established, $[\rho_1^P, \rho_2^P, \ldots, \rho_M^P]$.

P^2 : list of backup paths already established, $[\rho_1^B, \rho_2^B, \ldots, \rho_M^B]$.

P_1, P_2 : lightpaths.

$P_1^{payload}, P_2^{payload}$: payload of P_1, P_2.

$P_{mn}^{ij,c}$: number of lightpaths from i to j, routed through fiber $m \to n$ on channel c.

p_{mn} : the number of fibers from node m to node n.

q : number of entries in a traffic matrix T with nonzero values.

q : number of source/destination pairs in an instance for the RWA problem.

$\mathfrak{q}$: probability that a channel is used on a fiber.

Q_b : blocking probability when trying to establish a lightpath.

R_i : i^{th} router node in the network.

r_i : receiver tuned to the carrier wavelength corresponding to channel i.

r : number of hops in a broadcast and select network.

$\boldsymbol{R}$: the set of traffic requests.

$\boldsymbol{R_n}$: the bidirectional ring of n nodes.

$\mathfrak{R}$: list of MNH routes.

$\mathfrak{R} = [r_m]$: matrix of traffic requests.

r_u^k : number of receivers at end node u tuned to the wavelength corresponding to channel k.

ρ_i : the route through the physical topology of the lightpath corresponding to the ith logical edge, $1 \leq i \leq q$.

ρ_i : the route selected for the pair (s_i, d_i) in G, $1 \leq i \leq q$.

$\rho_i^1, \rho_i^2, \ldots, \rho_i^\aleph$: the set of routes from s_i to d_i in G.

ρ_i^P : the route of L_i^P through the physical topology.

$\mathcal{R}$: number of edge-disjoint routes over the physical topology, between every pair of end nodes.

$\boldsymbol{R}$: number of edge-disjoint routes over the physical topology, between every pair of end nodes.

$\mathfrak{R} = \{\rho_1, \rho_2, \ldots, \rho_q\}$: routing, i.e., set of selected routes, one for each element in $\mathcal{I}$.

ρ_{xy}^i : the ith precomputed route from E_x to E_y.

ρ_{xy} : a route from E_x to E_y.

ρ_j^i : the ith precomputed route from source(j) to destination(j) where j denotes the lightpath number.

ρ_j : the selected route from source(j) to destination(j) where j denotes the lightpath number.

ρ_i^B : the route of L_i^B through the physical topology.

$\boldsymbol{RR_i}$: the number of receivers at end node i.

$\mathcal{R}_{ij}$: receiver edge.

s^k : source of commodity k.

$\boldsymbol{s_r}$: the size of request $r \in R$, using the Optical Carrier level OC-n notation, used in the non-bifurcated traffic model.

s_i : the source of the ith logical edge, $1 \leq i \leq q$.

s_i : the source of the ith communication request, $1 \leq i \leq q$.

$\boldsymbol{S_n}$: the star of $n + 1$ nodes.

σ_{sd} : the string of length $2k - l_{sd}$ given by $s_k \ldots s_2 s_1 d_{k-l_{sD}} \ldots d_2 d_1$.

σ^i_{sd} : the substring of length k of σ_{sd} starting from the digit i.

src(k) : the source of the kth (source, destination) pair, $1 \leq k \leq \hat{q}$, having a nonzero value in T.

source(i) : the source of the logical edge i.

(S, D) : the pair of nodes for which there is a request for a connection.

$s(d)$: vertex in a de Bruijn graph having address $s_k s_{k-1} \ldots s_2 s_1$ $(d_k d_{k-1} \ldots d_2 d_1)$.

S_i : the set of all strings of k digits, $x_k x_{k-1} \ldots x_2 x_1$, taking each digit from Z_{d+1} such that the ith digit is d.

s^i_j : i^{th} input to an OXC using channel j.

$T = [t(i, j)]$: a $N_E \times N_E$ matrix, called the traffic matrix. The entry in row i and column j of T gives the traffic from end node i to end node j.

t_i : transmitter tuned to the carrier wavelength corresponding to channel i.

τ_{sd} : maximum traffic that can be sent from s to d.

$\theta^{kp}_{e_1 e_2}$: continuous variable which is only permitted to have a value of either 0 or 1.

t^k_u : number of transmitters at end node u tuned to the wavelength corresponding to channel k.

TR_i : the number of transmitters at end node i.

$\mathcal{T}_{ij}$: transmitter edge.

to(r) : destination of request r.

u : a variable denoting end node.

u_{ij} : capacity of edge $i \rightarrow j$ in a capacitated network.

$\vec{u_r} = [u_{v,r}]_{v \in V}$: the source-destination column vector for $r \in R$, where
$$u_{v,r} = \begin{cases} 1 & \text{if } v \text{ is the starting end node for request } r, \\ -1 & \text{if } v \text{ is the terminating end node for request } r, \\ 0 & \text{otherwise.} \end{cases}$$

$u_k u_{k-1} \ldots u_2 u_1$: the address of an end node.

Υ^k : the traffic $t(src(k), dest(k))$ for the kth commodity, $1 \leq k \leq q$.

$\hat{v}$: amount of commodity to be shipped.

v : a variable denoting end node.

V_E : set of end nodes.

V_S : set of vertices in a scalable topology.

V : set of nodes (end nodes or router nodes).

$\boldsymbol{V}$: set of vertices in a graph G.

$|\boldsymbol{V}|$: number of vertices in a graph G.

$\boldsymbol{V(G)}$: the set of vertices of a graph G.

$\boldsymbol{v_k v_{k-1} \ldots v_2 v_1}$: the address of an end node.

$\boldsymbol{w}$: node in a scalable topology.

$\boldsymbol{W}$: a walk in a graph.

$\boldsymbol{w^1, w^2}$: node in a scalable topology with edge to u.

$\boldsymbol{\mathcal{V}_E}$: $\{v : 1 \leq v \leq \mathcal{N}_E\}$, the set of all numbers identifying the end nodes in the network.

$\mathfrak{w}$: new vertex to be added to a scalable topology.

$\mathfrak{w}_i$: i^{th} digit of $\mathfrak{w}$.

$\boldsymbol{w_{kp}}$: a binary variable relevant when lightpath ℓ_p is selected.

$\boldsymbol{w_k}$: a binary variable, for all k, $1 \leq k \leq n_{ch}$, denoting whether channel number k has been used for the primary lightpath.

$\wp^i_{sd}$: the ith path from s to d in the logical topology.

$\wp_{sd}$: a shortest path from s to d in the logical topology.

$\mathfrak{w}(\boldsymbol{G^{\mathcal{I}}_{\Re}})$: chromatic number of graph $G^{\mathcal{I}}_{\Re}$.

$\mathfrak{w}(\boldsymbol{G^{\mathcal{I}}_{\Re_{min}}})$: minimum value of $\mathfrak{w}(G^{\mathcal{I}}_{\Re})$ considering all possible routings $\Re$.

$\boldsymbol{x^i}$: the ith element of vector $\overrightarrow{x}$.

$\boldsymbol{x}$: a vertex in G_S.

$\boldsymbol{x_i}$: a digit in Z_{d+1}.

$\boldsymbol{x^s_i}$: the ith slack variable.

$\overrightarrow{\boldsymbol{x^s}}$: a column vector of m slack variables so that the elements of $\overrightarrow{x^s}$ are x^s_1, $x^s_2, \ldots, x^s_m$.

$\overrightarrow{\boldsymbol{x_B}}$: the vector corresponding to the variables in the basis.

$\boldsymbol{x^i_B}$: the ith element of $\overrightarrow{x_B}$.

$\boldsymbol{x^{sd}_{ij}}$: a continuous variable to denote the portion of traffic $t(s, d) : t(s, d) > 0$, that is routed through the logical edge $E_i \Rightarrow E_j$.

$\overrightarrow{\boldsymbol{x_r}} = [x_{\ell,r}]_{\ell \in \mathcal{L}}$: the column vector containing lightpath routing variables for $r \in R$, where
$$x_{\ell,r} = \begin{cases} 1 \text{ if request } r \text{ is allotted to lightpath } \ell, \\ 0 \text{ otherwise.} \end{cases}$$

$\boldsymbol{x^k_j}$: a binary variable denoting the flow in the jth chain of the kth commodity.

$\widehat{x}_{pr}$: a binary variable for the primary path, used if lightpath ℓ_p is selected

$\overrightarrow{x^k}$: the vector $[x_1^k, x_2^k, \ldots x_{\hat{n}^k}^k]$ of variables.

x_{ij} : a binary variable for finding the (primary) path.

x_{ij}^k : a binary variable for finding the primary path for the kth commodity.

$\widehat{x}_{ij}^k$: a continuous variable denoting the amount of traffic for commodity k flowing on the logical edge $E_i \Rightarrow E_j$.

X_r : 1 if the rth request has been successfully allocated to a lightpath, 0 otherwise.

X_{ij} : amount of flow on edge (i, j),

$\mathcal{X}_{ijk}$: converter edge.

X_{ij}^k : an integer variable having a value of 0 or 1 where
$$X_{ij}^k = \begin{cases} 1 \text{ if the } k^{\text{th}} \text{ lightpath is routed through the edge } i \to j, \ (i,j) \in E \\ \quad \text{in the physical topology,} \\ 0 \text{ otherwise.} \end{cases}$$

X_{ij}^{wk} : an integer variable having a value of 0 or 1 where
$$X_{ij}^{wk} = \begin{cases} 1 \text{ if the } k^{\text{th}} \text{ lightpath is routed through the edge } i \to j, \\ \quad (i,j) \in E \text{ in the physical topology and is assigned channel } c_w, \\ 0 \text{ otherwise.} \end{cases}$$

$\overrightarrow{y}$: a vector of simplex multipliers.

y : a vertex in G_S.

$\widehat{y}_{pr}$: a binary variable for the backup path, used if ℓ_p is selected

Y_j^k : an integer variable having a value of 0 or 1 where
$$Y_j^k = \begin{cases} 1 \text{ if the } k^{\text{th}} \text{ lightpath is assigned channel } c_j, \\ 0 \text{ otherwise.} \end{cases}$$

$\overrightarrow{y}_c = [y_{\ell,c}]_{\ell \in \mathcal{L}}$: the column vector containing channel assignment variables for channel $c, 1 \leq c \leq n_{ch}$, where
$$y_{\ell,c} = \begin{cases} 1 \text{ if channel } c \text{ is assigned to lightpath } \ell, \\ 0 \text{ otherwise.} \end{cases}$$

y_{ij} : a binary variable for finding the backup path, used for dynamic wavelength allocation with wavelength continuity constraint.

y_{ij}^k : a binary variable for finding the backup path, used for 1:1 path protection in wavelength-convertible networks using static allocation.

z_k : a binary variable, for all k, $1 \leq k \leq n_{ch}$, denoting whether channel number k has been used for the backup lightpath.

z_{kp} : a binary variable relevant when ℓ_p is selected.

z_{ij}^k : continuous variable denoting the amount of flow of commodity number k on the edge $(i,j) \in E$.

Z_d : set of digits $\{0, 1, \ldots d - 1\}$.

Z_d^k : the set of all k-digit strings from Z_d.

Index